Basic Fracture Mechanics and its Applications

This textbook provides a comprehensive guide to fracture mechanics and its applications, providing an in-depth discussion of linear elastic fracture mechanics and a brief introduction to nonlinear fracture mechanics. It is an essential companion to the study of several disciplines such as aerospace, biomedical, civil, materials and mechanical engineering. This interdisciplinary textbook is also useful for professionals in several industries dealing with design and manufacturing of engineering materials and structures.

Beginning with four foundational chapters, discussing the theory in depth, the book also presents specific aspects of how fracture mechanics is used to address fatigue crack growth, environment assisted cracking, and creep and creep-fatigue crack growth. Other topics include mixed-mode fracture and materials testing and selection for damage tolerant design, alongside in-depth discussions of ensuring structural integrity of components through real-world examples. There is a strong focus throughout the book on the practical applications of fracture mechanics. It provides a clear description of the theoretical aspects of fracture mechanics and also its limitations. Appendices provide additional background to ensure a comprehensive understanding and every chapter includes solved example problems and unsolved end of chapter problems. Additional instructor support materials are also available.

Basic Fracture Mechanics and its Applications

Ashok Saxena

CRC Press
Taylor & Francis Group
Boca Raton London New York

CRC Press is an imprint of the
Taylor & Francis Group, an **informa** business

First edition published 2023
by CRC Press
6000 Broken Sound Parkway NW, Suite 300, Boca Raton, FL 33487-2742

and by CRC Press
4 Park Square, Milton Park, Abingdon, Oxon, OX14 4RN

CRC Press is an imprint of Taylor & Francis Group, LLC

Library of Congress Cataloging-in-Publication Data
Names: Saxena, A. (Ashok), author.
Title: Basic fracture mechanics and its applications / Ashok Saxena.
Description: First edition. | Boca Raton : CRC Press, [2023] | Includes
bibliographical references and index. |
Identifiers: LCCN 2022034759 (print) | LCCN 2022034760 (ebook) |
ISBN 9781032267197 (hardback) | ISBN 9781032273259 (paperback) |
ISBN 9781003292296 (ebook)
Subjects: LCSH: Fracture mechanics.
Classification: LCC TA409 .S2896 2023 (print) | LCC TA409 (ebook) |
DDC 620.1/126–dc23/eng/20221004
LC record available at https://lccn.loc.gov/2022034759
LC ebook record available at https://lccn.loc.gov/2022034760

ISBN: 9781032267197 (hbk)
ISBN: 9781032273259 (pbk)
ISBN: 9781003292296 (ebk)

DOI: 10.1201/9781003292296

Typeset in Times
by codeMantra

Access the Support Material: https://www.routledge.com/9781032267197

Contents

Preface

Albert Einstein once said,

> "it can scarcely be denied that the supreme goal of all theory is to make the irreducible basic elements as simple and as few as possible without having to surrender the adequate representation of a single datum of experience". Elliott Sober further elaborated "The search for simple theories, then, is a requirement of the scientific enterprise. When theories get too complex, scientists reach for Ockham's Razor, the principle of parsimony, to do the trimming."[1]

The theories of fracture mechanics are due for such trimming! Fracture in structural components is a very old problem with documented references in history dating back to several centuries BC. The phenomenon of fracture is a vast and complex subject that relies upon knowledge from several disciplines and is about bringing that knowledge to solve real technological problems for promoting safety in engineering structures and components. Because of its importance, the field has engaged some of the brightest minds among scientists and engineers. This book is an attempt to make these developments accessible to students and young engineers.

I was once asked about my definition of wisdom, to which I responded: *"wisdom is about having knowledge but even more importantly the ability to process that knowledge to make good decisions and, one without the other is not wisdom."* Fracture mechanics has benefited over the past 75 years from the engagement of several **wise men and women** from whose work we can put together a logical and cohesive approach to fracture problems based upon concepts taken from physics, mathematics, chemistry, and biology, and from aerospace, biomedical, civil, materials, and mechanical engineering. The primary objective of this book is to bring this development into classrooms around the world.

Risk of fracture is a major consideration in several industries such as transportation (air, ground, and sea), power-generation (from fossil, nuclear, and renewable energy sources), civil infrastructure (highways and bridges), recreation (speed boats, race cars, skis, sports protective gear such as helmets, tennis rackets, and golf clubs), biomedical (prosthetic implants), chemical (pressure vessels and reactors), and electronics packaging (mobile phones and lap top computers).

Engineers in basic materials industry such as steel, aluminum, plastics, and composites use fracture mechanics-based test methods to evaluate and characterize materials for potential use in products. Engineers in aerospace, automotive, ship building, power generation, petrochemical, and construction industries use fracture mechanics-based methods for design and life prediction, for remaining life prediction, for establishment of safe inspection criteria and intervals, and last but not least, for understanding field failures in components. Thus, a comprehensive course on fracture mechanics can equip students to address a broad range of fracture issues.

---

[1] Elliott Sober, Hans Reichenbach Professor and William F. Vilas Research Professor in the department of philosophy at the University of Wisconsin, Madison, USA, https://aeon.co/essays/are-scientific-theories-really-better-when-they-are-simpler.

Fracture mechanics has evolved over the past 75 years as an important engineering topic and is now already taught within several engineering disciplines such as mechanical, materials, civil, and aerospace engineering at the graduate level. It is often offered within the confines of a single discipline/department. This book aims to make the material accessible to upper-level undergraduate students as an elective, and to first year graduate students from the named four disciplines. The material-depth and topics in this book are organized such that the students from all four disciplines can enroll in a single course taught by one instructor. This approach has been successfully tried at the Georgia Institute of Technology and at the University of Arkansas for several years and has attracted a large number of students to the course each year. It is also intended for use in online professional development courses.

Basic Fracture Mechanics and its Applications is divided into 11 chapters and includes many Example Problems and End-of-Chapter problems. The first four chapters are foundational while the next six address specific phenomena involving crack growth and fracture in engineering materials and structural components under a variety of realistic loading and environmental conditions. The last chapter contains examples that are meant to be integrative and synthesize concepts described in the previous chapters using real-world problems related to structural integrity. This book attempts to bring together concepts of fracture mechanics from the perspective of (a) materials testing and selection for damage-tolerant design that ensures structural integrity of components and (b) specifying inspection intervals and criteria for making run, repair, retire decisions for components already in service. The latter is frequently used to justify extension of useful life of capital-intensive equipment such as rotors of steam and gas turbines, pressure vessels, and aircraft structures, resulting in considerable cost-savings. Despite the book's applications-oriented approach, the treatment of topics does not shy away from the important and rigorous mechanics and materials underpinnings of the concepts. There are appendices that provide the reader with additional needed knowledge that they might otherwise be lacking because of their disciplinary background.

When I first started in academia at Georgia Institute of Technology in 1985 after spending several years in industry, I immediately embarked on offering graduate courses on fracture mechanics. Initially, my objective for a two-semester long sequence of courses was to teach fracture mechanics to those materials science and engineering graduate students who would be conducting research in this field. Soon, the number of students in the course grew, and it began attracting students from other disciplines such as mechanical, engineering mechanics, aerospace, civil, and biomedical engineering. Catering to such a diverse student group required some changes to be made in my approach to teaching the subject. For example, the course had to include broader coverage of the component design and life prediction-related issues, in addition to materials testing and evaluation. Over the years, I have come to realize that these changes have helped in holding the interest of all students in the subject. Exposing materials science students to life prediction methods and design and structural integrity issues, and mechanics students to mechanistic aspects of fracture has been very useful to all. With assistance and cooperation from my colleagues, we were able to cross-list the courses in multiple engineering disciplines, but taught by a single instructor to a class consisting of students from the various disciplines.

This book has evolved from teaching fracture mechanics over the past 35 years. Several example problems are included to reinforce the concepts being discussed. Homework problems are provided for each chapter to further improve the understanding of students. Several universities have basic fracture courses available as part of their undergraduate and graduate curricula. This book can serve as a textbook for those classes. It is also a useful book for teaching fracture mechanics concepts to practicing engineers in continuing education courses.

The field of fracture continues to evolve as new engineered structural materials are introduced in energy-efficient, cost-effective, and light-weight components and systems, making it necessary to understand the limitations of current fracture theories. I have attempted to emphasize the simplifying assumptions used in the analytical framework for deriving the various fracture theories, and the limitations they place on the use of the theory.

As mentioned before, the field of fracture mechanics has been extremely fortunate to have attracted some of the most creative engineering minds. Achieving a high degree of rigor in their approaches to complex problems has always been a priority with these people. This is the reason why such rapid progress has occurred in the field and its simultaneous acceptance as a practical engineering tool. In the last 75 years, the field has progressed from its very elementary roots to essentially a well-developed and widely accepted approach for addressing structural integrity issues.

I have had the good fortune of having personally known several of the pioneers in this field who have provided the critical breakthroughs to solve a new class of fracture problems. I am indebted to these colleagues because I feel that they have contributed to the book by shaping my personal views on this complex subject. My involvement with fracture mechanics began in 1970 when I took my first graduate course at University of Cincinnati on the subject from my PhD thesis mentor, Prof. Stephen D. Antolovich. He inspired me to get involved in the subject for my masters and doctoral research, that started a relationship which is now in its sixth decade. Please see our recent article on the history of fracture mechanics research.[2] Following graduate school, I had the privilege of working for almost a decade at the Westinghouse Research and Development Center in a group established by late Edward T. Wessel, and by the side of some of the best researchers in the field. This affiliation also gave me access to a network of other researchers from all over the world who have made strong contributions to the field and have influenced my thinking. During my long affiliation with Georgia Tech (1985–2003) and with University of Arkansas (2003–2020), I worked with several colleagues and numerous bright graduate students, and post-doctoral fellows that have motivated me to write this book. I would like to gratefully acknowledge their contributions. In addition to Georgia Tech and University of Arkansas, I have used the contents of this book to teach fracture mechanics as part of professional development short courses to practicing engineers and to students at universities around the world.

This project could never be completed without the cooperation and inspiration from my family. My deepest gratitude is reserved for my wife of almost 50 years,

---

[2] S. D. Antolovich, A. Saxena, and W. W. Gerberich, "Fracture Mechanics—An Interpretive Technical History", *Mechanics Research Communications*, Vol. 91, July 2018, pp. 46–86.

Madhu, and our two children, Rahul, and Anjali, who have been very supportive and understanding of the long hours I have had to spend in completing the manuscript and in pursuit of other professional endeavors throughout my career from which this book has benefited. Also, my now deceased parents, Ram Kishore, and Shanti Saxena deserve credit for giving me the opportunity to have the best education that was available at the time and instilling values such as hard work and pursuing ambitious goals. To these members of my loving family, I dedicate this book!

Ashok Saxena
July 2022

Author

Dr. Ashok Saxena currently serves as President and CEO of WireTough Cylinders, a position he has held since January of 2018. Dr. Saxena also serves as Dean Emeritus and Distinguished Professor (Retired) in the Department of Mechanical Engineering at the University of Arkansas, Fayetteville and as Adjunct Professor in the School of Materials Science and Engineering at Georgia Institute of Technology. At University of Arkansas, he had previously served as the Provost and Vice-Chancellor of Academic Affairs, the Dean of Engineering, the Raymond and Irma Giffels' Chair, the Head of Biomedical Engineering, and the Billingsley Endowed Chair. At Georgia Tech, he held the position of Regents' Professor and Chair of the School of Materials Science and Engineering. Prior to that he was a Fellow Scientist at the Westinghouse Research and Development Center in Pittsburgh. He also served as the Vice-Chancellor of Galgotias University in Greater Noida, India from 2012 to 2014. Dr. Saxena served the American Board of Engineering and Technology (ABET) as Program Evaluator, member of the Engineering Accreditation Commission, and member of the ABET Board. He was a Director and the Vice-President of the International Congress on Fracture and the Executive Chair of the Fifteenth International Conference on Fracture in Atlanta, Georgia, during June 11–14, 2023. He is one of the founders of the Indian Structural Integrity Society (InSIS) and served as its President from 2015 to 2018. He has also been a Visiting Professor/Visiting Scientist in several institutions/research organizations around the world over his professional career spanning five decades.

Dr. Saxena received his MS and PhD degrees from University of Cincinnati in 1972 and 1974, respectively, in Materials Science and Metallurgical Engineering and his BTech degree from the Indian Institute of Technology, Kanpur in 1970 in Mechanical Engineering. Dr. Saxena's area of expertise is mechanical behavior of materials. He has published over 250 research papers, authored/edited 10 books. He is the recipient of numerous national and international awards and recognitions in the field of fracture research that include the George Irwin Medal, the Fracture Mechanics Medal from ASTM, the Wohler Fatigue Medal from the European Structural Integrity Society, Outstanding Research Author Award from Georgia Tech, and Paul Paris Gold Medal from the International Congress on Fracture. He is a Fellow of American Society for Testing and Materials, ASM International, International Congress of Fracture, and the Indian Structural Integrity Society. He also served as an elected member of the European Academy of Sciences.

1 Fracture in Structural Components

The economic impact of fracture in engineering materials and components in the United States was once estimated at about 4.4% of its gross domestic product (GDP) [1]. By the size of the economy in 2021, this would be approximately one-trillion US dollars which is high even if one allows for uncertainties associated with such estimates. Thus, fracture in structural components must be prevented across industries such as ground, air, and sea transportation, chemical, power-generation, civil infrastructure, sports equipment, and biomedical prosthetics where stakes are high. The potential risk of fracture also cuts across all types of structural materials used in construction such as metals, ceramics, polymers, and composites. Consequently, fracture has been studied extensively since industrialization in the mid-nineteenth century and continues to be an important area of research in the twenty-first century, especially as new engineering materials evolve, and designs become more efficient. This chapter explores the significance of fracture and why it has been and continues to be an important engineering topic for academic and industrial research. The chapter also presents the early history of developments in understanding of fracture and recognizes the contributions of pioneers who led these developments.

1.1 FRACTURE IN ENGINEERING MATERIALS AND STRUCTURES: SOCIETAL RELEVANCE

1.1.1 SAFETY ASSESSMENTS

Ensuring safety of people is one of the primary drivers of the need to prevent fractures in engineering structures. Fracture can lead to serious injuries and is threat to human life and are caused by failures in critical load-bearing components of automobiles, highways and bridges, trains, aircraft structures and engine components, pressure vessels used to store compressed combustible and non-combustible gases, and in host of other machinery and equipment used in our daily lives. Engineers are always aiming for fracture-safe design of components while designing them to be less bulky, using more economical materials, requiring less maintenance during service, and recyclable at the end of their design life. However, ensuring safety is paramount and any compromises cannot be justified by any of the above considerations. Thus, a thorough understanding of why components fail is an important aspect of safe designs. Quantifying risk of fracture is important in building adequate safety margins and redundancy in design of load-bearing components.

DOI: 10.1201/9781003292296-1

1.1.2 ENVIRONMENT AND HEALTH HAZARDS

Integrity of offshore oil rigs and leaks in oil and gas pipelines that run through populated areas are potential environmental hazards and threats to public safety and health. These risks must be mitigated with full understanding of potential risk and consequences of fracture. Similarly, the safe operation of nuclear power plants is essential, so nuclear regulatory agencies around the world employ and maintain a wide network of fracture experts to ensure safety of critical components in nuclear power plants.

Hydrogen is gaining momentum for use as fuel for generating electricity that can be used for running electric vehicles such as fork-lifts, cars, trucks, and potentially in trains, ships, and even aircrafts in the future. The process of generating electricity from hydrogen using fuel cells yields zero carbon emission that can greatly contribute to the goal of reducing carbon released into environment while safely, economically, and efficiently meeting the transportation needs. Hydrogen is also being used for storing excess energy generated from renewable sources such as wind and sun during peak periods to be made available during the lean periods. These applications require hydrogen ground storage capacity and onboard fuel storage capacity in vehicles. Being the lightest element in the periodic table, hydrogen must be compressed to very high pressures ranging from 500 to 900 bar to increase its volumetric energy density and become competitive with other conventional fuels. If not compressed to such high pressures, the storage tank sizes would become prohibitively large. Economically storing hydrogen at these high pressures presents significant technological challenges due to enhanced risk of fracture. The phenomenon of hydrogen embrittlement of steels used in the construction of storage tanks is a major design concern.

1.1.3 OPTIMIZING COSTS (FUEL ECONOMY, MATERIAL COSTS, OPPORTUNITY COSTS)

Cost of engineering machinery is determined by the cost of materials used in their constructions, the fabrication methods used in their manufacture and the techniques used in nondestructively inspecting them prior to placing them in service to avoid fracture. Operating costs depend on fuel efficiency, the design life, and the inspection intervals recommended for use during service to avoid sudden failures. Fracture also negatively impacts opportunity costs because revenue depends directly on the fraction of time the machine is available for use during service. Thus, detailed understanding of the factors that affect the risk of fracture during service are important in optimizing the overall life-cycle costs of operating expensive equipment.

1.1.4 PRODUCT LIABILITY

Fracture depends on many complex factors thus establishing the precise reasons of fracture in legal proceedings requires deep expertise in this interdisciplinary field. Fracture experts engage as expert witnesses in cases involving catastrophic failures that cause human casualties, harm to equipment, and economic losses in the form opportunity costs.

As an example, let us consider the failure (Figure 1.1) of a longitudinally welded pipe that carries steam at high pressure and high temperature to a steam turbine in

FIGURE 1.1   Picture of ruptured reheat pipe carrying supersaturated steam at a high pressure in an accident that occurred in June of 1985.

an electricity generation plant. This fracture occurred in June of 1985 and resulted in loss of lives, injuries, and prolonged shutdown of the plant. Thus, the losses can potentially be far more than the cost to replace the pipe and will also include:

- Cost of repair/replacement of other plant equipment that may have been in the vicinity of the ruptured pipe and suffered damage from the incident
- Lost opportunity costs for the period the plant will not be generating electricity
- Compensation costs paid to plant workers and/or their families that were injured or killed in the accident

Establishing potential liability for the losses can be complicated and will be distributed among the manufacturer of the pipe, the firm that designed the pipe and prepared its manufacturing specifications including pipe size, and the operators of the plant that had the responsibility for maintenance of the piping. The cause of fracture was inadequate pipe thickness and flaws in the longitudinal seam welds of the pipe.

1.2   EXAMPLES OF PROMINENT FRACTURES AND THE UNDERLYING CAUSES

1.2.1   FAILURES IN LIBERTY SHIPS

Responding to the demands of World War II to expedite production of cargo ships, United States launched the production of all welded ships called Liberty Ships to

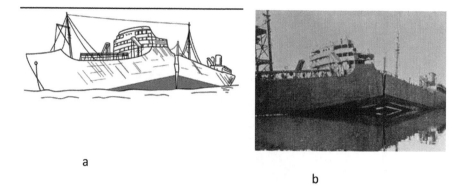

a

b

FIGURE 1.2 (a) A schematic drawing of the fractured all welded US Navy Liberty Ship Schenectady that fractured in a harbor at a welded joint, http://www.shippai.org/fkd/en/cfen/ CB1011020.html. (b) An actual picture of a fractured Liberty Ship available on the World Wide Web. https://metallurgyandmaterials.wordpress.com/2015/12/25/liberty-ship-failures/

transport military cargo. Use of welding in place of riveting to join preformed pieces of the ship reduced the production time very significantly. In all, 2,700 such ships were built in the early to mid-1940s of which 233 were either severely damaged or sunk in the ocean, some after catastrophic brittle fractures that propagated along the welds. Figure 1.2a shows a schematic and Figure 1.2b an actual picture of the fractured ship USS Schenectady as an example of such failures [2]. The failures were attributed to low fracture toughness of the welds, especially under cold temperatures. Fracture toughness, simply stated, refers to energy absorbed in the form of plastic deformation during the fracture process. Ductile fractures, providing high fracture resistance, occur when large amount of energy is absorbed during fracture. Brittle fracture, on the other hand, occurs, when the energy absorbed is relatively small. A much more formal definition of fracture toughness will be provided in chapters that follow. The welds, besides having low fracture toughness, were also ineffective in arresting fractures compared to riveted structures that can effectively arrest cracks and avoid catastrophic fractures.

Several improvements in welding techniques and in the materials used for construction of ships were implemented in future ship construction from this experience. The early efforts of George Irwin, who pioneered the developments in fracture mechanics at the US Naval Research Laboratory, can be traced to his work to understand fractures in Liberty Ships.

1.2.2 FAILURES OF COMET AIRCRAFT

The Comet aircraft went into service in the early 1950s was the world's first jet aircraft built in England and considered revolutionary by the airline industry. The first Comet crash occurred in March of 1953, but it was followed by two more crashes soon after that. The air worthiness certificate for the aircraft was withdrawn after those accidents. Presence of metal fatigue in the fuselage structure was blamed as the main reason for the crashes. The windows of the aircraft had sharp corners that raised local stress levels exacerbating the conditions that contributed to metal fatigue. In addition, the material used in the construction of the fuselage of the Comets that crashed was of very high strength, precipitate hardened aluminum alloy 7075-T6. The T6 temper led to strength levels in the range of 450 MPa which we now know is in the range in which the alloy becomes highly susceptible to environment assisted cracking in humid environments. Thus, cracks that originated due to metal fatigue grew rapidly by environment assisted cracking. This mechanism of cracking was not known at the time, therefore, not considered in selecting the material and its heat treatment. The redesign consisted of making the aircraft windows round or oval shaped to reduce the stress concentrations, and to use a heat treatment that reduces the strength levels to 350 MPa range to eliminate susceptibility to environment assisted cracking. The Al alloys used today to build aircrafts are not sensitive to environment assisted cracking. Because of the accidents, only 90 redesigned Comets were sold, and further production of planes was discontinued in the early 1980s.

1.2.3   CRACKS IN A380 AIRCRAFTS

In the Spring of 2012, cracks were found in the wing structure of Airbus A380 aircrafts within months after they were introduced into commercial service. The plane was designed to carry up to 800 passengers, so the stakes were high. It took over $20B to develop the plane so it would be several years before the investment would pay-off. The cost of each plane was $390M so the opportunity cost of grounding the planes was high. There were 69 planes that were operating at the time of which 20 had seen about 300+ landings and take-offs. At the time when cracks were discovered, a total of 253 aircrafts were on order from 19 airlines.

Cracks were discovered in the interior of the wings of an aircraft. The recommended inspection interval was 4 years, but these cracks appeared very quickly and were found during an unscheduled inspection. Most cracks were reported to be of micro-meter size, but some were as big as 1 cm in size. The cracks were found in a member that has an "L" shape and is used to attach the wing skins to the frame of the wing. Air Bus assessed that the cracks were unlikely to progress to the point where they would cause a catastrophic accident but, was considered serious enough that it required their full attention. The cracks were traced back to a problem during manufacturing that was quickly fixed, and the aircrafts were recertified for service. The aircraft have now been in service for several years without an incident and a potentially dangerous situation has been averted by timely inspections of a newly designed aircraft that did not have a long service history.

1.2.4   CRACK IN A STRUCTURAL MEMBER OF AN INTERSTATE HIGHWAY BRIDGE

Figure 1.3 shows a crack in a weld in a structural member of a highway bridge across the Mississippi river between Arkansas and Tennessee. This crack was discovered during a routine visual inspection of the bridge and was several inches in length. Bridges are designed with redundancy, so the crack was able to grow to that size, while the bridge continued to operate normally. Nevertheless, after the crack was discovered, the bridge was shut down for further inspections and repair for a long period

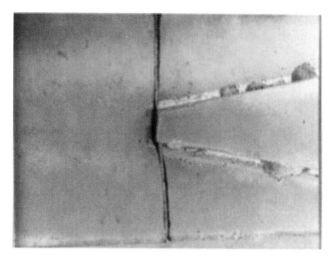

FIGURE 1.3 Cracking in a welded structural member in a highway bridge across the Mississippi river.

causing major disruptions to traffic. The crack formed due to a lack of penetration defect left there during welding and grew under the action of residual stresses that are locked during cooling of the molten metal. The differential thermal contraction during cooling in the various regions of the weldment from temperature gradients can leave behind high residual stresses in the local region. Welds are subjected to post weld heat treatment to relieve the residual stresses and to rejuvenate the microstructure of the heat-affected-zone in the base metal. If this process is not followed correctly, cracks can easily form in welds.

1.2.5 CRACKS IN HUMAN BONES

If one looks at cracks in the human bones from several people at high magnification, they will find the bone density of younger people is typically higher compared to an older person. Consequently, the pore sizes are small along the crack path in the fracture of a young person compared to an older person with higher levels of porosity making it more susceptible to fracture.

1.2.6 ANEURYSMS IN HUMAN ABDOMINAL AORTAS

In previous examples, the fractures resulted from dominant cracks. In some cases, the dominant crack developed from a pre-existing defect introduced during manufacturing. In other cases, early crack formation occurred during service and subsequently the crack became dominant and grew to a larger size. However, not all failures result from a dominant crack. Imagine a slender but long specimen of uniform cross-section being subjected to a tensile force from its two ends. When the

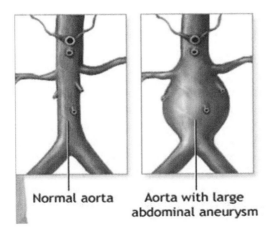

Normal aorta   Aorta with large
                  abdominal aneurysm

FIGURE 1.4   A schematic of human aorta with an acute aneurysm prior to rupture. https://medlineplus.gov/ency/article/000162.htm

applied stress approaches the ultimate strength of the material, the specimen develops a neck and ultimately fails. Thus, failures can also occur because the applied stress in a significant region of the component exceeds the ultimate strength of the material. The component may also have a local weak region in which strains concentrate, creating either a bulge or a neck that ultimately fractures. Such failures, also referred to as rupture, are found in pipes carrying fluids under pressure. The human body is full of blood vessels with circulating fluid (blood) in the vessels, so there is potential for such failures to occur.

Aneurysms (or bulges) can develop in regions of the blood vessels that are weak and can ultimately rupture when the strain exceeds a critical value. The largest blood vessel in the body is the aorta that supplies blood to several other smaller arteries. Figure 1.4 shows a model of a human aorta that has a weak region in the abdominal area where irreversible deformation or strain accumulates with time. With every systolic and diastolic event resulting from heart beats, strain accumulates incrementally until the day when the aorta ruptures if the aneurysm remains undiagnosed and/or unchecked. A very high percentage of such ruptures are fatal.

Such incremental growth in the aneurysm has been computationally simulated using finite element method, typically used in studying fractures in engineering structures. Thus, the understanding of fracture processes in structural materials and analytical tools developed to perform the analyses are now finding their way into assisting biomedical research. It is important to note that the stress-strain behavior of human tissue is quite different from the stress-strain behavior in metals and other structural materials, but there are common principles and techniques involved in studying and understanding their behavior. For example, degradation of the properties of human tissue with time (or age) is analogous to the degradation of material properties in metals that occurs during service, especially at high temperatures. The kinetics of material degradation must be built into the stress-strain models to have good predictive capability.

1.3   DEGRADATION PHENOMENA AND FRACTURE IN ENGINEERING MATERIALS AND STRUCTURES

Failures seldom occur when a structural component is first thrust into service because potential causes of failures have generally been anticipated, and design analysis and prototype testing take those into account. Materials used in construction of load-bearing components are thoroughly tested to demonstrate their worthiness for the application. Further, several design codes that address allowable stresses and specify suitable materials, are also available to ensure safe designs. Most fractures occur after the component has been in service for some time and are caused by (a) hidden defects that are left behind during manufacturing and escape detection, (b) defects that form due to negligence during operation, (c) an unknown degradation mechanism reveals itself that was not foreseen during accelerated testing of the material. An example of the latter was found in the failure of the Comet aircrafts where environment assisted cracking was not anticipated during design but revealed itself during service.

A typical sequence of events that lead to fracture involve the following steps:

• Presence of crack-like defects left during manufacturing and escaped detection during inspection or initiation of cracks during service.
• Growth of cracks during service due to fatigue loading, due to creep deformation, due to environmental degradation, or a synergistic interaction between two and more of these degradation mechanisms, such as environment-fatigue or creep-fatigue-environment interactions.
• Crack grows to a critical size and fracture suddenly occurs, and in several cases under normal operating conditions without much warning.
• Occasionally, upset conditions occur during service resulting in situations not considered during design, or due to a domino effect when a failure in one component leads to failure in others.

1.3.1   CRACK INITIATION/FORMATION AND GROWTH

The service time required for crack of a detectable size to form in a component/specimen is labelled as the crack initiation/formation life. Locations where such cracks are most likely to form are called the "hot spots" and can be identified by looking at the stress or strain distributions in the body. For example, in the case of the Comet aircrafts, it was the sharp corners associated with the rectangular windows in the body of the planes. In the case of components that operate at high temperatures, they are the points at which the stresses are high and/or the service temperatures are high. The time to initiate a crack is correlated with stress amplitudes if cyclic loading is involved or simply the stress, if only sustained stresses are present. The time or fatigue cycles necessary to form cracks of detectable size is used as criterion for quantifying crack formation/initiation life. This is then followed by the crack growth phase. Service life prediction models must account for both crack initiation/formation and crack growth, and an estimate of the critical size at which sudden fracture is expected to occur under normal loading conditions must also be made.

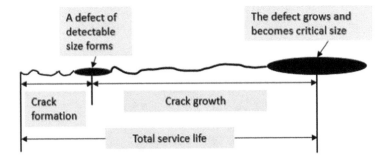

FIGURE 1.5 Schematic of the various phases of degradation eventually causing failure in structural components.

Fracture mechanics is an analytical approach which accounts for the reduction in fracture strength of structures and structural materials due to presence of cracks or crack like defects. The crack growth life of components is estimated using fracture mechanics as summarized in Figure 1.5.

1.4 HISTORY OF DEVELOPMENTS IN UNDERSTANDING FATIGUE AND FRACTURE

Developments in understanding of the phenomena of fatigue and fracture began in the late nineteenth and early twentieth century with ushering in of the age of mechanization and the industrial era. Machines for large-scale production, and trains, automobiles, and aircrafts for transportation of people and goods were introduced as part of daily life in the society. Failures in machinery and rail-road parts necessitated the need to understand fracture. The quest for understanding the reasons for fracture continued through the twentieth and into the twenty-first century to keep up with advancements in the auto, aerospace, and the power generation industries [3]. The initial focus on fracture, for understandable reasons, was on metals, especially steels, and later shifted to other more advanced metals and alloys and eventually to non-metals and composite materials. Rather than discussing these developments chronologically, we have chosen to address them phenomenologically.

1.4.1 DEVELOPMENTS IN UNDERSTANDING OF FATIGUE

In 1870, a German scientist and a railway engineer, A. Wohler (Figure 1.6), discovered the phenomenon of fatigue in response to unexpected failures in wheels and axles of train wagons. He discovered the relationship between cyclic life, N, and the cyclic stress, S, now known as the S-N curve, or the Wohler-Diagram. According to this relationship, the cyclic life of a component decreases as the stress amplitude increases. Later in 1899, Goodman modified this relationship by introducing mean stress during the fatigue cycle as a secondary variable that also affects the fatigue life. This relationship became known as the Goodman-Diagram and to this day is used in design of structural components for cyclic duty. These studies provided the

FIGURE 1.6   August Wohler (1819–1914), a German rail-road engineer who was the first to systematically study the phenomenon of fatigue in metals. He invented the Wohler Diagram. https://en.wikipedia.org/wiki/August_W%C3%B6hler

a   b

FIGURE 1.7   (a) S. S. Manson (1919–2013), https://sem.org/Files/about/ETHistJA04.pdf and (b) Louis F. Coffin (1917–2008), the inventors of the Coffin-Manson Law relating fatigue life to plastic strain range.

much-needed breakthrough in understanding of fatigue failures and led to the development of the local strain approach championed by Louis Coffin and Stan Manson (Figure 1.7) in 1950s, known as the Coffin-Manson low cycle fatigue law. This approach was further developed in late 1950s and 1960s by JoDean Morrow and his group at the University of Illinois and the automotive industry worldwide. It remains an active area of research with focus shifting from metals to advanced materials such as structural composites and additively manufactured components.

FIGURE 1.8 Paul C. Paris (1930–2017), the inventor of Paris-Law that correlates fatigue crack growth rate to the magnitude of the stress intensity factor range associated with the fatigue cycle. (This photograph is from 2013 when Prof. Paris attended the 13th International Conference on Fracture in Beijing, China.)

In early 1960s, Paul C. Paris (Figure 1.8) working at Boeing and later at Lehigh University, developed the first approach to systematically characterize the rates at which cracks grow under cyclic loading. This relationship, known as the Paris-law, states that the rate at which cracks grow during fatigue loading is determined by the cyclic stress intensity parameter, ΔK. This break-through development in modeling how fatigue damage progresses led to the damage tolerant approach in design of structural components. In this approach, a crack of the size that could escape detection is assumed to be present in the component, and the useful design life is estimated as the cycles needed for the crack to grow from this initial size to the critical size that can cause fracture, as illustrated in Figure 1.5. Design life estimated in this way is often used as the inspection interval during service, so life of the component can be extended if cracks are not found during reinspection.

1.4.2 Understanding Brittle and Ductile Fracture

In the early 1900s, a French Metallurgical Engineer by the name Georges Charpy (Figure 1.9) discovered that brittle fracture can occur in an otherwise ductile material in the presence of notches that promote a triaxial state of stress. He went on to develop what is known as the Charpy impact test. The test measures the fracture energy absorbed by a specimen of standard dimension containing a standard notch and impacted by a hammer of standard weight dropped from a standard height. The kinetic energy of the hammer at the bottom of its swing fractures the specimen and its residual energy can be measured by the height it attains following the impact. The difference in the initial potential energy of the hammer and its residual potential energy after impact is the energy absorbed by specimen during fracture and is termed as the Charpy impact energy and is a material property. This simple test, even

FIGURE 1.9 Georges Charpy (1865–1945), a French Metallurgist, who first proposed the Charpy impact test in the early 1900s that is used as a material qualification test even today.

today, is used for material screening and for quality assurance purposes during metal production and qualification. Brittle materials are those that fracture at low Charpy energy and ductile materials are those that absorb high amounts of Charpy energy prior to fracture.

1.4.3   EARLY DEVELOPMENTS IN FRACTURE MECHANICS

In 1913, C. E. Inglis (Figure 1.10a) successfully derived the relationship between stress concentration at the root of an elliptical notch and the remote applied uniform stress in a semi-infinite plate. His results were used to show the importance of the notch root radius on the intensity of stress concentration. The results allowed one to compute the elastic strain energy associated with the body containing the elliptical notch. By changing the ratio of the minor to major axes of the ellipse, the results can be applied to bodies containing sharp cracks. In 1921, A. A. Griffith (Figure 1.10b), using the results from Inglis' derivation, proposed a theory of brittle fracture emanating from cracks based on energy exchanges taking place during the fracture process. He noted that the energy for creating new surfaces associated with crack growth must account for the difference between the work done by external force and the change in strain energy of the body due to crack growth. He omitted the consideration of energy absorbed in the form of plastic deformation in metals that accompanies crack growth. Thus, his theory successfully described the behavior of very brittle materials such as glass but could not describe the behavior of metals and consequently lay dormant for three decades. It is important to note that the concept of line defects in crystals or dislocations that enable plastic deformation had not been proposed at the time Griffith advanced his theory.

In 1934, Taylor and Polanyi (Figure 1.11a and b) working independently proposed the presence of atomic scale line defects now also known as dislocations to explain why materials are not able to achieve their theoretical strength. This provided insight into the role of plastic deformation during fracture, especially in metals. Recognizing

a　　　　　　　　　b

FIGURE 1.10　Two British scientists (a) C. E. Inglis (1875–1952) who derived the stress distribution in front of an elliptical notch in a semi-infinite plate loaded with a remote uniform stress, and (b) A. A. Griffith (1893–1963), who advanced the early theory of fracture based on energy exchanges that take place when a crack grows in a brittle elastic solid. Griffith's theory, as proposed, was limited to extremely brittle materials such as glass and ceramics but was later modified to include metals.

a　　　　　　　　　b

FIGURE 1.11　(a) G. I. Taylor (1886–1975) and (b) Michael Polanyi (1891–1976) who independently proposed the presence of line defects in crystalline solids known as dislocations that enable plastic deformation in crystalline solids.

the importance of including the energy needed for plastic deformation accompanying fracture, in 1948, E. Orowan (Figure 1.12) and G. R. Irwin (Figure 1.13) independently proposed a modification to Griffith's theory to account for plastic deformation accompanying fracture. This extended the original Griffith's theory to metals.

In 1956, G. R. Irwin proposed an approach based on stress analysis of cracks utilizing the results from Westergaard's (Figure 1.14) analysis of stress distributions ahead of sharp cracks, first published in 1939. It is worthwhile to note that while Inglis and Westergaard were mathematicians working on classical elasticity

FIGURE 1.12   E. Orowan (1902–1989) proposed the modified Griffith's theory that extended the earlier version of the theory to include plastic deformation that accompanies brittle fracture.

FIGURE 1.13   George R. Irwin (1907–1998) popularly known as the father of modern fracture mechanics for his pioneering research that established the concept of stress intensity parameter, K, and for showing the equivalence between K and the Griffith's crack extension force, G and for numerous other developments.

FIGURE 1.14   H. M. Westergaard (1888–1950), derived the Airy's stress function for an elastic body with a sharp crack loaded in the crack opening mode and used it to determine the stress and displacement fields in the vicinity of the crack.

problems and made seminal contributions to the field of elasticity, it took people such as Griffith and Irwin, respectively, who were materials physicists/engineers to highlight the true value of their work. Irwin also demonstrated the equivalence between the stress and energy-based approaches in 1957 that allowed the field of fracture mechanics to be built upon a rigorous mechanics foundation. During 1960s and the 1970s, rapid developments in linear elastic fracture mechanics occurred but were limited to fracture and crack growth under the conditions of small-scale-yielding when plasticity was restricted to a small zone ahead of the crack tip in an otherwise elastic body.

1.4.4   DEVELOPMENTS IN ELASTIC-PLASTIC FRACTURE MECHANICS

In the late 1960s, the field of elastic-plastic fracture mechanics began to take root to meet the needs of the power generation industry, particularly nuclear power plants where safety considerations and fracture resistant designs were paramount and use of very ductile and fracture resistant materials was necessary. This required that fracture mechanics methods to be extended to conditions of gross plasticity. In 1968, James Rice put forward the concept of J-integral that was equivalent to the Griffith's crack extension force for nonlinear elastic materials. In the same year, J. W. Hutchinson, and J. Rice and G. Rosengren independently derived a relationship between the crack tip stress, strain, and displacements and the J-integral. J. A. Begley and J. D. Landes in 1971 proposed a fracture criterion for elastic-plastic materials using J-integral and produced experimental evidence in its support. During the same period, A. A. Wells proposed a fracture criterion based on crack tip opening displacement and produced

FIGURE 1.15 Cracked cartridge cases made from brass exposed to ammonia that led to environment assisted cracking in areas containing a high residual stress from metal forming.

experimental evidence in its support. C. F. Shih in 1975 showed analytically that J-integral and the crack tip opening displacement were uniquely related to each other, so the two seemingly different approaches converged into a single cohesive theory. Rapid progress occurred in elastic-plastic fracture mechanics in the 1970s and 1980s.

1.4.5 ENVIRONMENT ASSISTED CRACKING

Environment assisted cracking or also known as stress corrosion cracking of metals has been a subject of interest for over 150 years. The first documented case of stress corrosion failures is found in brass cartridge cases in the mid-nineteenth century in British India. The cause of cracking was not understood until much later (1921) and was labelled as Season Cracking. Figure 1.15 shows pictures of cracked brass cartridge cases in ammonia environment. The brass cartridge cases had high retained residual stresses from the plastic deformation during forming of the cups and were stored in horse stables where horse urine was the source of ammonia that caused the cracking. The requisites for environment assisted cracking or stress corrosion cracking are stress and a corrosive environment. Not all metals exhibit stress corrosion cracking in any given environment, and not all environments cause cracking in any given metal. Corrosive environments can be aqueous solutions such as salt-water, or they can be gaseous such as oxygen, hydrogen, and hydrogen sulfide. The fracture mechanics approach to what is known as environment assisted cracking developed in the early 1960s. This brought about many interdisciplinary collaborations between surface chemists, electro-chemists, materials scientists and physicists, and fracture mechanics experts to this field. This remains a field of active research because the problems are complex and multi-faceted.

1.4.6 DEVELOPMENTS IN TIME DEPENDENT FRACTURE MECHANICS

Static loading at high temperatures also promotes extensive plasticity as a function of time by a phenomenon known as creep. Creep and creep-fatigue considerations are important for components operating in high temperature environments such as steam turbines and gas turbines for land, sea, and aerospace applications. Metals and their alloys are known to deform by creep at temperatures that are higher than 35%–40%

of their melting point on the absolute scale. As an example, for lead because of its low melting point, room temperature is in the range of temperatures at which creep can occur. Lead is also a heavy metal so, if unsupported, it can creep under its own weight at room temperature. Thus, high temperature applications require use of refractory metals and alloys with high melting points. Resistance to deformation and fracture at high temperature of the alloys used in the application determines the operating temperatures of steam and gas turbines.

In early 1950s, Larson and Miller proposed a very useful parameter that allows one to consider creep rupture behavior of metals and its relationship with temperature and stress. This approach is still widely used in design of high temperature components.

Time-dependent deformation associated with creep alters the crack tip stress fields established as part of initial loading. This evolution of the stress field with time must be addressed in any viable theory for creep crack growth. Some of the earliest studies involving the use of fracture mechanics for predicting crack growth rates at elevated temperatures are found in the work of Lee James who in 1972 extended the idea of using Paris' cyclic stress intensity parameter, ΔK, to correlate fatigue crack growth behavior of 304 stainless steel at various loading frequencies. About the same time, Siverns and Price unsuccessfully attempted to correlate creep crack growth rate with stress intensity parameter, K and thus identified the need for time-dependent fracture mechanics parameters. Under static loading, K uniquely characterizes the crack tip stress field only upon loading at time, $t = 0^+$. With time, stress relaxation at the crack tip occurs due to creep deformation and there is no longer uniqueness in the relationship between K and the crack tip stress fields as pointed out by Saxena, and by Riedel and Rice in 1980.

Landes and Begley in 1976 proposed C^*-integral for characterizing creep crack growth rates under steady-state creep conditions that occur after the stress redistribution due to creep is complete. C^* is based on the mathematical analogy with Rice's J-integral for characterizing creep crack growth. The attention in the early 1980s then turned to search for parameters that were suitable for use under small-scale creep conditions which would become identical to C^* when widespread steady-state creep conditions developed in the cracked body. The work of Riedel and Rice, Ohji, Ogura and Kubo and Bassani and McClintock was key to unlocking the mechanics of this problem. Riedel and Rice and Ohji et al. analytically derived the equations for stress fields ahead of the crack tip under small-scale creep conditions while Bassani and McClintock numerically verified these analytical results. Saxena proposed the C_t parameter, that is measurable at the load-line of the cracked body and thus includes contributions from all the different creep deformation mechanisms present at the crack tip. This was then extended to characterize creep-fatigue crack growth, a type of loading experienced frequently by components in high temperature service.

1.5   SUMMARY

Over the past century, fracture research has evolved into a coherent interdisciplinary field motivated by the need for structural integrity assurance and safe operation of components that are part of air, ground and water transportation systems and power generation systems, as well as others. The design, material selection and choice of inspection techniques, inspection criteria and intervals are seldom considered concurrently leading to designs that often take more time from start to finish than afforded by product design cycle times. This is particularly the case when new and innovative materials are used in a competitive marketplace. The interdisciplinary field of *Fra*cture *C*ontrol and *S*tructural *I*ntegrity (FraCSI) has evolved over the past 60 years to respond to this need. FraCSI analyses focus on design evaluation and assessment of margins between safe and normal operation and the possibility of fracture. This is sometimes considered in a probabilistic framework and expressed as probability of failure.

FraCSI has evolved as a field at the interface of several disciplines including Mechanical, Materials, Aerospace, and Civil Engineering, Mechanics, Physics, and Mathematical and Computational Sciences and is now also touching Biomedical Engineering and Geology. The field has continued to prosper over the past six decades while continuously expanding its capabilities to address new classes of deformation, fracture and crack initiation and crack growth problems.

Progress has been made in experimental capabilities for materials characterization, molecular and atomistic scale computational modelling, nondestructive inspection techniques. Progress has also occurred in scientific understanding of the fracture phenomenon within the crack tip process zone via damage mechanics. Better understanding of the limitations of single parameter linear and nonlinear fracture mechanics also exists. Similarly, there has been progress in understanding of crack initiation phenomena related to fatigue, creep and environmental effects and their synergistic effects. However, a question arises that deserves in-depth probing. Have we maximized the potential of these developments in the design of next generation structural materials and/or highly efficient structural components? For example, we can now estimate and measure as well as compute the concentration of hydrogen in the crack tip region in ferritic and austenitic stainless steels because of progress in molecular dynamics and atomic scale ab initio calculations but there is more work needed to bring these developments into engineering practice.

There is a need for various disciplines within the umbrella of FraCSI to coalesce more effectively to deliver implementable methods and analyses tools that can be inserted in the direct path of design of future engineering systems; this will accelerate the design process by minimizing the number of iterations needed and the time between iterations. Similarly, there is an issue on how long it takes to develop and qualify new materials for use in primary structural components. Can this new material development cycle be compressed in time to match with the typical product development cycle? Thus, fracture related topics will remain important in engineering applications for the foreseeable future.

REFERENCES

1. J.J. Dugan, W.H. Fisher, R.W. Buxbaum, A.R. Rosenfield, A.R. Buhr, E.J. Honton, and S.C. McMillan, *Economic Effects of Fracture in the United States: Part 2. A report to NBS by Battelle Columbus Laboratories*, United States, 1983.
2. Failure Knowledge Data Base/100 Selected Cases, "Brittle Fracture of Liberty Ships", http://www.shippai.org/fkd/en/cfen/CB1011020.html
3. S.D. Antolovich, A. Saxena, and W.W. Gerberich, "Fracture Mechanics: An Interpretive Technical History", *Mechanics Research Communications*, Vol. 91, 2018, pp. 46–86.

2 Early Theories of Fracture

Fracture in structural components made from metals has attracted great deal of interest among scientists and engineers for over a century for reasons discussed in the previous chapter. In this chapter, we will consider the early fracture theories that are the foundation on which the contemporary theories of fracture are built. To understand these theories fully, it is important to recall concepts of stress, strain, and elasticity theory. These are briefly reviewed in Appendix 2A at the end of this chapter.

2.1 MICROSCOPIC ASPECTS OF BRITTLE FRACTURE

Prior to launching into the theories of fracture, it is useful to review the physical phenomena that are being modeled. Engineering materials typically consist of numerous crystals with crystals ranging in size from approximately a $100\,nm$, $\sim 10^{-7}\,m$, to several $100\,\mu m$, or $\sim 10^{-4}\,m$.[1] These crystals form aggregates to make up bulk materials used to fabricate engineering components as seen in Figure 2.1 [1]. There is a distribution of crystal sizes in bulk materials; in other words, all crystals in an aggregate are not the same size or shape. Also referred to as grains, individual crystals consist of billions and trillions of atoms, depending on their size, that are arranged in regular crystallographic arrays.

Neighboring crystals have different crystallographic orientations and grain boundaries are regions that accommodate the change in orientation between adjoining grains. Thus, the periodicity of atomic arrangements in the grain boundary region is disrupted. These regions contain stretched and compressed bonds that are necessary to maintain continuity in the metal aggregate. Since atoms in the grain boundary regions are separated by distances that are nonequilibrium, there is strain energy associated with them known as grain boundary energy. The energy state of atoms in the grain boundaries is higher than the energy states in the grain interior. The dislocation density in the grain boundary regions is similarly higher than in the grain interior to accommodate the higher lattice distortions. Such dislocations are known as geometrically necessary dislocations, necessary for maintaining continuity in the material.

2.1.1 Intergranular and Transgranular Fracture

Fracture in metals proceeds along grain boundaries if the grain boundaries are the weak link. Alternatively, fracture proceeds within crystals along crystallographic planes of low atomic density that are relatively weak because they require breaking the fewest primary atomic bonds per unit area of fracture surface. The former type of fracture is called intergranular fracture and the latter transgranular fracture shown schematically in Figure 2.1. In metals at low temperatures, fracture along crystallographic planes is far more common, but that is not always the case as seen in antimony-copper alloys in Figure 2.2a.

DOI: 10.1201/9781003292296-2

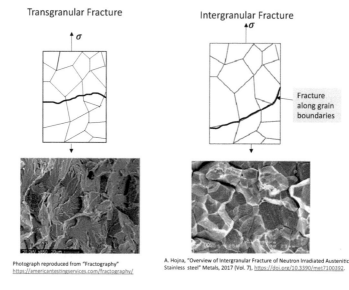

Transgranular Fracture   Intergranular Fracture

Photograph reproduced from "Fractography"
https://americantestingservices.com/fractography/

A. Hojna, "Overview of Intergranular Fracture of Neutron Irradiated Austenitic
Stainless steel" Metals, 2017 (Vol. 7), https://doi.org/10.3390/met7100392.

FIGURE 2.1 Schematics of (a) transgranular fracture occurring along weak crystallographic planes in metals and (b) grain boundary fracture. (Reproduced from A. P. Morowitz, "Fracture Processes in Aerospace Materials", in Introduction to Aerospace Materials, Chapter 18, Science Direct, 2012, pp. 428–453.)

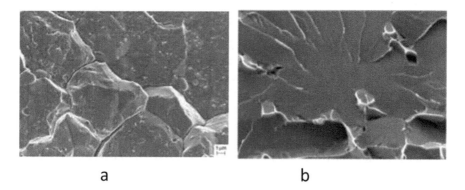

a   b

FIGURE 2.2 Photomicrographs of (a) intergranular fracture in an antimony-copper alloy and (b) transgranular fracture in steel. (Reproduced from J. H. Chen and R. Cao, *Micromechanism of Cleavage Fracture in Metals*, Elsevier Inc., New York, NY, 2014.)

Figure 2.2a shows a high magnification (approximately $5,000x$) picture of intergranular fracture in an antimony-Cu alloy at a low temperature [2]. In these alloys, antimony diffuses and seggregates preferentially to regions near the grain boundaries and makes them weaker in relation to the interior of the grain. Therefore, fracture occurs dominantly along the grain boundaries in these alloys.

Figure 2.2b shows an example of transgrannular fracture along a crystallographic plane in a ferritic steel [2]. Ferritic steels have a body centered cubic structure containing a family of planes, <100>,[2] with low atomic density so the fracture prefers to propagate along those planes. The very flat fracture mode along a crystallographic plane is known as cleavage fracture and can have a mirror-like shiny appearance if it occurs in single crystals or in metals with grains that are large enough in size to be visible to an unaided eye.

2.1.2   EQUI-COHESIVE TEMPERATURE

Intergranular fractures are more common at higher temperatures because in crystalline materials, grain boundary strength degrades at a faster rate with temperature than the interior of the grain. Conversely, transgranular fracture is more common at lower temperatures. Equi-cohesive temperature is defined as the temperature where the grain boundary strength and the strength of the crystal interior are equal. Thus, below equi-cohesive temperature, predominantly transgranular fracture occurs and above that temperature, predominantly intergranular fracture is observed.

2.1.3   DUCTILE AND BRITTLE FRACTURE

The classification of the fracture as intergranular and transgranular was based on atomic scale processes. At a length scale in the range of 0.1–100 mm, fractures are characterized as brittle or ductile. Brittle fracture involves fracture with little plastic deformation of the material and ductile fractures involve large plastic deformation, see Figure 2.3 [3]. Brittle fractures are undesirable because they lead to rapid and catastrophic failure of the structure when a critical load is reached. At magnifications of about 5,000x, brittle fracture surface often shows cleavage facets. While all brittle fractures are not caused by cleavage, the presence of cleavage facets on the fracture surface are often indicative of brittle fracture.

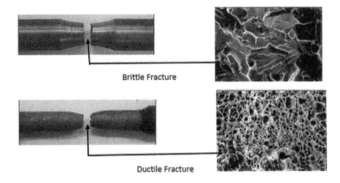

FIGURE 2.3   Appearance of brittle and ductile fractures at macroscopic and microscopic levels [3].

Ductile fractures are accompanied by significant plastic deformation of the material in the fracture region. This typically results in stable fracture and is therefore the preferred failure mode for structures. Continued propagation of fracture requires that the applied load increase. If the load is reduced, or not increased, the fracture stops propagating. At microscopic levels, ductile fracture shows dimples on the fracture surfaces formed by nucleation and growth of voids on second phase particles. It is possible for ductile fracture to end by cleavage, leading to sudden fracture after sustaining some stable crack growth. In these situations, fracture becomes unstable like brittle fracture and the fracture surface will show evidence of both ductile void growth and cleavage facets.

2.2 MODELS OF FRACTURE AT THE ATOMIC SCALE

As mentioned in the previous section, cleavage fracture in metals occurs by breaking of atomic bonds along crystallographic planes as schematically illustrated in Figure 2.4. Intergranular fracture also requires severing atomic bonds in the grain boundary region. Thus, early theories of fracture focused on estimating forces and energies required to break atomic bonds along the fracture plane.

In Figure 2.5, we consider two adjacent atoms in a crystalline solid that are initially separated by an equilibrium distance x_0 between their centers. We designate P as the applied force tending to separate these atoms and P_c as the cohesive or the critical force needed to break the atomic bond. If N is the number of atoms in a surface area A of the crystalline solid, N bonds must be broken for the fracture to spread over the surface area A. Let x be the separation distance between the centers of neighboring atoms, $\sigma = $ stress, $\varepsilon = $ strain, and E the elastic modulus of the material. The top half of Figure 2.5 schematically shows the relationship between the binding potential energy that has a minimum at the equilibrium distance between the neighboring atoms, x_0, and it increases as the two atoms are displaced by the force P, away from x_0 in either direction. The bottom half of Figure 2.5 shows the relationship between the applied force P and the separation distance x. λ is the separation distance at which there is no longer a binding force between the atoms.

If the relationship between P and x is modeled as being sinusoidal, we can write the following relationships:

$$P = P_c \sin \frac{\pi(x - x_0)}{\lambda} \tag{2.1}$$

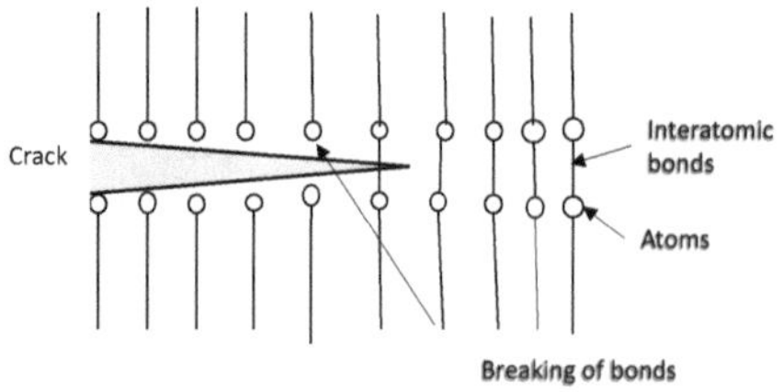

FIGURE 2.4 A schematic showing how atomic bonds are severed as the fracture progresses.

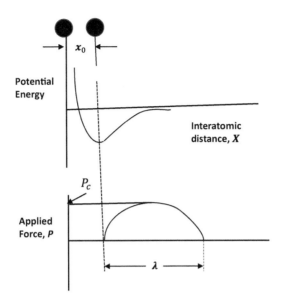

FIGURE 2.5 (a) Potential energy as a function of separation distance between two neighboring atoms and (b) force required to separate the neighboring atoms from their equilibrium distance.

$$E_b = \int_{x_0}^{\lambda+x_0} P dx \qquad (2.2)$$

where E_b is the binding energy between atoms.

For small displacements, $P \approx P_c \dfrac{\pi(x - x_0)}{\lambda}$ and then multiplying and dividing by x_0 we get,

$$\frac{PN}{A} = \frac{P_c N}{A} \frac{\pi(x - x_0)x_0}{\lambda x_0}$$

Noting that $\varepsilon = \dfrac{x - x_0}{x_0}$, we can write $\sigma = \sigma_c \dfrac{\pi x_0}{\lambda} \varepsilon$, where σ_c = cohesive strength of the solid and $\dfrac{\sigma}{\varepsilon} = E = \sigma_c \dfrac{\pi x_0}{\lambda}$ or $\sigma_c = \dfrac{\lambda E}{\pi x_0}$, and recognizing that $\dfrac{\lambda}{x_0} \sim 1$, we get,

$$\sigma_c \sim \frac{E}{\pi} \qquad (2.3)$$

Thus, according to equation (2.3) the theoretical cohesive strength is approximately one-third of the elastic modulus. However, this estimate of cohesive strength is several orders of magnitude higher than the observed strength of materials. For example, for steel E is approximately 200 GPa so the calculated cohesive strength is 66 GPa; but even in the case of strongest steels, the strength is no more than 3 GPa which is approximately 20 times less than the theoretically estimated value from equation (2.3). So, by itself, this model is insufficient for explaining fracture.

An alternate estimate of theoretical strength can be based on the energy expended in creating new surfaces. This energy must be supplied by the external forces. Thus, if γ_s is the energy required to create a unit area of new surface, the energy required for extending the crack by the unit area must be $2\gamma_s$ because two new surfaces of unit area are created during crack formation. The energy balance can be written as:

$$2\gamma_s = \int_{x_0}^{\lambda+x_0} \sigma_c \, \sin\frac{\pi(x-x_0)}{\lambda} dx = \frac{2\lambda}{\pi} \sigma_c \tag{2.4}$$

From the previous derivation, $\sigma_c = \dfrac{E\lambda}{\pi x_0}$

$$\text{Thus,} \quad \sigma_c = \sqrt{\frac{E\gamma_s}{x_0}} \tag{2.5}$$

For illustration if we use $E = 200$ GPa for steels, $x_0 \sim 10^{-9}$ m, and $\gamma_s \approx 10$ J/m^2

$$\sigma_c \sim \sqrt{\frac{200 \times 10^9 \left(\text{N/m}^2\right) \times 10\left(\text{Nm/m}^2\right)}{10^{-9}\,(\text{m})}} \approx 45\,\text{GPa}$$

Thus, the energy balance approach also yields much higher estimates of strength than the observed strength of materials. Both approaches based on force required to break atomic bonds and energy required to create new surfaces yield essentially equivalent results, and both predict much higher than the observed values of strength. Thus, the search for a good theory for fracture continued and led to the notion that defects that are known to exist in materials, must cause reduction in their strength. The theory of stress concentrations around material discontinuities and notches was envisioned to play a significant role in fracture of materials as discussed in the next section.

2.3   STRESS CONCENTRATION EFFECTS OF FLAWS

In 1913, Inglis [4] presented an important analysis recognizing the role of flaws in locally raising the stresses that can potentially cause fracture. He considered an elliptical shaped defect with a major axis of $2a$ and a minor axis of $2b$ and a root radius of ρ in a plate subjected to a remote stress, σ, Figure 2.6. His analysis led to the following relationships:

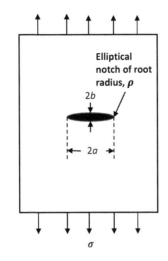

FIGURE 2.6 A semi-infinite plate loaded with a uniform remote stress, σ, with an elliptical notch in its center analyzed by Inglis [5].

$$\rho = \frac{b^2}{a} \quad \text{or} \quad \frac{a}{b} = \sqrt{\frac{a}{\rho}}$$

$$\sigma_A = \sigma\left(1 + 2\sqrt{\frac{a}{\rho}}\right) \tag{2.6}$$

where σ_A = stress at the notch root.

For $a \gg \rho$ such as for sharp cracks or crack-like defects, $\sigma_A \approx 2\sigma\sqrt{\dfrac{a}{\rho}}$

As $\rho \rightarrow 0$, $\sigma_A \rightarrow \infty$. Thus, sharp cracks can raise the local stress level considerably. Theoretically, ρ cannot be smaller than the interatomic distance, x_0, so, $\sigma_A \approx 2\sigma\sqrt{\dfrac{a}{x_0}}$

Accordingly, fracture occurs when $\sigma_A = \sigma_c = \sqrt{\dfrac{E\gamma_s}{x_0}}$. Thus, the remote stress $\sigma = \sigma_f$ the stress at fracture, is given by,

$$\sqrt{\frac{E\gamma_s}{x_0}} = 2\sigma_f\sqrt{\frac{a}{x_0}} \quad \text{or} \quad \sigma_f = \sqrt{\frac{E\gamma_s}{4a}} \tag{2.7}$$

Equation (2.7) provides an estimate of how fracture stress changes with crack size and is based upon the local stress at the crack tip approaching the theoretical cohesive strength of the material rationalizing the previously observed discrepancy between the observed and estimated strengths. However, many questions remain about reliably predicting fracture so, the search for a fracture theory continued.

2.4   GRIFFITH'S THEORY OF BRITTLE FRACTURE

Griffith [5] sought to model the energy exchanges during propagation of brittle fracture by stating that during brittle fracture, the external forces perform the work to provide the surface energy needed to create new surfaces and the changes in elastic strain energy of the cracked body, Figure 2.7a. He pointed out that in planar bodies as the crack forms, strain energy must be released in the regions above and below the crack faces because no stresses can be transmitted across the crack. This allowed him to estimate, using Inglis' analysis, the change in strain energy as the crack size increases. The energy balance during incremental crack growth, Δa, can be written as:

$$\Delta F = \Delta \Pi + \Delta W_s$$

where ΔF = work done by the external force, $\Delta \Pi$ = change in potential energy of the system which for isothermal bodies is the same as the change in elastic strain energy, and ΔW_s = the surface energy needed to create new surfaces associated with the crack size increment. If the thickness of the planar cracked body is B and the area of the new crack surface is ΔA. Then,

$$\frac{\Delta F}{\Delta A} = \frac{\Delta \Pi}{\Delta A} + \frac{\Delta W_s}{\Delta A}; \quad \Delta A = 2B\Delta a \tag{2.8}$$

where a = crack size, Δa = increment in crack size and the factor 2 appears because there are two crack tips, and each grows by Δa.

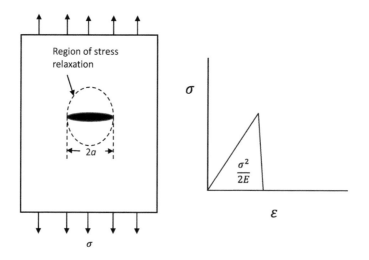

FIGURE 2.7   (a) Griffith's formulation of a crack problem in a semi-infinite plate loaded with a remote stress, σ, showing a region of stress relaxation due to the presence of the crack and (b) strain energy per unit volume in a linear-elastic body represented by the area under the stress–strain diagram.

If Π is the potential energy of the cracked body with a crack size $2a$ and Π_0 is the initial potential energy of the un-cracked body, then

$$\Delta\Pi = \Pi - \Pi_0$$

When fracture occurs under isothermal conditions, the potential energy change, $\Delta\Pi$, as mentioned before, is the same as the change in strain energy, ΔU, because other forms of potential energy such as thermal energy remain constant. Therefore, moving forward, we will use change in strain energy ΔU in place of change in potential energy. The strain energy per unit volume, U, is given by the area under the stress–strain curve and is equal to $\dfrac{\sigma^2}{2E}$ as seen in Figure 2.7b. Also, see Appendix A2 for a more detailed discussion of strain energy. After formation of the crack, the region of the cracked body directly above and below the crack becomes unstressed and consequently is unable to store any strain energy. Griffith, based on Inglis' analysis, estimated the strain energy reduction, ΔU, associated with the presence of the crack of length $2a$ is given by equation (2.9):

$$\Delta U = -\frac{\sigma^2 \pi a^2}{E} B \tag{2.9}$$

In other words, a semi-infinite planar body of thickness B, stressed to a remote uniform stress, σ, containing a center crack of length $2a$ stores less strain energy given by equation (2.9) compared to an uncracked body stressed similarly.

Rewriting equation (2.8) replacing Π with U, for a small increment in crack length,

$$\lim_{\Delta a \to 0} \frac{\Delta F}{\Delta A} = \frac{1}{2B}\frac{dF}{da} = \frac{1}{2B}\left(\frac{dU}{da} + \frac{dW_s}{da}\right)$$

$$\frac{d\Delta U}{2Bda} = \frac{dU}{2Bda} = -\frac{\sigma^2}{E}\pi a \tag{2.10}$$

On the other hand,

$$W_s = 2(\text{surfaces}) \times 2aB\gamma_s = 4B\gamma_s a$$

$$\text{and} \quad \frac{1}{B}\frac{dW_s}{da} = 4\gamma_s$$

Griffith further postulated that if the crack growth occurred under the conditions of constant displacement, the external work done, $dF = 0$, then:

$$-\frac{dU}{Bda} = \frac{2\sigma_f^2}{E}(\pi a) = \frac{dW_s}{Bda} = 4\gamma_s$$

Under these conditions, the energy needed to create new surfaces during fracture must be compensated by the reduction in the overall strain energy of the cracked body. We get, fracture stress, σ_f:

$$\sigma_f = \sqrt{\frac{2E\gamma_s}{\pi a}} \tag{2.11}$$

Equation (2.11) yields fracture stress levels that are similar in magnitude to the stress-based approach, equation (2.7). The two equations yield estimates of fracture stress that are similar in form and comparable in magnitudes. Griffith performed his verification experiments on glass because at the time glass was one of the most brittle materials known and showed that his hypothesis was indeed correct. However, several attempts to extend the analysis to metals, even brittle metals, were unsuccessful and Griffith's work remained dormant for more than two decades until modifications to the theory were proposed.

2.5 OROWAN'S MODIFICATION TO GRIFFITH'S THEORY

The modification proposed by Orowan [6] to Griffith's theory was simple but elegant and made a big impact on the acceptance of Griffith's theory by extending its applicability to metals. Orowan's proposal stated that fracture in metals is always accompanied by plastic deformation which is an energy dissipative process much like energy for creation of new surfaces and therefore, it must be included when accounting for energy exchanges taking place during fracture. External forces must then also provide energy for plastic deformation accompanying fracture in addition to the surface energy needed to produce new surfaces. In the case of fracture occurring under fixed grip conditions when the external work done is zero, the release of stored elastic energy must be sufficient to also supply the energy needed for the accompanying plastic deformation during fracture along with the energy for creating new surfaces.

The energy used to cause plastic deformation is dissipated in the form of heat and cannot be recovered. In that sense, it is like surface energy because that too is dissipated and is irrecoverable. Thus, Orowan proposed that the total dissipated energy during the fracture process consists of (a) energy needed to form new surfaces and (b) energy needed for plastic deformation in the surrounding region accompanying fracture. This can be expressed as:

$$W_f = W_s + W_p$$

where W_f is the total energy for fracture, W_s is the surface energy, and W_p is the energy for plastic deformation. Expressing the above equation in terms of unit areas, we can write:

$$\gamma_f = \gamma_s + \gamma_p$$

where γ_f is the total energy per unit area of crack surface dissipated during fracture, γ_s = surface energy as defined before, and γ_p is the energy per unit area of crack

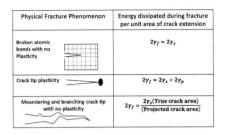

Physical Fracture Phenomenon	Energy dissipated during fracture per unit area of crack extension
Broken atomic bonds with no Plasticity	$2\gamma_f \approx 2\gamma_s$
Crack tip plasticity	$2\gamma_f \approx 2\gamma_s + 2\gamma_p$
Meandering and branching crack tip with no plasticity	$2\gamma_f = \dfrac{2\gamma_s (\text{True crack area})}{(\text{Projected crack area})}$

FIGURE 2.8 Physical crack tip phenomena at the crack tip and the associated energy dissipation [7].

surface associated with the accompanying plastic deformation. Thus, equation (2.11) can be rewritten as:

$$\sigma_f = \sqrt{\frac{2E\left(\gamma_s + \gamma_p\right)}{\pi a}} \tag{2.12}$$

In metals, it was noted that the plastic energy dissipated is much larger than the surface energy and in very brittle materials such as ceramics and glasses the reverse is true. This is schematically shown in Figure 2.8.

Figure 2.8 [7] also includes the case of a branched crack that meanders as it propagates through different grains of varying orientations. In such cases, the actual crack area consists of the sum of the areas of the various segments which is larger than the projected area of the crack on the nominal crack plane. Thus, a meandering crack can absorb more energy than energy needed to grow a flat crack that is confined to the nominal crack plane, and "appear" to have a higher toughness.

2.6   THE CONCEPT OF CRACK EXTENSION FORCE, G

To generalize the Griffith's fracture theory, we introduce the concept of crack extension force, G, which in the thermodynamic sense, is the free energy available for crack extension. It is also known as the Griffith's Crack Extension Force. When the available energy becomes equal to or exceeds the energy needed for fracture, spontaneous crack growth is expected to occur. This condition is both necessary and sufficient for fracture. As already mentioned, the energy needed to advance fracture by a unit area is $2\gamma_f$. From rearranging equation (2.8), we get for fracture under isothermal conditions, when $\Delta\Pi = \Delta U$:

$$G = \frac{1}{B}\left(\frac{dF}{da} - \frac{d\Pi}{da}\right) = \frac{1}{B}\left(\frac{dF}{da} - \frac{dU}{da}\right) \tag{2.13}$$

In equation (2.13), all the terms that contribute to the free energy for fracture are collected as G. Until fracture conditions are attained, G represents the potential for fracture. Also note that if strain energy change (ΔU) is negative, as in the case of fracture under constant displacement conditions, the strain energy is released to act as the crack driving force. However, if there is a net increase in strain energy of the body as crack extends, the work done by the external force must provide for that increase and also the energy needed for fracture. The critical value of G is also called G_c that is equal to $2\gamma_f$ because two surfaces are created when the crack grows by a unit area. This is analogous to stress and yield strength; until stress exceeds the yield strength, plastic yielding does not occur, and stress only represents the tendency for yielding. Similarly, until G exceeds G_c fracture does not occur and G represents the tendency for fracture. For the Griffith's problem including Orowan's modification, we can write

$$G_c = \frac{\sigma_f^2 \pi a}{E} = 2\gamma_f \qquad (2.14)$$

Thus, the crack extension force for this problem in which $\dfrac{dF}{Bda} = 0$, is given by:

$$G = -\frac{dU}{Bda} = \frac{\sigma^2 \pi a}{E} \qquad (2.15)$$

The formula for calculating G such as equation (2.15) will be different for problems other than the Griffith's problem. This will be discussed next. The Griffith's problem had simplifying assumptions that the crack was in the center of a uniformly stressed semi-infinite plate and that the fracture occurred under the fixed displacement conditions. This is a very special case and in applications we will need a much more general formulation for G, as presented next.

2.6.1 ESTIMATION OF GRIFFITH'S CRACK EXTENSION FORCE FOR AN ARBITRARY SHAPED BODY

A planar cracked body with a uniform thickness, B, subjected to a point force P and a load-line displacement, Δ is shown in Figure 2.9. The relationship between the total displacement, Δ, the displacement due to the crack, Δ_c, and the displacement of an uncracked body, $\Delta_{no\ crack}$, is given by equation (2.16) below:

$$\Delta = \Delta_c + \Delta_{no\ crack} \qquad (2.16)$$

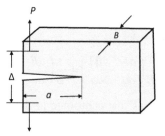

FIGURE 2.9   An arbitrary shaped cracked body loaded by point loads normal to the crack surface.

Since the region above and below the crack is mostly stress free, $\Delta_{\text{no crack}}$ is negligible and $\Delta_c \approx \Delta$. The relationship between the load, P, and the displacement, Δ, for linear-elastic body is shown in Figure 2.10 for varying crack sizes. The inverse of the slope is called compliance, C, that is constant for a fixed crack size but increases as the crack size increases and is given by:

$$C = \frac{\Delta}{P}$$

If we increment the crack by an amount equal to Δa under the conditions of fixed load P, the load-displacement relationships for the two crack sizes can be given by Figure 2.11. The strain energy designated by U is given by the area under the respective load-displacement diagram, and

$$\Delta U = \left(U_{a+\Delta a} - U_a\right) \tag{2.17}$$

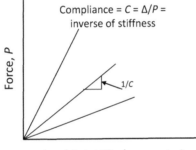

FIGURE 2.10 Relationship between load and load-line displacement for bodies of various crack sizes.

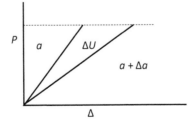

FIGURE 2.11 Load versus load-line displacement for a crack of length a and for a crack of length, $a + \Delta a$ while the increment in the crack length occurs under the conditions of constant load.

In this case, ΔU is given by the area enclosed between the two $P-\Delta$ lines corresponding to crack sizes a and $a+\Delta a$ in Figure 2.11. We also know that the external work done ΔF is given by,

$$\Delta F = P\left(\Delta_{a+\Delta a} - \Delta_a\right)$$

$$\Delta U = \frac{P}{2}\left(\Delta_{a+\Delta a} - \Delta_a\right)$$

$$G = \frac{1}{B}\left(\frac{dF}{da} - \frac{dU}{da}\right)$$

$$G = \frac{P}{2B}\left(\frac{\partial\Delta}{\partial a}\right)_P = \frac{P^2}{2B}\frac{dC}{da} \tag{2.18}$$

Next, we consider crack growth under the conditions of fixed displacements as shown in Figure 2.12.

In the case of constant displacement,

$$\frac{dF}{da} = 0, \quad \text{so}$$

$$G = -\frac{1}{B}\left(\frac{dU}{da}\right)_\Delta = -\frac{\Delta}{2B}\frac{\partial P}{\partial a} = \frac{P^2}{2B}\frac{dC}{da}$$

Note that, compliance is only a function of crack size for a planar body with constant thickness, therefore, $\frac{\partial C}{\partial a} = dC/da$. We see that under both conditions of crack extension under fixed displacement and fixed load,

$$G = \frac{P^2}{2B}\frac{dC}{da} \tag{2.19}$$

The full derivations of equations (2.18) and (2.19) are left as a homework exercise.

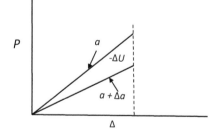

FIGURE 2.12 Load versus load-line displacement when the crack extension occurs under the conditions of fixed displacement.

It also follows that for the conditions of fracture occurring under constant load and constant displacement and recognizing that potential energy, Π, for isothermal conditions is equal to the strain energy, U

$$G = \left(\frac{d\Pi}{Bda}\right)_P = \left(\frac{dU}{Bda}\right)_P = -\left(\frac{d\Pi}{Bda}\right)_\Delta = -\left(\frac{dU}{Bda}\right)_\Delta \qquad (2.20)$$

Equation (2.20) can be used to compute the value of G using finite element analysis by a technique called the virtual crack extension. In this technique, the strain energy states of bodies with crack length a, and that of crack length $a + \Delta a$ are compared to determine the value of the crack extension force. Because of these relationships, the crack extension force is also often known as the strain energy release rate. It also leads to an experimental method for determining the value of G, as illustrated in Figure 2.13. The steps consist of taking several specimens that are identical in all respects except have different crack lengths, $a_1, a_2, a_3.....a_i$. Each specimen is loaded with a force of sufficient magnitude to characterize the $P - \Delta$ trends but not sufficient to cause the cracks to grow or cause a large plastic zone in front of the crack tip. Next, the compliance, C, of each specimen is calculated and plotted against crack size. This relationship can then be expressed in a nondimensional form by plotting CBE against a/W. Then G is calculated using equation (2.21) below:

$$G = \frac{P^2}{2B}\frac{dC}{da} = \frac{P^2}{2B}\frac{1}{BEW}\frac{dCBE}{d\left(\frac{a}{W}\right)} = \frac{1}{2}\left(\frac{P}{B}\right)^2 \frac{1}{EW}\left(\frac{d(CBE)}{d\left(\frac{a}{W}\right)}\right) \qquad (2.21)$$

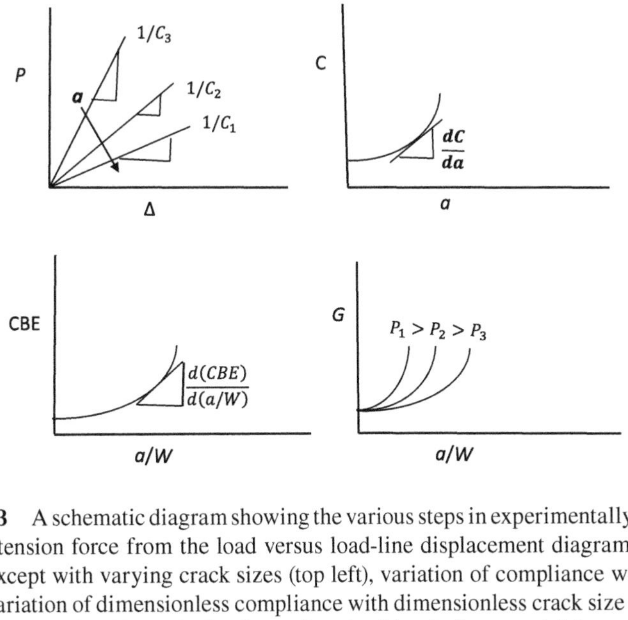

FIGURE 2.13   A schematic diagram showing the various steps in experimentally determining the crack extension force from the load versus load-line displacement diagram for identical specimens except with varying crack sizes (top left), variation of compliance with crack size (top right), variation of dimensionless compliance with dimensionless crack size (bottom left), and the variation of G with crack size for various load levels (bottom right).

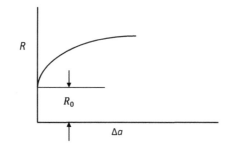

FIGURE 2.14 Crack growth resistance curve for ductile materials that exhibit stable crack growth with the resistance rising with crack growth.

Note that by presenting the results in a nondimensional form, the relationship formulated in equation (2.21) can be applied to cracked bodies of a given geometry made from any linear-elastic material and of any thickness, and size, W, that maintains proportionality among the other in-plane dimensions. In other words, we can independently choose the thickness, B, and the width W of the body but all the other in-plane dimensions, except the crack size, must be proportioned to W. In other words, equation (2.21) applies to small and large bodies of the same in-plane geometry.

2.7 CRACK GROWTH RESISTANCE, *R*

Crack growth resistance, R, is a material property related to the energy needed to extend the crack by a unit area and is equal to $2\gamma_f$. Crack growth resistance for ductile and brittle materials can be contrasted as follows.

If a is the current crack size, and a_0 the original crack size, then $\Delta a = a - a_0$. In ductile materials, the resistance to crack growth rises with crack extension and can be written as $R = R(\Delta a)$ and schematically shown in Figure 2.14. Equation (2.22) is used to describe the crack growth resistance:

$$R = R_0 + C_1 (\Delta a)^{C_2} \tag{2.22}$$

where R_0, C_1, and C_2 are regression constants derived from experimental data.

In brittle materials, fracture occurs suddenly when the applied value of G becomes equal to G_c or $2\gamma_f$ as seen in Figure 2.15. The fracture resistance in brittle materials is characterized by a single parameter, G_c because once the crack begins to grow, it immediately becomes unstable.

2.8 PREDICTING INSTABILITY IN CRACKED STRUCTURES

Unstable crack growth in a structure occurs when fast-fracture begins and the crack propagation is uncontrolled. In brittle materials with a flat R-curve, fast fracture occurs as soon as the critical value of the crack extension force is reached regardless of the size and geometry of the structure. In ductile materials, the fracture behavior of structures made of the same material but of different geometry can vary considerably

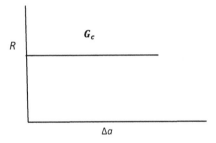

FIGURE 2.15 Unstable crack growth behavior typical of brittle materials in which the crack growth resistance remains flat with crack extension.

depending on the size, geometry, and the loading characteristics of the structure. For example, if we consider a structure being subjected to a monotonically increasing load, fast fracture is expected to occur soon after the crack begins to grow, with a small amount of stable crack growth. On the other hand, if the structure is loaded by slowly increasing displacement, we could get substantial amounts of stable crack growth prior to the fast fracture. Thus, a generalized theory for predicting stable crack growth and unstable fracture is needed. The conditions for stable and unstable fracture are stated as follows.

For stable crack growth to occur, the following two conditions must be satisfied:

$$G = R \tag{2.23a}$$

$$\left(\partial G/\partial a\right)_{P,\Delta} < \left(dR/d(\Delta a)\right) \tag{2.23b}$$

The right hand of equation (2.23b) is a total derivative because the crack growth resistance is a material property that is only a function of the amount of crack extension, Δa. The fracture becomes unstable when:

$$G \geq R \tag{2.24a}$$

$$\left(\partial G/\partial a\right)_{P,\Delta} \geq \left(dR/d(\Delta a)\right) \tag{2.24b}$$

Next, we will apply these conditions for stable/unstable fracture under the conditions of fixed load and fixed displacement, respectively. We will see how stable crack growth is achieved much more readily under displacement-controlled conditions than under load-controlled conditions as illustrated in Figures 2.16 and 2.17, respectively. Figure 2.16 shows how the crack extension force changes with crack size for different levels of applied loads. Superimposed on those curves is the R-curve of the material that shows ductile behavior. For load P_1, the crack grows by an amount equal to Δa_1 and then stops growing at that point because G becomes less than R, so the fracture is stabilized. At load P_2, the crack grows in a stable fashion by Δa_2 but when it gets

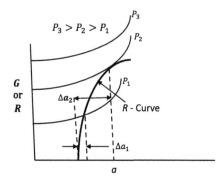

FIGURE 2.16 Crack driving force curves for various load levels and the crack growth resistance curves.

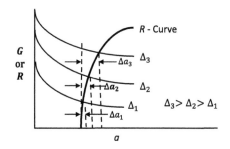

FIGURE 2.17 Crack driving force as a function of crack size for constant deflection levels and the crack growth resistance curve.

there, the condition of equation (2.24b) is satisfied and the fracture is expected to become unstable at that point. For load P_3, unstable fracture occurs right from the beginning because G always exceeds R and $\partial G/\partial a$ is greater than dR/da before the G versus a curve even meets the R-curve.

Figure 2.17 shows the relationship between G and the crack size for different values of fixed displacements, Δ_1, Δ_2, Δ_3, and the superimposed R-curve. In each case, it is expected that the cracks will grow in a stable fashion by increments of Δa_1, Δa_2, and Δa_3 respectively, and then stop growing when G becomes less than R. Thus, in general, displacement-controlled loading provides more stability than load-controlled conditions. This is illustrated mathematically by the following example.

Example Problem 2.1

A material exhibits the following crack growth resistance behavior:

$$R = 50 + 200(a - a_0)^{0.5}$$

where a = crack size and a_0 = initial crack size. R is in units of kJ/m² and crack size is in mm. The elastic modulus of the material is 207,000 MPa. Consider a semi-infinite plate subjected to remote uniform stress σ and a crack of length $2a$ that is located at the center of the plate; thus, a represents half crack size.

a. If the plate fractures at 500 MPa, compute the following:
  I. Half crack size, a_f at failure.
  II. The amount of stable crack growth (at each crack tip) that precedes failure.
b. If this plate has an initial crack length ($2a_0$) of 50.8 mm and the plate is loaded to failure, compute the following:

  I. The stress at failure.
  II. The half crack size at failure.
  III. The stable crack growth at each crack tip.

To be consistent in units, we take stress and E in N/m² or in Pa, R in J/m² and crack size in m, the crack growth resistance equation can be re-written as:

$$R = \left(50 + 200(1,000)^{0.5}(a - a_0)^{0.5}\right)1,000$$

$$= 5 \times 10^4 + 6.324 \times 10^6 (a - a_0)^{0.5}$$

The critical conditions for unstable fracture to occur are,

$$G = R \quad \text{and} \quad \frac{dG}{da} = \frac{dR}{da}$$

Thus, G at fracture is given by,

$$G = R = \frac{\sigma_f^2 \pi a_f}{E}$$

where σ_f = stress at fracture. Thus,

$$\frac{(500 \times 10^6)^2 \pi a_f}{207 \times 10^9} = 5 \times 10^4 + 6.324 \times 10^6 (a_f - a_0)^{0.5}$$

$$a_f = \frac{5 \times 10^4 + 6.324 \times 10^6 (a_f - a_0)^{0.5}}{\dfrac{(500 \times 10^6)^2 \pi}{207 \times 10^9}}$$

$$a_f = 0.01318 + 1.671(a - a_0)^{0.5}$$

Also,
$$\frac{dG}{da} = \frac{dR}{da}$$

$$\frac{\sigma_f^2 \pi}{E} = 3.792 \times 10^6 = 3.162 \times 10^6 \left(a_f - a_0\right)^{-0.5}$$

$$\left(a_f - a_0\right)^{-0.5} = 1.1992$$

$$\left(a_f - a_0\right) = 0.695 \text{ m}$$

or
$$a_f = 0.01318 + 1.67(0.695)^{0.5} = 1.4 \text{ m}$$

and
$$a_0 = 0.70 \text{ m}$$

Thus, the amount of stable crack extension=0.695 m. These numbers are high indicating that fracture in the component at a stress of 500 MPa is unlikely.

b. If $2a_0 = 50.8$ mm $= 0.0508$ m or $a_0 = 0.0254$ m

Invoking the two conditions for fracture for this situation,

$$\frac{\left(\sigma_f\right)^2 \pi a_f}{207 \times 10^9} = 5 \times 10^4 + 6.324 \times 10^6 \left(a_f - a_0\right)^{0.5}$$

and $\dfrac{\sigma_f^2 \pi}{E} = 3.162 \times 10^6 \left(a_f - a_0\right)^{-0.5}$

thus, $3.162 \times 10^6 \left(a_f - a_0\right)^{-0.5} a_f = 5 \times 10^4 + 6.324 \times 10^6 \left(a_f - a_0\right)^{0.5}$

or $a_f = 0.0158\left(a_f - a_0\right)^{0.5} + 2\left(a_f - a_0\right)$

or $-a_f = 0.0158\left(a_f - a_0\right)^{0.5} - 2a_0$

or $0.0158\left(a_f - 0.0254\right)^{0.5} + a_f - 0.0508 = F\left(a_f\right) = 0$

We can solve this equation iteratively as follows:

a_f	$F(a_f)$
0.03	−0.0197
0.04	−0.0089
0.045	−0.003588
0.048	−0.00042
0.0485	0.000101

Form the above table, $F(a_f)$ is approximately 0 when, $a_f \cong 0.0484$ m

Also, $\sigma_f = \sqrt{\left(\dfrac{E}{\pi} 3.162 \times 10^6 \times (0.0484 - 0.0254)^{0.5}\right)} = 1172.4 \,\text{MPa}$

The amount of stable crack growth at each crack tip $= (0.0484 - 0.0254) = 0.023$ m $= 23$ mm.

Example Problem 2.2

A double cantilever beam specimen is shown in Figure 2.18 is loaded with a point load, P. Find an expression for estimating G for this configuration.

Assuming the crack tip, because of symmetry about the horizontal axis, will remain fixed upon application of the load. From elementary mechanics of materials [8]:

For a cantilever beam, $\quad \Delta = \dfrac{2Pl^3}{3EI} = \dfrac{2Pa^3}{3EI}; \quad I = \dfrac{BH^3}{12}$

$$C = \frac{\Delta}{P} = \frac{2a^3}{3EI}$$

$$\Delta = \frac{8P}{EB}\left[\frac{a}{H}\right]^3$$

$$\frac{dC}{da} = \frac{6a^2}{3EI} = \frac{24a^2}{EBH^3}$$

$$G = \frac{P^2}{2B}\frac{dC}{da} = \left[\frac{P}{B}\right]^2 \frac{12a^2}{EH^3} = \frac{3EH^3}{16}\frac{\Delta^2}{a^4}$$

From the above equation, for constant load conditions, the value of G increases rapidly with crack size, leading to conditions that will result in unstable fracture. On the other hand, for constant displacement conditions, the value of G decreases rapidly with crack size, leading to conditions that will stabilize fracture.

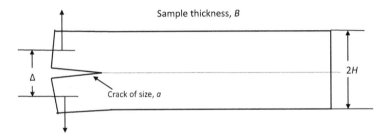

FIGURE 2.18 Double cantilever beam specimen.

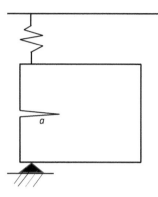

FIGURE 2.19 A cracked body in a loading train between two fixed ends with varying stiffness simulated by varying the spring constant.

2.8.1 PREDICTING INSTABILITY CONDITIONS FOR A GENERAL CASE

The conditions derived in Example 2.2 for a simple specimen geometry can be generalized for any geometry as shown in the derivation that follows.

To illustrate how conditions between the extremes of monotonically increasing load and those of gradually rising displacement rates affect fracture instability, we consider the configuration shown in Figure 2.19 where a cracked body is attached to a long loading train between two fixed ends. To vary the stiffness of the loading train we insert a spring. The stiffness of the spring can be made very high to simulate displacement-controlled conditions to one of low stiffness that will simulate the load-controlled conditions.

$$\Delta_T = C_M P + \Delta(aP)$$

where C_M=Machine compliance, Δ_T=total displacement and Δ=displacement due to crack

$$d\Delta_T = 0 = C_M dP + \left(\frac{\partial \Delta}{\partial a}\right)_P da + \left(\frac{\partial \Delta}{\partial P}\right)_a dP$$

$$\left(\frac{dP}{da}\right)_{\Delta_T} = -\frac{\left(\dfrac{\partial \Delta}{\partial a}\right)_P}{C_M + \left(\dfrac{\partial \Delta}{\partial P}\right)_a}$$

$$G = G(a,P)$$

$$dG = \left(\frac{\partial G}{\partial a}\right)_P da + \left(\frac{\partial G}{\partial P}\right)_a dP$$

$$\frac{dG}{da} = \left(\frac{\partial G}{\partial a}\right)_P - \left(\frac{\partial G}{\partial P}\right)_a \frac{\left(\frac{\partial \Delta}{\partial a}\right)_P}{C_M + \left(\frac{\partial \Delta}{\partial P}\right)_a}$$

$$\frac{dG}{da} = \left(\frac{\partial G}{\partial a}\right)_P - \left(\frac{\partial G}{\partial P}\right)_a \frac{\left(\frac{\partial \Delta}{\partial a}\right)_P}{C_M + C} \tag{2.25}$$

For very long structures, $C_M \to \infty$ so $\dfrac{dG}{da} = \left(\dfrac{\partial G}{\partial a}\right)_P$ which is the same as constant load conditions. For very stiff structures, $\dfrac{dG}{da} = \left(\dfrac{\partial G}{\partial a}\right)_P - \left(\dfrac{\partial G}{\partial P}\right)_a \dfrac{1}{C}\left(\dfrac{\partial \Delta}{\partial a}\right)_P$; the first term on the right-hand side is positive and all quantities in the second term are positive and increase rapidly with crack size, therefore G is expected to decrease with crack size and stabilize the fracture.

2.9 SUMMARY

In this chapter, the mechanisms of fracture and the early theories upon which the contemporary theories of fracture mechanics are based, are discussed. We defined phenomena such as intergranular and transgranular fracture at the atomic and microscopic levels and the meaning of terms such as brittle and ductile fracture at the macro level. We also observed that transgranular cleavage fracture leads to brittle behavior. Ductile behavior, on the other hand, is associated with the formation and growth of microvoids and fracture occurs by coalescence of voids.

We approached fracture by estimating the stress required to break atomic bonds along the fracture plane called the cleavage plane. The fracture stress thus derived overestimated the measured fracture stress in engineering materials by approximately one to two orders of magnitude. Fracture of atomic bonds was also modeled as the energy required to overcome the binding energy between neighboring atoms. This too yielded estimates of fracture stress that were high and in fact comparable to those derived from the stress-based approach. To rationalize this large discrepancy between theoretical estimates of fracture stress and the experimentally measured values, it was necessary to invoke the presence of crack like defects that greatly intensify the stress at the crack tip allowing for stresses to build up to the level necessary to break atomic bonds.

The Griffith–Orowan theory of brittle fracture that accounts for the role of defects in the fracture process was described in its original and its modified form. It was shown to quantitatively and realistically describe the brittle fracture phenomenon based on energy exchanges that take place during the fracture process. The concept of the crack extension force, G, to generalize the Griffith's theory to cracked bodies of arbitrary shape was described and was shown to have the capability to predict fracture in components containing cracks. It was shown that G in all instances was related to the change in strain energy of the cracked body and therefore can be determined by comparing the strain energy of identical bodies but incrementally differing crack sizes. Thus, a parameter that is computed using global stresses, strains, and strain energy was shown to predict fracture from crack tips in brittle materials. The latter restriction on the theory was based on the consideration that the underlying theory used for estimating stress, strain and strain energy was based on the theory of linear elasticity. The theory predicts that as the magnitude of crack extension force, G, equals or exceeds the critical value G_c, fracture is expected to occur.

The last part of the chapter was dedicated to establishing the conditions for stable and unstable fracture. It was shown that the onset of unstable fracture depends on the loading configuration such as monotonically increasing load or displacement and on the size and geometry of the cracked body.

REFERENCES

1. A.P. Morowitz, "Fracture Processes in Aerospace Materials", in *Introduction to Aerospace Materials, Chapter 18*, 2012, Science Direct, Elsevier BV, 2022, pp. 428–453.
2. J.H. Chen and R. Cao, *Micromechanism of Cleavage Fracture in Metals*, Elsevier Inc., New York, NY, 2014.
3. D. Francois, "Brittle Fracture", in *Handbook of Materials Behavior Models*, 2001, Elsevier Inc., New York, NY, pp. 56–576.
4. C.E. Inglis, "Stresses in a Plate Due to Presence of Cracks and Sharp Corners", *Transactions of the Institute of Naval Architecture*, Vol. 55, 1913, pp. 219–241.
5. A.A. Griffith, "The Phenomena of Flow and Rupture in Solids", *Philosophical Transactions, Series A*, Vol. 221, 1921, pp. 163–197.
6. E. Orowan, "Fracture and Strength of Solids", *Reports on Progress in Physics*, Vol. XII, 1948, pp. 185–232.
7. T. L. Anderson, *Fundamentals of Fracture Mechanics*, 2nd Edition, CRC Press, London, England, 1996.
8. F.P. Beer, E.R. Johnston, Jr., J.T. Wolf, and D.F. Mazurek, *Mechanics of Materials*, 7th Edition, McGraw-Hill Publishers, New York, NY, 2015.

HOMEWORK PROBLEMS

1. According to the Griffith's theory for brittle fracture, the energy required to increase the crack area by a unit amount is equal to twice the energy per unit area required for creating new surfaces. Why is the factor of 2 needed?
2. What is the Griffith's crack extension force, G?
3. Show on your own that the Griffith's Crack Extension Force G, for crack extension under both displacement-control and under load-control:

$$G = \frac{P^2}{2B} \frac{dC}{da}$$

4. A material exhibits the following crack growth resistance curve; $R = 10 + 10(a - a_0)^{0.5}$ where a_0 = initial half crack size and a = current half crack size. In the above equation, the crack size is in mm and R in kJ/m^2.

If a semi-infinite plate with a central crack of length $2a$ fractures at 200 MPa, calculate (a) half crack size at fracture and (b) amount of stable crack growth at fracture. The Young's modulus, E, of the material is 200 GPa. The Griffith's crack extension force, G, for this configuration is given by:

$$G = \frac{\sigma^2 \pi a}{E}$$

5. A crack is found in the middle region of the long femur bone in an adult. Based on what you have learnt about stable and unstable fracture, discuss the merits of surgically implanting a stainless-steel plate across the fracture.
6. Explain why the terms crack extension force and energy release rates are used interchangeably in the literature?
7. Concrete by itself is brittle. Why is it more suitable for applications involving compressive loading but not for tensile loading? What approaches would you recommend for building its fracture resistance under tensile loads?
8. A large, uniformly stressed plate containing a circular hole of radius 25 mm in the middle has straight slot on one side of the hole that is normal to primary stress direction. If the overall length of the slot is 5 mm and its root radius is 1 mm, estimate the stress concentration factor at the root of the slot. If the slot is long relative to the diameter of the hole, what is the estimated stress concentration factor?

APPENDIX 2A: REVIEW OF SOLID MECHANICS

To understand the material covered in the book, a good working knowledge of solid mechanics is essential. The brief overview provided here is to introduce the reader to conventions, symbols, and the specific concepts of solid mechanics used in this book. It is expected that the reader is already familiar with mechanics of deformable bodies so for brevity, the important relationships in solid mechanics are merely stated without providing proofs. Those that desire to see the proofs are referred to other books on the subject. An example of such a book is by Beer, Johnston, DeWolf, and Mazurek [8].

2A.1 STRESS

Stress is defined as force-per-unit area. Imagine a force ΔF acting at a point, P, which lies on an area ΔA as shown in Figure 2A.1a. Reducing the force into components that are normal and tangential to ΔA defines the normal stress (σ) and shear stress (τ) as:

$$\lim_{\Delta A \to 0} \sigma = \frac{\Delta F_n}{\Delta A} \quad \text{and} \quad \tau = \frac{\Delta F_t}{\Delta A} \tag{2A.1}$$

where ΔF_n and ΔF_t are normal and tangential components of force as shown in Figure 2A.1b. In the most general case, each face of a cube may be subjected to an

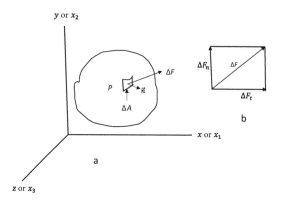

FIGURE 2A.1 Elemental area ΔA subjected to (a) force, ΔF and (b) normal and tangential components of the force.

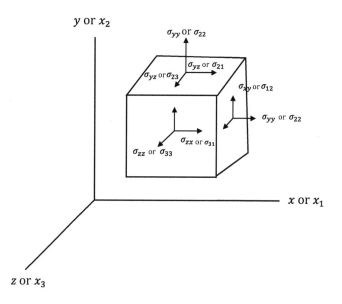

FIGURE 2A.2 Components of stress acting on an element. Also shown are the positive directions of the stress components.

arbitrary force. However, each of these forces may be resolved into components that are parallel to the three coordinate axes. One of these on each face will be normal and the other two will be tangential resulting in nine components. If each of these nine components are divided by the area of the face upon which they act, nine components of stress will result as shown in Figure 2A.2. The first subscript in describing the stress component refers to the direction normal to the plane and the second subscript to the direction of the stress component. These are shown for the (x, y, z) or the (x_1, x_2, x_3) designations of the coordinate system. σ_{xx} (or σ_{11}) arises from a force acting in the positive x-direction (or x_1) on a plane with a normal also in the positive x direction. When both subscripts are either positive or negative, the stress is designated to be positive or tensile. A positive-negative (or vice-versa) suffix would indicate a negative (or compressive) stress. Also, if the stress state is uniform, a normal stress of the same magnitude must act on the positive x-plane as on the negative x-plane. Since the direction of the stress on the negative x-plane is negative, the stress will be positive (or tensile). This collection of the nine components of stresses is called the stress tensor, designated as σ_{ij}, and can be expressed as in equation (2A.2).

$$\sigma_{ij} = \begin{pmatrix} \sigma_{xx} & \sigma_{yx} & \sigma_{zx} \\ \sigma_{xy} & \sigma_{yy} & \sigma_{zy} \\ \sigma_{xz} & \sigma_{yz} & \sigma_{zz} \end{pmatrix} \qquad (2A.2)$$

where i, j are iterated over x, y, z, respectively. Two identical subscripts indicate a normal stress while differing subscripts indicate a shear stress. Normal stresses are often designated by a single subscript and shear stress by τ, with both subscripts:

$$\sigma_{xx} \equiv \sigma_x \qquad (2A.3a)$$

$$\sigma_{xy} \equiv \tau_{xy} \qquad (2A.3b)$$

If the element being considered is in equilibrium, it leads to the result that:

$$\sigma_{ij} = \sigma_{ji} \quad \text{or} \quad \sigma_{xy} = \sigma_{yx} \qquad (2A.4)$$

Thus, only six of the nine stress components are independent. Also, to maintain equilibrium, the following equations must be satisfied:

$$\frac{\partial \sigma_{xx}}{\partial x} + \frac{\partial \sigma_{yx}}{\partial y} + \frac{\partial \sigma_{zx}}{\partial z} = 0 \qquad (2A.5a)$$

$$\frac{\partial \sigma_{xy}}{\partial x} + \frac{\partial \sigma_{yy}}{\partial y} + \frac{\partial \sigma_{zy}}{\partial z} = 0 \qquad (2A.5b)$$

$$\frac{\partial \sigma_{xz}}{\partial x} + \frac{\partial \sigma_{yz}}{\partial y} + \frac{\partial \sigma_{zz}}{\partial z} = 0 \qquad (2A.5c)$$

In the contracted tensor notations, the above equations can be written as:

$$\frac{\partial \sigma_{ij}}{\partial x_j} = 0 \qquad (2A.5d)$$

where i and j vary as 1, 2, 3.

The resultant force-per-unit area is called the traction vector T that differs from stress in that it is a vector with a defined direction. The three components of T in the x, y, and z directions are T_1, (or T_x), T_2 (or T_y), and T_3 (or T_z) as follows:

$$T = T_1 i + T_2 j + T_3 k \tag{2A.6}$$

where i, j, k are unit vectors in the x, y, z directions. The various components of the traction vector can be related to the components of stress as:

$$T_1 = \sigma_{11} n_1 + \sigma_{21} n_2 + \sigma_{31} n_3 \tag{2A.7a}$$

$$T_2 = \sigma_{12} n_1 + \sigma_{22} n_2 + \sigma_{32} n_3 \tag{2A.7b}$$

$$T_3 = \sigma_{13} n_1 + \sigma_{23} n_2 + \sigma_{33} n_3 \tag{2A.7c}$$

where n_1, n_2, and n_3 are the three direction cosines associated with the outward normal n for the plane of the traction vector. In the short index notation, equation (2A.7) can be expressed as:

$$T_i = \sigma_{ij} n_j \tag{2A.8}$$

At every point in a body, there exists a set of three orthogonal planes called the principal planes on which the traction vector lies normal to the plane and thus, no shear stresses exist on these planes. The three principal stresses can be determined by solving the following cubic equation (2A.9):

$$\sigma^3 - I_1 \sigma^2 + I_2 \sigma - I_3 = 0 \tag{2A.9}$$

$$I_1 = \sigma_{11} + \sigma_{22} + \sigma_{33}$$

$$I_2 = \left(\sigma_{11}\sigma_{22} + \sigma_{22}\sigma_{33} + \sigma_{33}\sigma_{11} - \sigma_{12}^2 - \sigma_{23}^2 - \sigma_{31}^2 \right)$$

$$I_3 = \begin{vmatrix} \sigma_{11} & \sigma_{12} & \sigma_{13} \\ \sigma_{21} & \sigma_{22} & \sigma_{23} \\ \sigma_{31} & \sigma_{32} & \sigma_{33} \end{vmatrix}$$

I_1, I_2, and I_3 are also known as stress invariants because their values do not vary with the choice of coordinate system. The state of plane stress is defined when the nonzero components of stress are restricted to a plane; in other words, when $\sigma_z = \sigma_{xz} = \sigma_{yz} = 0$. Thus, the nonzero components of stress are σ_x, σ_y, and τ_{xy}.

2A.2 STRAIN

Let two points located in a solid being deformed be displaced relative to their original positions. Strain is defined in terms of such displacements in a manner as to exclude the rigid body translation and rotation. For example, if the distance l_0 between two points A and B in a solid refers to an initial undeformed condition, and A moves to A' and B to B' after application of the load, a state of strain exists if the distance between A' and B', $l \neq l_0$. The strain, ε, is defined as:

$$\varepsilon = \frac{l - l_0}{l_0} = \frac{\Delta l}{l_0} \tag{2A.10}$$

If the change in length is large, we may define strain as follows:

$$\varepsilon = \int_0^e d\varepsilon = \int_{l_0}^l \frac{dl}{l} = ln\frac{l}{l_0} \tag{2A.11}$$

If we consider displacements of the four corners of a two-dimensional element ABCD in Figure 2A.3 instead of a line as in equation (2A.10), the two-dimensional strain tensor can be written as:

$$\varepsilon_{11} = \varepsilon_{xx} = \frac{\partial u}{\partial x} \tag{2A.12a}$$

$$\varepsilon_{22} = \varepsilon_{yy} = \frac{\partial v}{\partial y} \tag{2A.12b}$$

$$\gamma_{xy} = 2\varepsilon_{xy} = 2\varepsilon_{12} = \frac{\partial u}{\partial y} + \frac{\partial v}{\partial x} \tag{2A.12c}$$

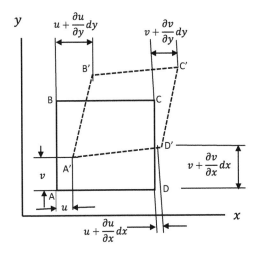

FIGURE 2A.3   Displacements in a two-dimensional deformed element.

If we consider a three-dimensional elemental cube, the following additional components of strain can be defined:

$$\varepsilon_{33} = \varepsilon_{zz} = \frac{\partial w}{\partial z} \qquad (2A.12d)$$

$$\gamma_{xz} = 2\varepsilon_{xz} = 2\varepsilon_{13} = \frac{\partial u}{\partial z} + \frac{\partial w}{\partial x} \qquad (2A.12e)$$

$$\gamma_{yz} = 2\varepsilon_{yz} = 2\varepsilon_{23} = \frac{\partial v}{\partial z} + \frac{\partial w}{\partial y} \qquad (2A.12f)$$

where the u and v are displacements as shown in Figure 2A.3 and w is the corresponding displacement in the z-direction. Equations (2A.12) can be collectively written in contracted or index notations as:

$$\varepsilon_{ij} = \frac{1}{2}\left(\frac{\partial u_i}{\partial x_j} + \frac{\partial u_j}{\partial x_i}\right) \qquad (2A.13)$$

From equation (2A.13), it also follows that:

$$\varepsilon_{ij} = \varepsilon_{ji} \qquad (2A.14)$$

Also note the relationship in equations (2A.12) between the engineering shear strain and the tensorial shear strain components. The tensorial shear strains are half of the values of the corresponding engineering shear strain components. Thus, all components of the strain tensor may be written as:

$$\varepsilon_{ij} = \begin{pmatrix} \varepsilon_{11} & \varepsilon_{21} & \varepsilon_{31} \\ \varepsilon_{12} & \varepsilon_{22} & \varepsilon_{32} \\ \varepsilon_{13} & \varepsilon_{23} & \varepsilon_{33} \end{pmatrix} \qquad (2A.15)$$

Like principal stresses, principal strains are defined as the normal strains on the planes with zero shear strains and are represented by ε_1, ε_2, ε_3. It can also be rigorously shown that the planes of principal stresses and principal strains are the same.

The state of plane strain is defined as the condition when the strains are limited to a single plane; in other words, when $\varepsilon_{zz} = \gamma_{xz} = \gamma_{yz} = 0$.

2A.3   ELASTICITY

The relationships between stress and strain are known as the constitutive equations. For isotropic, homogeneous, and elastic materials, these relationships are defined by the Hooke's Law. For uniaxial tension, the normal strain in the direction of loading is given by:

$$\varepsilon_1 = \frac{\sigma_1}{E} \qquad (2A.16a)$$

and the transverse strains (or the other principal strains are given by):

$$\varepsilon_2 = \varepsilon_3 = -v\varepsilon_1 \qquad (2A.16b)$$

where E = elastic modulus (or Young's modulus) and v = Poisson's ratio. For a three-dimensional stress state, the relationships between stress and elastic strains are given by:

$$\varepsilon_x = \frac{1}{E}\left[\sigma_x - v(\sigma_y + \sigma_z)\right] \qquad (2A.17a)$$

$$\varepsilon_y = \frac{1}{E}\left[\sigma_y - v(\sigma_x + \sigma_z)\right] \qquad (2A.17b)$$

$$\varepsilon_z = \frac{1}{E}\left[\sigma_z - v(\sigma_x + \sigma_y)\right] \qquad (2A.17c)$$

$$\gamma_{xy} = 2\varepsilon_{12} = \frac{\tau_{xy}}{G} \qquad (2A.17d)$$

$$\gamma_{yz} = 2\varepsilon_{23} = \frac{\tau_{yz}}{G} \qquad (2A.17e)$$

$$\gamma_{zx} = 2\varepsilon_{31} = \frac{\tau_{zx}}{G} \qquad (2A.17f)$$

where G is the shear modulus and for an isotropic, homogeneous material, it is given by:

$$G = \frac{E}{2(1+v)} \qquad (2A.18)$$

For the state of plane strain:

$$\sigma_z = v(\sigma_x + \sigma_y) \qquad (2A.19)$$

2A.4 ELASTIC STRAIN ENERGY

If a bar of length x subjected to a force, F, elongates by a distance dx, the work done by the force, F is given by Fdx. Thus, the work per unit volume is given by:

$$dU = \frac{Fdx}{Ax} = \sigma_1 d\varepsilon_1 \qquad (2A.20)$$

For uniaxial tension, $\sigma_1 = E\varepsilon_1$. Thus:

$$dU = E\varepsilon_1 d\varepsilon_1 \tag{2A.20a}$$

$$\text{or} \quad U = \frac{E\varepsilon_1^2}{2E} = \frac{\sigma_1^2}{2E} \tag{2A.21}$$

Since for elastic materials, W is also the stored strain energy-per-unit volume in the solid, it is also known as the strain energy density. For a three-dimensional state of stress, W is given by:

$$U = \frac{1}{2}\left(\sigma_{xx}\varepsilon_{xx} + \sigma_{yy}\varepsilon_{yy} + \sigma_{zz}\varepsilon_{zz} + \tau_{xy}\gamma_{xy} + \tau_{yz}\gamma_{yz} + \tau_{xz}\gamma_{xz}\right) \tag{2A.22a}$$

n in the form of index notation can be written as:

$$U = \frac{1}{2}\sigma_{ij}\varepsilon_{ij} \tag{2A.22b}$$

For elastic materials that do not obey the Hooke's law (also known as nonlinear elastic materials), we can write the equivalent of equation (2A.20a) as:

$$dU = \sigma_{ij}d\varepsilon_{ij}$$

$$U = \int_0^{\varepsilon_{ij}}\sigma_{ij}d\varepsilon_{ij} \tag{2A.23}$$

Thus, if the strain energy density distribution $U(x, y, z)$ is known, we can also write:

$$\sigma_{ij} = \frac{\partial U}{\partial \varepsilon_{ij}} \tag{2A.24}$$

2A.5   STRESS TRANSFORMATION EQUATIONS

Let the stress state of a point be referenced to a coordinate system by x and y by components σ_x, σ_y, and τ_{xy}. If the same stress state is referenced to another coordinate system x', y' with its origin located at the same point but the axes rotated by an angle θ, the normal and shear stress components with respect to the new coordinate axes, $\sigma_{x'}$, $\sigma_{y'}$, and $\tau_{x'y'}$ are given by the stress transformation equations below. These equations are also represented by the Mohr's circle.

$$\sigma_{x'} = \frac{\sigma_x + \sigma_y}{2} + \frac{\sigma_x - \sigma_y}{2}\cos 2\theta + \tau_{xy}\sin 2\theta \tag{2A.25a}$$

$$\tau_{x'y'} = -\frac{\sigma_x - \sigma_y}{2}\sin 2\theta + \tau_{xy}\cos 2\theta \tag{2A.25b}$$

$$\sigma_{x'} = \frac{\sigma_x + \sigma_y}{2} - \frac{\sigma_x - \sigma_y}{2}\cos 2\theta - \tau_{xy}\sin 2\theta \qquad (2A.25c)$$

We also notice that $\sigma_{x'} + \sigma_{y'} = \sigma_x + \sigma_y$ and the sum of the normal stresses is known as stress invariant.

Principal stresses are normal stresses that act on planes that have 0 shear stresses. If σ_1 is the maximum normal stress and σ_2 is the minimum normal stress, then those values are given by:

$$\sigma_1, \sigma_2 = \frac{\sigma_x + \sigma_y}{2} \pm \sqrt{\left(\frac{\sigma_x - \sigma_y}{2}\right)^2 + \tau_{xy}^2} \qquad (2A.26)$$

2A.6   STRESS–STRAIN BEHAVIOR

Frequently, fracture in metals is accompanied by plastic deformation. It is therefore important to understand the stress–strain behavior from when the force is first applied to when fracture occurs in a specimen. The specimens used for characterizing the stress–strain behavior over the full range of elastic-plastic behavior are long with a length, l, and uniform cross-section area A throughout the length. Typically, l/A ratio is the equal to or greater than 4. When force is applied and increased gradually until fracture, stress and strain rise and produce a relationship given by Figures 2A.4 and 2A.5 for ductile and brittle materials, respectively.

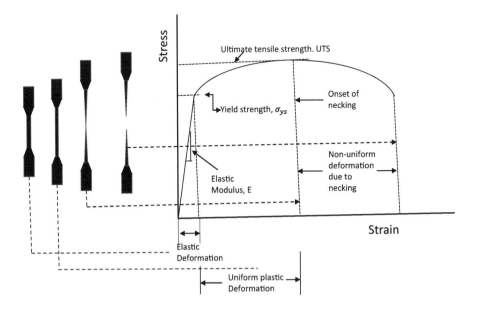

FIGURE 2A.4   Tested specimens of a ductile material along with the accompanying stress–strain curve.

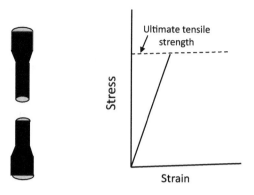

FIGURE 2A.5   Stress–strain behavior of a brittle material showing little plastic deformation prior to rupture (or fracture).

From these tests, properties such as elastic modulus, E, the yield strength, σ_Y or σ_{ys}, ultimate tensile strength, σ_U, the percent elongation prior to onset of rupture, the reduction in area prior to fracture are derived. For additional information on the tensile test, refer to excellent resources available on the internet.

NOTES

1 Technically, crystal sizes can range from a few nanometers to several centimeters in single crystals.

2 The designation <100> refers Miller indices that are used to designate crystallographic planes.

3 Theoretical Basis for Linear Elastic Fracture Mechanics

Before discussing the mechanics of how cracks degrade the ability of structures to withstand loads, it is important to understand the deformation behavior of structural materials. Except in the case of very brittle materials, fracture is preceded by inelastic deformation that leads to concentration of strains in the region where fracture initiates. Microcracks first form in microstructurally weak regions or in regions that have the highest stresses and then fracture spreads by coalescence of microcracks and leads to separation. An example of this phenomenon is the onset of necking during a tensile test in metallic materials where strain concentration occurs in vicinity of nonmetallic inclusions or a cluster of inclusions that are present as impurities at some location along the length of the specimen. Fracture occurs at that location because it is microstructurally weaker than the rest of the material. In the case where macro-cracks are already present, strain concentration occurs at the tip of the crack and fracture emanates from that pre-existing defect.

As discussed in Chapter 2, if cracks exist in a load-bearing component, its resistance to fracture depends on (a) the material's ability to absorb energy through irreversible processes such as plastic deformation in the crack tip region, (b) meandering of the crack, or (c) by fiber pull out in composite materials etc. If these energy absorbing processes are absent, such as during fracture in glasses, all available energy is directed toward forming new surfaces leading to a low energy, brittle fracture.

3.1 ENGINEERING MATERIALS AND DEFECTS

Engineering materials for structural applications are classified as metals and alloys, ceramics, polymeric materials, and composites. Composites are combinations of two or more types of materials mixed in pre-determined proportions to create tailored properties. Examples of composites include brittle concrete that is reinforced with steel rods/wires to increase the toughness. Because the fibers are aligned in one direction, the properties of composites vary with direction and are therefore anisotropic.

The stress–strain behavior of materials can be as simple as linear elastic or as complex as nonlinear and time-dependent. In this chapter, we will be concerned only with materials that behave as linear elastic solids and are homogeneous and have isotropic properties. In other words, the stress–strain behavior in these materials is consistent everywhere and, in all directions, and can be described by linear elasticity. Despite these simplifications, the theory of elasticity is applicable to large number

DOI: 10.1201/9781003292296-3

of engineering materials. None-the-less, it is important to always remember these limitations of this theory, so it is not used where these conditions cannot be assured.

Our ability to accurately describe fracture also depends on how well we understand the deformation mechanisms and damage processes in structural materials. Defects in materials determine the deformation and the damage progression so we begin with briefly describing defects in crystalline materials. For a more complete understanding of defects, the readers are referred to other sources [1,2].

Point defects consist of missing atoms or the presence of foreign atoms at lattice points of crystalline solids. Missing atoms are known as vacancies and foreign atoms occupying a lattice point in place of a regular atom is known as a substitutional point defect. A cluster of vacancies can form cavities especially at the grain boundaries and can significantly affect fracture behavior. This is particularly important at high temperatures where point defects can migrate at a faster rate to form clusters that are classified as volume defects and degrade the fracture strength. Therefore, grain boundaries degrade in strength at a faster rate at high temperatures than the grain interior, making intergranular fracture the dominant failure mode.

Line defects or dislocations are associated with a missing half-plane of atoms in crystalline materials. The direction of atomic displacements due to missing row of atoms is normal to the edge of the missing half plane or the so-called dislocation line. These dislocations are known as edge dislocations. Another form of line defects are ones in which the atomic displacements are parallel in direction to the dislocation line. These dislocations are called screw dislocations. Dislocations in a crystal either end on themselves to form a loop or they end at the crystal boundary. When dislocations move under the action of applied forces, materials deform plastically or permanently, and fracture is avoided or at least delayed. Inability of dislocations to move within the crystal causes brittle behavior such as in ceramics. The density of line defects in materials can be expressed in the number of intersections per unit area and can vary between 10^6 and $10^{12}/cm^2$. This is equivalent to 10^6–10^{12} cm of dislocation line length per cm^3. Dislocations move most easily on closed-packed planes where the resistance to their movement is the least and in closed-packed directions where the jump distance to an equivalent atomic position in the crystal is the shortest.

Planar defects consist of grain boundaries between adjoining crystals or stacking faults within crystals. Both have substantial roles in plastic deformation processes. In ceramics, adjoining grains are often joined by amorphous layers of glassy compounds that can flow in a viscous fashion causing the more rigid grains to slide relative to each other. Grain boundary sliding accommodates plastic deformation in this case and can be beneficial in forming of shapes in these otherwise very difficult to form materials.

Volume defects consist of discontinuities such as large inclusions, a large cluster of vacancies in the form of cavities, and cracks or crack-like defects. Presence of such defects significantly reduce the strength of solids and are the focus of this book.

3.2 STRESS ANALYSIS OF CRACKS

In this section, we will derive the stress fields in the vicinity of crack tips using theory of elasticity that applies to linear-elastic, homogeneous and isotropic materials. We will only consider cracks in planar bodies so the theory of planar elasticity (2-D elasticity theory) can be used to derive the equations. Despite these simplifying assumptions, the results are highly useful and applicable to predicting fracture in wide range of engineering materials such as metals and ceramics and provide guidance on how to approach fracture problems in complex materials such as composites, that are neither homogeneous, nor isotropic. The mechanics of fracture in these cases is complex and will be covered in a later chapter of this book.

3.2.1 EQUATIONS OF ELASTICITY

The equations of elasticity consisting of equilibrium, equations (2A.5), strain displacement relationships, equations (2A.12), and the stress–strain relations, equations (2A.17), were reviewed in Appendix 2A as part of the previous chapter, so they are not repeated here. In plane elasticity, there are three independent stress components, two independent displacement components and three independent strain components for a total of eight unknowns. We also have two equilibrium equations, three stress–strain relations, and three strain-displacement relations, thus we can solve for all eight unknowns. In 3-D elasticity, there are six independent components of stress, six independent components of strain and three independent components of displacement for a total of 15 unknowns. Thus, there are three equilibrium equations, six strain-displacement relationships and six stress–strain relationships adding to 15 available equations. Since we will be considering cracks in planar bodies for most part in this book, we will be using methods applicable to planar elasticity to arrive at those solutions.

3.2.2 COMPATIBILITY EQUATIONS

Compatibility equations below are a way to eliminate strain-displacement relationships to reduce the number of equations in plane elasticity which, in 2D elasticity and can be written as:

$$\frac{\partial^2 \varepsilon_{xx}}{\partial y^2} = \frac{\partial^3 u}{\partial x \partial y^2} \tag{3.1a}$$

$$\frac{\partial^2 \varepsilon_{yy}}{\partial x^2} = \frac{\partial^3 v}{\partial x^2 \partial y} \tag{3.1b}$$

$$\frac{\partial^2 \gamma_{xy}}{\partial x \partial y} = \frac{\partial^3 u}{\partial x \partial y^2} + \frac{\partial^3 v}{\partial x^2 \partial y} \tag{3.1c}$$

The three equations above lead to the following single equation known as the compatibility condition:

$$\frac{\partial^2 \varepsilon_{xx}}{\partial y^2} + \frac{\partial^2 \varepsilon_{yy}}{\partial x^2} = \frac{\partial^2 \gamma_{xy}}{\partial x \partial y} \tag{3.2}$$

Substituting the stress–strain relationships, equations (2A.17), into the compatibility equations, we get:

$$\frac{\partial^2}{\partial y^2}\left[\frac{1}{E}\left(\sigma_{xx} - v\sigma_{yy}\right)\right] + \frac{\partial^2}{\partial x^2}\left[\frac{1}{E}\left(\sigma_{yy} - v\sigma_{xx}\right)\right] = \frac{\partial^2}{\partial x \partial y}\left(\frac{\tau_{xy}}{\mu}\right)$$

or,

$$\frac{\partial^2}{\partial y^2}\sigma_{xx} + \frac{\partial^2}{\partial x^2}\sigma_{yy} - v\left[\frac{\partial^2}{\partial x^2}\sigma_{xx} + \frac{\partial^2}{\partial y^2}\sigma_{yy}\right] = \frac{\partial^2}{\partial x \partial y}\left(\frac{E\tau_{xy}}{\mu}\right)$$

or

$$\frac{\partial^2}{\partial y^2}\sigma_{xx} + \frac{\partial^2}{\partial x^2}\sigma_{yy} - v\left[\frac{\partial^2}{\partial x^2}\sigma_{xx} + \frac{\partial^2}{\partial y^2}\sigma_{yy}\right] = 2(1+v)\frac{\partial^2}{\partial x \partial y}\left(\tau_{xy}\right)$$

$$= -(1+v)\left(\frac{\partial^2}{\partial x^2}\sigma_{xx} + \frac{\partial^2}{\partial y^2}\sigma_{yy}\right) \text{ (Use the equilibrium equations 2A.5)}$$

or

$$\frac{\partial^2}{\partial y^2}\sigma_{xx} + \frac{\partial^2}{\partial x^2}\sigma_{yy} + \frac{\partial^2}{\partial x^2}\sigma_{xx} + \frac{\partial^2}{\partial y^2}\sigma_{yy} = 0$$

or

$$\left(\frac{\partial^2}{\partial x^2} + \frac{\partial^2}{\partial y^2}\right)\sigma_{xx} + \left(\frac{\partial^2}{\partial x^2} + \frac{\partial^2}{\partial y^2}\right)\sigma_{yy} = 0$$

$$\text{or}\quad \nabla^2\left(\sigma_{xx} + \sigma_{yy}\right) = 0 \tag{3.3}$$

We see that any stress distribution that satisfies the Harmonic equation (3.3), automatically satisfies the equilibrium equations, stress–strain relations, and the compatibility conditions. Thus, instead of solving eight equations simultaneously, we can focus on a single equation to find admissible stress, strain, and displacement relations.

Airy's stress function

We define a function, Φ, also known as the Airy's stress function, such that,

$$\sigma_{xx} = \frac{\partial^2 \Phi}{\partial y^2} \tag{3.4a}$$

$$\sigma_{yy} = \frac{\partial^2 \Phi}{\partial x^2} \tag{3.4b}$$

$$\tau_{xy} = -\frac{\partial^2 \Phi}{\partial x \partial y} \tag{3.4c}$$

$$\text{Thus,}\quad \nabla^2\left(\frac{\partial^2 \Phi}{\partial x^2} + \frac{\partial^2 \Phi}{\partial y^2}\right) = \frac{\partial^4 \Phi}{\partial x^4} + 2\frac{\partial^4 \Phi}{\partial x^2 \partial y^2} + \frac{\partial^4 \Phi}{\partial y^4} = \nabla^4 \Phi = 0 \tag{3.5}$$

Equation (3.5) is called the biharmonic equation. If we define Φ such that it meets the boundary conditions of a problem and it also satisfies the biharmonic equation, we can use Φ to determine the stress distribution in a body corresponding to the boundary conditions it describes. This will be the correct stress distribution because it will automatically satisfy the equilibrium equations and the compatibility conditions everywhere in the body. If function Φ is defined such that,

$$\Phi = \Psi_1 + x\Psi_2 + y\Psi_3 \qquad (3.6)$$

where Ψ_i are all harmonic functions, then it can be readily shown that Φ satisfies the biharmonic equation.

3.2.3 APPLICATION OF AIRY'S STRESS FUNCTION TO CRACK PROBLEMS

Cracked bodies can be loaded in three distinct modes, Mode I, Mode II, and Mode III shown in Figure 3.1. Mode I is most common and is known as the crack opening mode. Mode II is the in-plane shear mode and Mode III the anti-plane shear mode. The crack opening mode presents the largest risk of fracture and therefore has been most extensively explored in fracture-mechanics research. We will derive the stress field at the tip of a crack loaded in Mode I.

Westergaard [3] was able to identify several Airy's functions of the complex variable $z = x + iy$ including one that is useful for solving crack problems. Here, $i =$ imaginary number, $\sqrt{-1}$. A function Z of the complex variable z is defined as follows:

$$Z(z) = ReZ + iImZ$$

$$z = x + iy = re^{i\theta} = r(\cos\theta + i\sin\theta)$$

If Z is analytic, it must have continuous derivatives in its domain and it must meet the Cauchy–Reiman conditions. See Appendix 3A to learn more about Cauchy–Reiman conditions. Applied to the Z-function, we get the following equations:

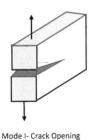

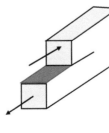

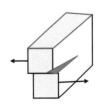

Mode III – Out-of-Plane Shear

Mode I- Crack Opening Mode

Mode - II In-Plane Shear

FIGURE 3.1   The three modes of fracture in cracked bodies.

$$\frac{\partial Rez}{\partial x} = \frac{\partial Im\ z}{\partial y} = Re\frac{dZ}{dz} \tag{3.7a}$$

$$\frac{\partial Imz}{\partial x} = -\frac{\partial Rez}{\partial y} = Im\frac{dZ}{dz} \tag{3.7b}$$

If we also require that Z be harmonic then,

$$\nabla^2 Rez = \nabla^2 ImZ = 0$$

We further specify that

$$Z' = \frac{dZ}{dz};\ Z = \frac{d\bar{Z}}{dz};\ \bar{Z} = \frac{d\bar{\bar{Z}}}{dz}\ \text{from which it follows that,}$$

$$\bar{\bar{Z}} = \int \bar{Z}dz;\quad \bar{Z} = \int Zdz;\quad Z = \int Z'dz$$

Let us choose an analytic function Φ such that,

$$\Phi = Re\bar{\bar{Z}} + yIm\bar{Z} \tag{3.8}$$

Then Φ satisfies the bi-harmonic equation and we can write the components of stress using equation (3.4a–c). Applying the Cauchy–Riemann conditions to the real and imaginary parts of Φ, we get in addition to equations (3.7) the following relationships:

$$\frac{\partial Re\bar{\bar{Z}}}{\partial x} = \frac{\partial Im\bar{\bar{Z}}}{\partial y} = Re\bar{Z} \tag{3.9}$$

$$\frac{\partial Im\bar{\bar{Z}}}{\partial x} = -\frac{\partial Re\bar{\bar{Z}}}{\partial y} = Im\bar{Z} \tag{3.10}$$

$$\frac{\partial Re\bar{Z}}{\partial x} = \frac{\partial Im\bar{Z}}{\partial y} = ReZ \tag{3.11}$$

$$\frac{\partial Im\bar{Z}}{\partial x} = -\frac{\partial Re\bar{Z}}{\partial y} = ImZ \tag{3.12}$$

Then,

$$\sigma_x = \frac{\partial^2 \Phi}{\partial y^2} = \frac{\partial}{\partial y}\left[\frac{\partial}{\partial y}Re\bar{\bar{Z}} + \frac{\partial}{\partial y}yIm\bar{z}\right] = \frac{\partial}{\partial y}\left[-Im\bar{Z} + yReZ + Im\bar{Z}\right]$$

$$\sigma_x = ReZ - yImZ' \tag{3.13a}$$

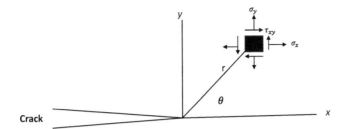

FIGURE 3.2   A Mode I crack and the crack tip coordinate system.

We can derive the equations for the other two components of stress as:

$$\sigma_y = \frac{\partial}{\partial x}\left[\frac{\partial}{\partial x}\left(\frac{\partial}{\partial x}Re\bar{Z} + \frac{\partial}{\partial x}y\,Im\,\bar{Z}\right)\right]$$

$$= \frac{\partial}{\partial x}\left[Re\bar{Z} + y ImZ\right] = ReZ + yImZ' \tag{3.13b}$$

$$\tau_{xy} = -\frac{\partial}{\partial y}\left[\left(\frac{\partial}{\partial x}Re\bar{Z} + \frac{\partial}{\partial x}y\,Im\,\bar{Z}\right)\right] = -\frac{\partial}{\partial y}\left[Re\bar{Z} + yImZ\right] = -yReZ \tag{3.13c}$$

Next, we define a Z-function that meets the boundary conditions of Mode I crack problem shown in Figure 3.2. The boundary conditions for a crack problem state that, σ_y and τ_{xy} are both 0 on the crack surface. Thus, for $y=0$ and $x<0$, σ_y, $\tau_{xy}=0$. Westergaard proposed the following Z-function that meets the boundary conditions for a Mode I crack:

$$Z = \frac{K}{\sqrt{2\pi z}} \tag{3.14}$$

K is a parameter that will be shown to be a function of geometry and size of the cracked body, crack size and a characteristic applied remote stress. For $y=0$ and $x<0$, Z is purely imaginary, thus from equations (3.13), $\sigma_y=0$ and $\tau_{xy}=0$. He also guessed correctly that the dominant stress singularity at the crack tip must be of the form of $r^{-\frac{1}{2}}$. This was later proven to always be the case by Williams in 1958 [4] for all crack problems.

$$\text{If } z = re^{i\theta} \text{ then } Z = \frac{K}{\sqrt{2\pi r}}e^{-\frac{i\theta}{2}} = \frac{K}{\sqrt{2\pi r}}\left(\cos\frac{\theta}{2} - i\,\sin\frac{\theta}{2}\right)$$

$$\sigma_x = ReZ - yImZ' = \frac{K}{\sqrt{2\pi r}}\cos\frac{\theta}{2} - rsin\theta\left(ImZ'\right)$$

$$Z' = -\frac{K}{2\sqrt{2\pi}}z^{-\frac{3}{2}} = -\frac{K}{2\sqrt{2\pi}}\left(r^{-\frac{3}{2}}\right)\left(\cos\frac{3\theta}{2} - i\,\sin\frac{3\theta}{2}\right)$$

$$y ImZ' = \frac{K}{2\sqrt{2\pi}}r^{-\frac{3}{2}}r\sin\theta\sin\frac{3\theta}{2} = \frac{K}{\sqrt{2\pi}}r^{-\frac{1}{2}}r\sin\frac{\theta}{2}\cos\frac{\theta}{2}\sin\frac{3\theta}{2}$$

Thus, it is seen that,

$$\sigma_x = \frac{K}{\sqrt{2\pi r}}\cos\frac{\theta}{2}\left[1 - \sin\frac{\theta}{2}\sin\frac{3\theta}{2}\right] \tag{3.15a}$$

$$\sigma_y = \frac{K}{\sqrt{2\pi r}}\cos\frac{\theta}{2}\left[1 + \sin\frac{\theta}{2}\sin\frac{3\theta}{2}\right] \tag{3.15b}$$

$$\tau_{xy} = \frac{K}{\sqrt{2\pi r}}\sin\frac{\theta}{2}\cos\frac{\theta}{2}\cos\frac{3\theta}{2} \tag{3.15c}$$

Equations (3.15) are known as the crack tip field equations. For $\theta = 0$,

$$\tau_{xy} = 0 \quad \text{and} \quad \sigma_x = \sigma_y = \frac{K}{\sqrt{2\pi x}}$$

Since the crack tip fields are expected to dominate the stresses near the crack tip, the stress functions for all Mode I cracks, regardless of their geometry must reduce to the form,

for $z \to 0$, $Z = \dfrac{K}{\sqrt{2\pi z}}$ hence,

$$K = \lim_{z \to 0}\sqrt{2\pi z}\,Z \tag{3.16}$$

This method can be used to determine the K-expressions for Mode I crack geometries for which a Z-function can be found. The following examples illustrate this method for determining K in few configurations.

3.3   STRESS INTENSITY PARAMETER, K, FOR VARIOUS CRACK GEOMETRIES AND LOADING CONFIGURATIONS BY THE WESTERGAARD METHOD

As mentioned in the previous section, if we can identify a Z-function that meets the boundary conditions of a crack problem that also reduces to the form in equation (3.14), as $z \to 0$, we can then determine the magnitude of the crack tip stresses and the K from equation (3.15).

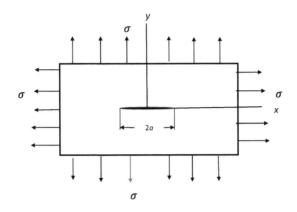

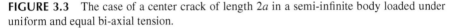

FIGURE 3.3 The case of a center crack of length $2a$ in a semi-infinite body loaded under uniform and equal bi-axial tension.

Example Problem 3.1

The case of a semi-infinite panel with a center crack subjected to uniform equi-biaxial tension

A semi-infinite panel containing a crack of length $2a$ in the center and loaded with remote and equal uniform stresses in the x and y directions is shown in Figure 3.3. The following Z-function meets all the boundary conditions of the problem:

$$Z = \frac{\sigma z}{\sqrt{z^2 - a^2}} \qquad (3.17)$$

Z is analytic everywhere except in the region, $-a \le z \le a$. For $z \gg a$, $Z = \sigma$, thus $\sigma_x = \sigma_y = \sigma$. We will use equation (3.16) to determine the expression for K for this crack configuration. Note that the coordinate systems in the crack problem that led to equation (3.16) was located at the crack tip and that in equation (3.17) is located in the middle of the crack. Thus, we need to make the following coordinate transformation.

We define $z' = z - a$, or $z = z' + a$. The new coordinate system will be located at the right tip of the crack. Since both cracks are identical, consideration of any one of them provides the conditions for the other by symmetry. Substituting for z in equation (3.17) gives:

$$Z(z') = \frac{\sigma(z' + a)}{\sqrt{(z' + a)^2 - a^2}} = \frac{\sigma(z' + a)}{\sqrt{z'(z' + 2a)}}$$

$$\lim_{z' \to 0} Z(z') = \frac{\sigma a}{\sqrt{z'(2a)}}$$

Thus, $K = \dfrac{\sigma a}{\sqrt{z'(2a)}} \sqrt{2\pi z'} = \sigma\sqrt{\pi a}$

$$K = \sigma\sqrt{\pi a} \qquad (3.18)$$

Substituting for K in the field equations (3.15) provides the stress distribution in front of the crack tip for this crack configuration. This solution also applies to the same configuration but under a remote uniaxial stress only in the y-direction. Since the remote stress in the x-direction does not contribute to opening the crack, it also does not contribute to the magnitude of K. Therefore, the above solution also applies to the configuration consisting of a semi-infinite body with a center crack of length $2a$ in the middle and loaded with a uniform stress, σ just in the y-direction. The latter is the same as the Griffith's problem of Chapter 2. Equation (3.18) also gives the value of K for the other crack tip from symmetry.

K is called the stress intensity parameter and was so labelled by George Irwin who put forward this concept in the mid-1950s [5]. He noted that K represents the amplitude of the stress singularity at the crack tip, equations (3.15). In other words, a higher value of K would imply higher levels of stress in the crack tip region. Also note that the value of K depends on the magnitude of the remote applied stress and the crack size; increasing either the stress or the crack size intuitively increases the risk of fracture. Thus, Irwin proposed that K may be used in a fracture criterion as follows. If K approaches the critical value, K_c, for a given material, fracture is expected to occur. This is analogous to the statement that when the applied stress, σ, reaches a critical value, σ_{ys}, plastic yielding (or plastic deformation) occurs. In this example, σ is the applied stress and σ_{ys} is a material property. Similarly, K is the applied stress intensity parameter and K_c is a material property that can be measured by testing the material. The test that is used to determine the value of K_c of a material is called the fracture toughness test that will be considered in a later chapter. Often, K is also written as K_I, where the subscript I refers to the crack opening mode or Mode I, loading. In the literature and here, we will simply use K and K_I interchangeably.

Example Problem 3.2

K-solution for an array of co-linear cracks in a semi-infinite plate

We will consider one more example of the estimation of K using the Westergaard's approach. Figure 3.4 shows an array of collinear cracks of equal

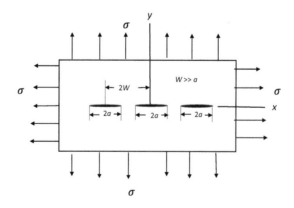

FIGURE 3.4   An array of collinear cracks of equal size in a semi-infinite body subjected to remote equi-biaxial stress.

size and equal spacing in a semi-infinite plate subjected to equi-biaxial tension. Westergaard provided the Z-function described by equation (3.19) for this case.

$$Z(z) = \cfrac{\sigma}{\left[1 - \left\{\cfrac{\sin\cfrac{\pi a}{2W}}{\sin\cfrac{\pi z}{2W}}\right\}^2\right]^{1/2}} \tag{3.19}$$

Z is only real for $z > a$, and is only imaginary for $z < a$. Therefore, the condition for cracks that $\sigma_y = \tau_{xy} = 0$ for $z < a$, and $y = 0$ are met. Also, for $a \ll W$ and $z \gg a$, but smaller than W, $\sigma_x = \sigma_y = \sigma$. Thus, all boundary conditions for the problem are met so the Z-function is the right function for this problem. These boundary conditions imply that there is no interaction between the stress fields of neighboring cracks.

Next, we transform the origin to the crack tip by stipulating that $z' = z - a$, or $z = z' + a$.

$$Z(z') = \cfrac{\sigma}{\left[1 - \left(\cfrac{\sin\cfrac{\pi a}{2W}}{\sin\cfrac{\pi(z'+a)}{2W}}\right)^2\right]^{1/2}}$$

$$K = \lim_{z' \to 0} \sqrt{2\pi z'}\, Z$$

$$\text{Let, } H = 1 - \left(\cfrac{\sin\cfrac{\pi a}{2W}}{\sin\cfrac{\pi(z'+a)}{2W}}\right)^2 = 1 - \left(\cfrac{\sin\cfrac{\pi a}{2W}}{\sin\cfrac{\pi z'}{2W}\cos\cfrac{\pi a}{2W} + \sin\cfrac{\pi a}{2W}\cos\cfrac{\pi z'}{2W}}\right)^2$$

$$\text{or } H = 1 - \left(\cfrac{1}{\sin\cfrac{\pi z'}{2W}\cfrac{1}{\tan\cfrac{\pi a}{2W}} + \cos\cfrac{\pi z'}{2W}}\right)^2$$

$$\text{As } \lim_{z' \to 0} H = 1 - \left(\cfrac{1}{\cfrac{\pi z'}{2W}\cfrac{1}{\tan\cfrac{\pi a}{2W}} + 1}\right)^2 = 1 - \left(\cfrac{\pi z'}{2W}\cfrac{1}{\tan\cfrac{\pi a}{2W}} + 1\right)^{-2}$$

Using the binomial theorem that states that for $x \ll 1$, $(1-x)^n \cong 1 - nx$, we get:

$$H = 1 - \left(1 - 2\frac{\pi z'}{2W}\frac{1}{\tan\dfrac{\pi a}{2W}}\right) = 2\frac{\pi z'}{2W}\frac{1}{\tan\dfrac{\pi a}{2W}}$$

Thus, $K = \lim_{z' \to 0}\sqrt{2\pi z'}\,Z = \dfrac{\sigma\sqrt{2\pi z'}}{\left(2\dfrac{\pi z'}{2W}\dfrac{1}{\tan\dfrac{\pi a}{2W}}\right)^{1/2}} = \dfrac{\sigma\sqrt{2\pi z'}}{\dfrac{\sqrt{2\pi z'}}{\sqrt{2W}\left(\tan\dfrac{\pi a}{2W}\right)^{1/2}}}$

or $K = \sigma\sqrt{\pi a}\left\{\dfrac{2W}{\pi a}\tan\dfrac{\pi a}{2W}\right\}^{1/2}$ (3.20)

The above expression was also used to estimate the value of K for center cracked specimens shown in Figure 3.5 with a finite width equal to $2W$ and a crack in the middle equal to a length of $2a$.

More accurate expressions for K for the center cracked panel were developed later. These are given as:

$$K = \sigma\sqrt{\pi a}\left[\sec\left(\frac{\pi a}{2W}\right)\right]^{1/2}\left[1 - 0.025\left(\frac{a}{W}\right)^2 + 0.06\left(\frac{a}{W}\right)^4\right]$$ (3.21)

Equation (3.21) is derived from finite element analysis and is most accurate. In the above figure, the various expressions for K are compared to equation (3.21) and it is seen that equation (3.20) is quite accurate up to a/W values of 0.5 or less. Beyond that, equation (3.20) begins to lose its accuracy, so equation (3.21) is recommended. An alternate expression below without the polynomial function in

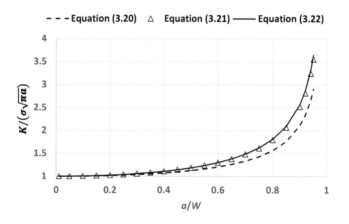

FIGURE 3.5   Solutions of stress intensity parameter from the various equations showing a good agreement among them up to a/W values of 0.5.

equation (3.21) is also frequently used and is shown to be quite accurate for a/W values of up to 0.8.

$$K = \sigma\sqrt{\pi a}\left[\sec\left(\frac{\pi a}{2W}\right)\right]^{1/2} \qquad (3.22)$$

Example Problem 3.3

K-solution for a wedge loaded crack in a semi-infinite body
    The Westergaard's Z-function for a wedge-loaded crack, Figure 3.6, in a semi-infinite plate of unit thickness is given by,

$$= \frac{Pa}{\pi(z-b)z}\sqrt{\frac{1-\left(\dfrac{b}{a}\right)^2}{1-\left(\dfrac{a}{z}\right)^2}}$$

We are asked to find the expressions for K at the crack tips A and B.
    Crack tips A and B are expected to have different values of K if $b \neq 0$. So let us first derive an expression for K for crack tip A. We first transform the coordinates from the center of the crack so the origin is now located at point A,
$z' = z - a$ or $z' = z + a$ and then substituting for z in the Z-function above,

$$Z(z') = \frac{Pa}{B\pi(z'+a-b)(z'+a)}\sqrt{\frac{1-(b/a)^2}{1-\left(\dfrac{a}{z'+a}\right)^2}}$$

$$\lim_{z'\to 0} Z(z') = \frac{Pa}{B\pi(a-b)(a)}\sqrt{\frac{1-(b/a)^2}{1-\left(\dfrac{1}{1+z'/a}\right)^2}}$$

$$= \frac{P}{B\pi(a-b)}\sqrt{\frac{(a-b)((a+b))}{a^2(1-2z'/a)}} = \frac{P}{B\pi}\sqrt{\left(\frac{a+b}{a-b}\right)/(2az')}$$

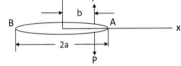

FIGURE 3.6 A wedge loaded crack in a semi-infinite plate.

$$K_A = \lim_{z' \to 0} Z\sqrt{2\pi z'} = \sqrt{2\pi z'}\sqrt{\left[\frac{a+b}{a-b}\frac{1}{2az'}\right]\frac{P}{B\pi}}$$

$$K_A = \frac{P}{B\sqrt{\pi a}}\sqrt{\frac{a+b}{a-b}}$$

K_B can be derived by substituting $-b$ for b in the equation for calculating K_A. Thus,

$$K_B = \frac{P}{B\sqrt{\pi a}}\sqrt{\frac{a-b}{a+b}}$$

3.4 CRACK TIP DISPLACEMENT FIELDS

We begin with the derivation of the crack tip displacement fields for plane stress conditions. From the strain-displacement relationships, we can write:

$$\varepsilon_x = \frac{\partial u}{\partial x} = \frac{1}{E}\left[\sigma_x - v\sigma_y\right] = \frac{1}{E}\left[(ReZ - yImZ') - v(ReZ + yImZ')\right]$$

$$= \frac{1}{E}\left[(1-v)ReZ - yImZ'(1+v)\right]$$

$$u = \int \varepsilon_x dx$$

Implementing the Cauchy–Reiman conditions and following steps detailed in Appendix 3A, we can show that:

$$u = \frac{K}{2\mu}\sqrt{\frac{r}{2\pi}}\cos\frac{\theta}{2}\left[\kappa - 1 + 2\sin^2\frac{\theta}{2}\right] \qquad (3.23a)$$

Similarly, $v = \int \varepsilon_y dy$ leads to

$$v = \frac{K}{2\mu}\sqrt{\frac{r}{2\pi}}\sin\frac{\theta}{2}\left[\kappa + 1 - 2\cos^2\frac{\theta}{2}\right] \qquad (3.23b)$$

It also turns out that equation (3.23a and b) can also be used to compute the displacements for plane strain by substituting $\kappa = (3-4v)$. The full derivations for equation (3.23a and b) are given in Appendix 3A.

3.5 THE RELATIONSHIP BETWEEN G AND K

The potential of K for characterizing fracture and the previous success of G in characterizing fracture, raises the question about the relationship between G and K. This was directly addressed by George Irwin [5] in 1957 through this elegant derivation.

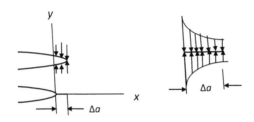

FIGURE 3.7 Estimation of the work required for closing a Mode I crack by a small distance Δa.

Recall that G is based on satisfying the thermodynamic considerations during fracture while the concept of K is based on stress considerations in the crack tip region. However, also recall from Chapter 2 that early fracture theories based on stress and energy approaches led to somewhat similar estimates of the critical stress required for fracture.

Irwin postulated that the work required to close the crack tip through a "pinching action" over an incremental portion of the crack of length equal to Δa should be the same as the elastic energy released if the crack grew by the same amount. This is schematically shown in Figure 3.7. Since the energy released is directly related to G, we should be able to use that to derive the relationship between G and K.

We assume that the "act of pinching" the crack tip occurs under the conditions of constant remote displacement for which $G = -\dfrac{1}{B}\left(\dfrac{\partial U}{\partial a}\right)_\Delta$, equation (2.20) and the work done in the process equals the change in strain energy. Let us also assume that the original crack length was $a + \Delta a$ and the crack over the length Δa is being closed by applying a stress equal to σ_y.

$$G = \lim_{\Delta a \to 0} -\frac{1}{B}\left(\frac{\Delta U}{\Delta a}\right)_\Delta$$

$$-\Delta U = \int_0^{\Delta a} \frac{1}{2}\sigma_y B dx (2v) = \int_0^{\Delta a} \sigma_y B v dx$$

$$G = \lim_{\Delta a \to 0} \frac{1}{\Delta a}\int_0^{\Delta a} \sigma_y v dx$$

$\sigma_y = \dfrac{K(a)}{\sqrt{2\pi x}}$ where $K(a)$ is the value of K corresponding to crack size a, and from equation (3.23b)

$$v = \frac{K}{2\mu}\sqrt{\frac{r}{2\pi}}\sin\frac{\theta}{2}\left[\kappa + 1 - 2\cos^2\frac{\theta}{2}\right]$$

Since v is referenced to a crack size of $a + \Delta a$, the value of K in the above equation should be for a crack size of $a + \Delta a$ also. Further, to estimate v along the crack plane going in a direction away from the crack tip, θ must be 180° and r must be referenced to the tip of the larger crack of $a + \Delta a$ in equation (3.23b). If the origin is referenced to the tip of the crack closed over a distance Δa, $r = \Delta a - x$. This yields the correct result that $r = \Delta a$ at the tip. For $\theta = \pi$, as explained above,

$$\text{Thus, } v = \frac{K(a + \Delta a)}{2\mu}\sqrt{\frac{\Delta a - x}{2\pi}}(\kappa + 1)$$

$$\text{and } G = \lim_{\Delta a \to 0} \frac{1}{\Delta a} \int_0^{\Delta a} \frac{K(a)K(a + \Delta a)}{\sqrt{2\pi x}\,(2\mu)}\sqrt{\frac{\Delta a - x}{2\pi}}(\kappa + 1)dx$$

$$\lim_{\Delta a \to 0} K(a + \Delta a) = K(a) = K$$

$$\text{Thus, } G = \lim_{\Delta a \to 0} \frac{1}{\Delta a} \int_0^{\Delta a} \frac{K^2}{4\pi\,\mu}\sqrt{\frac{\Delta a - x}{x}}(\kappa + 1)dx$$

Substituting $x = \Delta a \sin^2\theta;\quad dx = 2\Delta a(\sin\theta\cos\theta)d\theta$

$$G = \lim_{\Delta a \to 0} \frac{1}{\Delta a} \int_0^{\frac{\pi}{2}} \frac{K^2(\kappa + 1)}{4\pi\,\mu}\sqrt{\frac{\Delta a \cos^2\theta}{\Delta a \sin^2\theta}}(2\Delta a\,\sin\theta\cos\theta)d\theta$$

Recognizing that $2(\cos\theta)^2 = 1 + \cos 2\theta$

$$G = \int_0^{\frac{\pi}{2}} \frac{K^2(\kappa + 1)}{4\pi\,\mu}(1 + \cos 2\theta)d\theta = \frac{K^2(\kappa + 1)}{8\mu}$$

It can be shown by substituting the values of κ for plane stress and plane strain, that:

$$G = \frac{K^2}{E} \quad \text{for Plane Stress} \tag{3.24a}$$

$$G = \frac{K^2}{E}(1 - v^2) \quad \text{for Plane Strain} \tag{3.24b}$$

These relationships clearly demonstrate the equivalence of the two approaches and lend credibility to the hypothesis that K is a viable fracture parameter. It also provides the basis for an experimental method of determining K that is left as a homework exercise. It can now be claimed that the stress intensity parameter, K, not only uniquely characterizes the crack tip stresses, it is also uniquely related to the energy available for crack extension. It will be shown in the next chapter that K also uniquely characterizes the shape and size of the crack tip plastic zone if the conditions of small-scale plastic deformation exist in the cracked body, providing one more reason for why K is attractive as a fracture parameter.

3.6 DETERMINING K FOR OTHER LOADING AND CRACK GEOMETRIES

The Westergaard's method of determining K using the Airy's stress function is limited to a few simple geometries. Now that the potential of the stress intensity parameter, K, for characterizing fracture in materials and for predicting fracture in components appears promising, there is need for determining its value for crack shapes found in engineering components to fully exploit its potential. There are handbooks [6] that list the K-expressions for several crack configurations that have been determined either through Westergaard's Z-functions or from finite element analyses and other numerical techniques that have been conducted on specific geometries and loading conditions. In this discussion, we will learn about some methods that can be effectively used to derive K expressions for new situations using the expressions that are already available.

Proportioning the specimen or a component allows one to use the same expression for K for cracked bodies that are identical in planar geometry but of different sizes. For example, let us consider a planar cracked body with a crack size, a, thickness, B, and a single characteristic planar dimension width, W. The width W is independently chosen, and the other planar dimensions of the body are proportioned to W. The K-expression can be developed such that it will hold for any value of W.

K is expressed in terms of the point load P or a characteristic stress σ depending on how the external loads/stresses are applied.

$$K = \frac{P}{BW^{\frac{1}{2}}} F\left(\frac{a}{W}\right); \quad \text{or} \quad K = \sigma \sqrt{\pi a} f\left(\frac{a}{W}\right)$$

If there is a relationship between the applied load P and a characteristic stress σ, such as for center cracked tension specimen, where $P = 2BW\sigma$, the relationship between F and f can be obtained. F and f are known as boundary correction factors and are often written as polynomial expressions such as in equation (3.25).

$$F \text{ or } f = b_0 + b_1\left(\frac{a}{W}\right) + b_2\left(\frac{a}{W}\right)^2 + b_3\left(\frac{a}{W}\right)^3 + \ldots \qquad (3.25)$$

where b_i are regression constants determined from fitting the polynomial function to K values calculated for discrete crack sizes obtained from finite element analyses. Some examples of such K-solutions for commonly used specimens in fracture testing and for cracked configurations found in practice are shown in Table 3.1.

TABLE 3.1

Stress Intensity Parameter Expressions for Common Geometries Including Specimens and Components

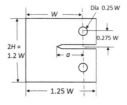

Compact type, C(T)

$$F\left(\frac{a}{W}\right) = \frac{K}{P} BW^{1/2} = F\left(\frac{a}{W}\right) = \frac{2 + a/W}{(1 - a/W)^{3/2}}$$

$$\left[0.886 + 4.64(a/W) - 13.32\left(\frac{a}{W}\right)^2 + 14.72\left(\frac{a}{W}\right)^3 - 5.6(a/W)^4 \right]$$

Center crack tension, CC(T)

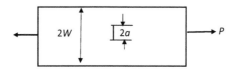

$$F\left(\frac{a}{W}\right) = \frac{K}{P} BW^{1/2} = \frac{1}{2}\sqrt{\pi\left(\frac{a}{W}\right) \sec\frac{\pi}{2}\frac{a}{W}} \left\{ 1 - 0.025\left(\frac{a}{W}\right)^2 + 0.06\left(\frac{a}{W}\right)^4 \right\}$$

Three-point bend specimen

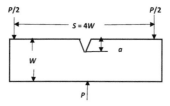

$$F\left(\frac{a}{W}\right) = \frac{K}{P} BW^{1/2}$$

$$= \frac{3\left(\dfrac{S}{W}\right)\left(\dfrac{a}{W}\right)^{1/2}\left[1.99 - \dfrac{a}{W}(1 - a/W)\left[2.15 - 3.93\left(\dfrac{a}{W}\right) + 2.7(a/W)^2 \right] \right]}{2(1 + 2a/W)(1 - a/W)^{3/2}}$$

(Continued)

TABLE 3.1 (*Continued*)
Stress Intensity Parameter Expressions for Common Geometries Including Specimens and Components

Single edge crack specimen subject to pure bending, SEC(B)

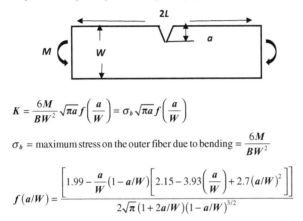

$$K = \frac{6M}{BW^2} \sqrt{\pi a}\, f\left(\frac{a}{W}\right) = \sigma_b \sqrt{\pi a}\, f\left(\frac{a}{W}\right)$$

$$\sigma_b = \text{maximum stress on the outer fiber due to bending} = \frac{6M}{BW^2}$$

$$f(a/W) = \frac{\left[1.99 - \frac{a}{W}(1 - a/W)\left[2.15 - 3.93\left(\frac{a}{W}\right) + 2.7(a/W)^2\right]\right]}{2\sqrt{\pi}\,(1 + 2a/W)(1 - a/W)^{3/2}}$$

Single edge crack tension SEC(T)

$$\sigma = \frac{P}{BW}$$

$$K = \sigma\sqrt{\pi a}\, f(a/W)$$

$$f(a/W)$$

$$= F(a/W)\sqrt{\frac{W}{\pi a}}$$

$$F\left(\frac{a}{W}\right) = \frac{K}{P} BW^{1/2} = \sqrt{2\tan\left(\frac{\pi a}{2W}\right)} \left\{ \frac{0.752 + 2.02(a/W) + 0.37\left(1 - \sin\frac{\pi a}{2W}\right)^3}{\cos\left(\frac{\pi a}{2W}\right)} \right\}$$

(Continued)

TABLE 3.1 (*Continued*)

Stress Intensity Parameter Expressions for Common Geometries Including Specimens and Components

Double edge crack tension DEC(T)

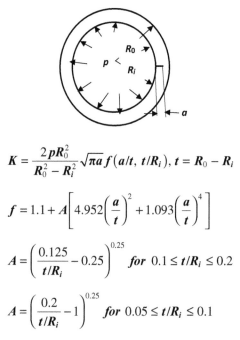

$$\sigma = \frac{P}{2BW}$$

$$K = \sigma\sqrt{\pi a}\,f(a/W)$$

$$f(a/W) = 2F(a/W)\sqrt{\frac{W}{\pi a}}$$

$$F\left(\frac{a}{W}\right) = \frac{K}{P}BW^{1/2}$$

$$= (1 - a/W)^{-1/2}\left\{1.122 - 0.561\left(\frac{a}{W}\right) - 0.205\left(\frac{a}{W}\right)^2 - 0.471\frac{a}{W}^3 - 0.7024(a/W)^4\right\}$$

Internally pressurized cylinder with an axial crack

$$K = \frac{2pR_0^2}{R_0^2 - R_i^2}\sqrt{\pi a}\,f(a/t,\, t/R_i),\ t = R_0 - R_i$$

$$f = 1.1 + A\left[4.952\left(\frac{a}{t}\right)^2 + 1.093\left(\frac{a}{t}\right)^4\right]$$

$$A = \left(\frac{0.125}{t/R_i} - 0.25\right)^{0.25}\quad \textit{for}\ 0.1 \le t/R_i \le 0.2$$

$$A = \left(\frac{0.2}{t/R_i} - 1\right)^{0.25}\quad \textit{for}\ 0.05 \le t/R_i \le 0.1$$

(*Continued*)

TABLE 3.1 (*Continued*)

Stress Intensity Parameter Expressions for Common Geometries Including Specimens and Components

Axially loaded cylinder with a circumferential crack

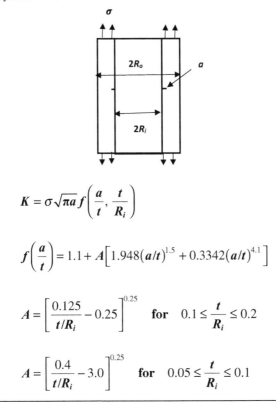

$$K = \sigma \sqrt{\pi a}\, f\left(\frac{a}{t}, \frac{t}{R_i}\right)$$

$$f\left(\frac{a}{t}\right) = 1.1 + A\left[1.948(a/t)^{1.5} + 0.3342(a/t)^{4.1}\right]$$

$$A = \left[\frac{0.125}{t/R_i} - 0.25\right]^{0.25} \quad \text{for} \quad 0.1 \le \frac{t}{R_i} \le 0.2$$

$$A = \left[\frac{0.4}{t/R_i} - 3.0\right]^{0.25} \quad \text{for} \quad 0.05 \le \frac{t}{R_i} \le 0.1$$

3.7 USE OF LINEAR SUPERPOSITION PRINCIPLE FOR DERIVING *K*-SOLUTIONS

In linear elasticity, individual components of stress caused by different applied loads can be summed to obtain the total magnitude of the applied stress on the body. For example, σ_x due to various forces applied on the body can be summed to calculate σ_x due to all forces combined. The same also holds for all other components of stress but σ_x cannot be combined with say τ_{xy}. In cracked bodies, the stress intensity parameter, K, associated with each external load that produces crack face displacements in Mode I can be linearly added as expressed below:

$$K_I^{\text{total}} = K_I^A + K_I^B + K_I^C + \ldots$$

where A, B, C, ... are separate loads that produce Mode I displacements in the cracked body. Note that K values from different modes cannot be added. An example of this procedure is illustrated in Figure 3.8.

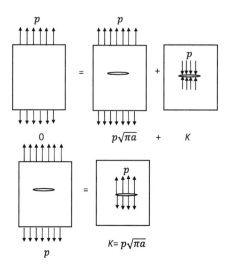

FIGURE 3.8   An illustration of the principle of linear superposition for obtaining K-solutions for a crack that is wedge-loaded by a pressure p using a known solution of a uniformly loaded center crack panel with the same crack and a remote stress equal to p.

The above example was a simple case consisting of a semi-infinite body under uniform remote stress and center crack that is much smaller than the width of the body. Let us consider a more general case of a structural component subjected to arbitrary loads and for which the stress distribution is completely defined without a crack, but we do not have a K-solution for a crack in that body under the applied external load. Such stress distributions that typically do not consider cracks are commonly available from analyses performed to support component design. The remote applied load consists of a nonuniformly distributed load on the boundary of the component described by a function $P(x)$ as seen in Figure 3.9. We are asked to find the stress intensity factor for a crack that is introduced along the x-axis. The normal stress distribution along the crack plane in an uncracked body is given by a function $\sigma(x)$ as also shown in Figure 3.9. From the principle of linear superposition, we can show that $K^a = K^b + K^c$ in Figure 3.10. Since $K^c = 0$, $K^a = K^b$.

The following example illustrates the use of linear superposition to assemble K expressions for a single edge crack specimen subjected to a combination of uniform tension and bending loads.

Example Problem 3.4

A single edge crack specimen with a width, $W = 10$ cm is subject to a stress in MPa that is described by $[300 - 20x]$ for $0 \leq x \leq 10$ cm where x is measured from the specimen edge containing the crack mouth. Calculate the value of K for a crack size to width ratio of 0.05, 0.1, 0.2, 0.5, and 0.6. The K-solutions for single edge crack specimen subject to pure bending and pure tension are given in Table 3.1.

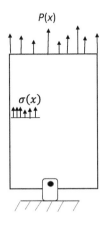

FIGURE 3.9 An uncracked structural component in which the load distribution at the boundary, $P(x)$, and the resultant stress distribution along the segment AB, $p(x)$, are both known from the stress analysis.

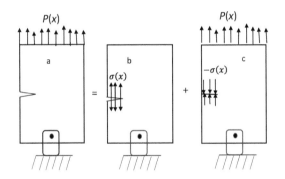

FIGURE 3.10 K for the geometry in (a) can be determined by summing the K from the two cases shown in (b) and (c). Note that the K for configuration in c is 0, leading to the result that the K for the geometry in (a) is equal to K for geometry in (b).

Solution

If σ_t and σ_b are the uniform tension and the maximum bending stress respectively,

$$\sigma_t = \frac{300+100}{2} = 200 \text{ MPa} \quad \text{and} \quad \sigma_b = \frac{300-100}{2} = 100 \text{ MPa},$$ where 300 is the

stress at $x=0$ and 100 is the stress at $x=10$ cm.

The total K is given by $K=K_t+K_b$ where K_t is the value of K contributed by the uniform stress σ_t and K_b is the value of K contributed by the bending stress σ_b.

From Table 3.1, the following expressions for K can be found for single edge crack geometries subjected to pure tension and pure bending:

$$K_t = \sigma_t \sqrt{\pi a} f_t(a/W)$$

$$f_t(a/W) = F(a/W)\sqrt{\frac{W}{\pi a}}$$

$$\sqrt{\frac{2W}{\pi a}}\tan\left(\frac{\pi a}{2W}\right) = \left\{\frac{0.752 + 2.02(a/W) + 0.37\left[1 - \sin\left(\frac{\pi a}{2W}\right)^3\right]}{\cos\left(\frac{\pi a}{2W}\right)}\right\}$$

$$K_b = \sigma_b\sqrt{\pi a}\,f_b(a/W),$$

where, $$f_b = \frac{\left[1.99 - \dfrac{a}{W}(1 - a/W)\left[2.15 - 3.93\left(\dfrac{a}{W}\right) + 2.7(a/W)^2\right]\right]}{2\sqrt{\pi}\,(1 + 2a/W)(1 - a/W)^{3/2}}$$

From the above equations,

$$K = K_t + K_b = \sqrt{\pi a}\left[\sigma_t f_t + \sigma_b f_b\right]$$

The results of the calculations are summarized in Table 3.2.

Note, how rapidly the K value increases with crack size. If this cracked member was made from a high strength low alloy steel, for which the critical K for fracture is about $95\,\mathrm{MPa}\sqrt{m}$, the component will fracture at a crack size that is approximately 0.02 m or 20 mm.

3.8   K-SOLUTIONS FOR 3-D CRACKS

Cracks in components are often of elliptical shape or half-elliptical shape if they emanate from the surface and therefore, have two characteristic length dimensions. In such instances, K varies along the crack front and depends on the shape of the crack. Examples of such cracks are seen in an aircraft propeller blade (Figure 3.11). Such cracks are known as 3-D cracks.

The K-solution for some 3-D cracks of common shapes is discussed next. The simplest form of a 3-D crack is a penny shaped (circular) crack in an infinite body

---

TABLE 3.2
K-Values for a Plate Containing an Edge Crack Subjected to Tension and Bending as a Function of Crack Size

a/W	a (m)	f_t	f_b	K (MPa(m))$^{1/2}$
0.05	0.005	1.225	0.525	51.5
0.1	0.01	1.33	0.504	75.4
0.2	0.02	1.56	0.494	117.33
0.5	0.05	2.369	0.708	276.53
0.6	0.06	2.73	0.916	362.78

FIGURE 3.11 Elliptical-shaped crack in an aircraft engine propeller blade (Example taken from published cases from Naval Surface Warfare Center).

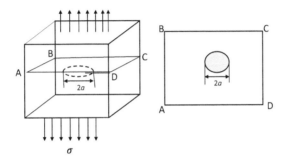

FIGURE 3.12 A penny shaped (circular) crack in an infinite body subjected to uniform tension, σ.

subjected to uniform tension shown in Figure 3.12 and the equation for estimating K is given in equation (3.26). This expression is derived from an exact stress solution.

$$K = \frac{2}{\pi}\sigma\sqrt{\pi a} = 0.64\ \sigma\sqrt{\pi a} \tag{3.26}$$

The K-solution for an elliptical crack in a semi-infinite body subject to uniform tension (Figure 3.13) is given by equation (3.27):

$$K = \frac{\sigma\sqrt{\pi a}}{\Psi}\left[\left(\frac{a}{c}\right)^2 \cos^2\phi + \sin^2\phi\right]^{1/4} \tag{3.27}$$

Ψ is an elliptical integral of the second kind and the angle ϕ is defined in Figure 3.14.

$$\psi = \int_{0}^{\pi/2}\left[1 - \left(\frac{c^2 - a^2}{c^2}\right)\sin^2\phi\right]d\phi$$

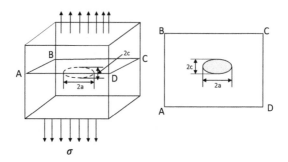

FIGURE 3.13   An elliptical crack in an infinite body subjected to uniform tension.

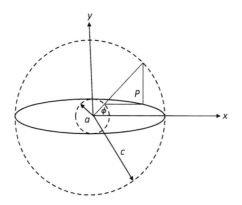

FIGURE 3.14   Representation of point P on the ellipse in terms of the angle ϕ.

The value of ψ depends on a/c and is given by the following equation:

$$\psi = \frac{\pi}{2}\left[1 - \frac{1}{1}\left(\frac{1}{2}\right)^2\left(\frac{c^2 - a^2}{c^2}\right) - \left(\frac{1}{3}\right)\left(\frac{1.3}{2.4}\right)^2\left(\frac{c^2 - a^2}{c^2}\right)^2\right.$$

$$\left. - \left(\frac{1}{5}\right)\left(\frac{1.3.5}{2.4.6}\right)^2\left(\frac{c^2 - a^2}{c^2}\right)^3 - \cdots\right]$$

(3.28)

We see that for elliptical cracks, with $c/a > 1$, the maximum K occurs at the deepest point when $\phi = \pi/2$ and the lowest value of K occurs when $\phi = 0$. Thus, the highest value of K is given by $K = \dfrac{\sigma\sqrt{\pi a}}{\psi}$. For a semi-elliptical surface flaw, $K = \dfrac{1.12\sigma\sqrt{\pi a}}{\psi}$ where 1.12 is the surface correction factor and can be written as in equation (3.29) below:

$$K = 1.12\sigma\sqrt{\frac{\pi a}{Q}}$$

(3.29)

where $Q = \psi^2$ is known as the shape correction factor. The value of Q can also be represented by the following approximate equations (3.30):

$$Q \approx 1 + 1.464\left(\frac{a}{c}\right)^{1.65} \quad \text{for} \quad a/c \leq 1 \tag{3.30a}$$

$$Q \approx 1 + 1.464\left(\frac{a}{c}\right)^{1.65} \quad \text{for} \quad c/a \leq 1 \tag{3.30b}$$

The K-expression for a surface elliptical flaw in a finite body subject to tension and bending (solutions from Newman and Raju [7]) as seen in Figure 3.15 is given in equation (3.31).

$$K = (\sigma_t + H\sigma_b)\sqrt{\frac{\pi a}{Q}} f\left(\frac{a}{t},\frac{a}{c},\frac{c}{W},\phi\right) \tag{3.31}$$

$$\text{where } f = \left[M_1 + M_2\left(\frac{a}{t}\right)^2 + M_3\left(\frac{a}{t}\right)^4\right] f_\phi \cdot g \cdot f_w$$

$$M_1 = 1.13 - 0.09\left(\frac{a}{c}\right)$$

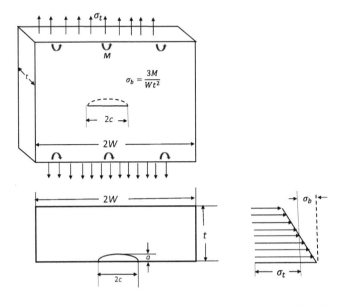

FIGURE 3.15 Geometry of a semi-infinite flaw subjected to tension and bending where σ_t is the stress due to uniform tension and σ_b is the bending stress on the outer fiber due to bending moment.

$$M_2 = -0.54 + \frac{0.89}{0.2 + a/c}$$

$$M_3 = 0.5 - \frac{1.0}{0.65 + \dfrac{a}{c}} + 14(1.0 - a/c)^{24}$$

$$f_\phi = \left[\left(\frac{a}{c} \right)^2 \cos^2\phi + \sin^2\phi \right]^{\frac{1}{4}}$$

$$g = 1 + \left[0.1 + 0.35(a/t)^2 \right](1 - \sin\phi)^2$$

$$f_w = \left[\sec\left(\frac{\pi c}{W} \sqrt{\frac{a}{t}} \right) \right]^{1/2} \quad \text{for} \quad \phi = 0$$

$$f_w = 1 \quad \text{for} \quad \phi = 90$$

$$H = H_1 + (H_2 - H_1)\sin^p\phi$$

$$p = 0.2 + \frac{a}{c} + 0.6\left(\frac{a}{t} \right)$$

$$H_1 = 1 - 0.34\left(\frac{a}{t} \right) - 0.11\left(\frac{a}{c} \right)\left(\frac{a}{t} \right)$$

$$H_2 = 1 + G_1\left(\frac{a}{t} \right) + G_2\left(\frac{a}{t} \right)^2$$

$$G_1 = -1.22 - 0.12\left(\frac{a}{c} \right)$$

$$G_2 = 0.55 - 1.05\left(\frac{a}{c} \right)^{3/4} + 0.47 \left(\frac{a}{c} \right)^{3/2}$$

The above equations can be used to predict the change in the shape of the crack profile as it grows under fatigue loading. To do that, we substitute $\phi = 0$ for tracking the growth in c and $\phi = \pi/2$ for tracking the growth in a. Thus, a/c changes continuously as the crack grows. An example of application for the above equations is illustrated in the following example.

Example Problem 3.5

Using ultrasonic testing, the area of a semi-elliptical surface crack in a long plate that is 20 cm wide and 5 cm thick and is subjected to uniform tension, σ, is determined as $\pi/2$ cm^2. Estimate the maximum value K/σ for a/c values of 1, 0.6, 0.4, and 0.2. These are the K values corresponding to a unit of applied stress to the plate.

Solution

The area, A, of a semi-elliptical crack is given by $A = \dfrac{\pi}{2}(ac)$; since $A = \dfrac{\pi}{2}$, $ac = 1$. Table 3.3 lists the values of a and c and various other parameters for calculating K for the desired a/c values given in the problem statement. These parameters correspond to those in equation (3.31). Also note that since the loading consists of pure tension, σ_b, the stress that relates to bending $= 0$.

Some observations from these results are as follows:

- The highest value of K along the crack profile occurs at the deepest point of the crack if $a/c < 1$.
- The highest values of K for a fixed crack area of the crack occur in the interval of $0.4 \le a/c \le 0.6$.
- If cracks grow due to fatigue loading or due to environment assisted cracking, the shape of the elliptical crack will continuously evolve. In other words, c/a is expected to continuously change, because of the differences in the crack growth rates at the deepest point and at the surface.

Semi-elliptical surface cracks are often a consideration in cylindrical members used in pressure vessel applications where the cracks on the radial-axial plane can be present either on the inside or outside surface of the pressure vessel. The hoop stress in the vessel acts normal to the cracks and gives rise to a K value that is a cause for concern. If the cracks are located on the inside surface, the pressure also acts to open the crack in Mode I. Then the value of K due to hoop stress and due to pressure acting on the crack surfaces must be added to estimate the total K.

3.9   SUMMARY

In this chapter, the concept of stress intensity parameter, K, that uniquely determines the crack tip stresses in the vicinity of cracks under Mode I loading was derived. The magnitude of K was shown to be related to the remotely applied stress, the crack size, the geometry of the body and the loading configuration. The theoretical basis of K is embedded in the theory of linear elasticity that assumes that the materials, besides being linear-elastic, are also homogeneous and isotropic. The equations relating K to the displacements in the vicinity to the crack tip were derived. It was also shown that K is uniquely related to the Griffith's crack extension force, G, making it a candidate for characterizing fracture.

Methods for determining K in different geometries containing cracks were discussed. The Westergaard's method based on Airy's stress function was used to determine the expressions for K for some simple crack geometries. For complex geometries, K is determined for discrete crack sizes by finite element techniques

TABLE 3.3

Estimation of K Values for a Semi-Elliptical Surface Crack at the Deepest Point and on the Surface in a Plate of Area $\pi/2$ cm² with Different Aspect Ratios

a/c	a (cm)	c (cm)	a/t	Q	M_1	M_2	M_3	f_w	g $\phi=\frac{\pi}{2}$	g $\phi=0$	$f\phi$ $\phi=\frac{\pi}{2}$	$f\phi$ $\phi=0$
1	1	1	0.2	2.464	1.04	0.202	−0.106	1.005	1.0	1.114	1.0	1.000
0.6	0.775	1.291	0.155	1.43	1.076	0.572	−0.300	1.006	1.0	1.108	1.0	0.775
0.4	0.632	1.581	0.126	1.22	1.09	0.943	−0.452	1.008	1.0	1.105	1.0	0.632
0.2	0.447	2.236	0.089	1.070	1.112	1.685	−0.610	1.011	1.0	1.103	1.0	0.447

a/c	$\left[M_1 + M_2\left(\dfrac{a}{t}\right)^2 + M_3\left(\dfrac{a}{t}\right)^4\right]$	$K/\sigma\,(\phi = \pi/2)$ MPa$\sqrt{m}$/MPa	$K/\sigma\,(\phi = 0)$ MPa$\sqrt{m}$/MPa
1.0	1.048	0.1189	0.1324
0.6	1.089	0.1429	0.1227
0.4	1.105	0.142	0.0992
0.2	1.125	0.1278	0.0632

and expressed analytically by fitting polynomial relationships through discrete data. The K-expressions for numerous cases are collected and reported in handbooks but expressions for some frequently encountered cases are listed in this chapter. The use of linear superposition principle for assembling K-expressions was demonstrated and so were expressions for determining K for elliptical and semi-elliptical cracks in semi-infinite and finite-size bodies.

REFERENCES

1. J. Weertman and J.R. Weertman, *Elementary Dislocation Theory*, MacMillan, New York, NY, 1964.
2. J.P. Hirth and J. Lothe, *Theory of Dislocations*, 2nd Edition, Krieger Publishing Company, Malabar, FL, 1992.
3. H.M. Westergaard, "Bearing Pressures and Cracks", *Journal of Applied Mechanics*, Vol. 6, 1939, pp. 49–53.
4. M.L. Williams, "On the Stress Distribution at the Base of a Stationery Crack", *Journal of Applied Mechanics*, Vol. 24, 1957, pp. 109–114.
5. G.R. Irwin, "Analysis of Stress and Strains Near the End of a Crack Traversing a Plate", *Journal of Applied Mechanics*, Vol. 24, 1957, pp. 109–114.
6. H. Tada, P.C. Paris, and G.R. Irwin, *The Stress Analysis of Cracks Handbook*, 3rd Edition, American Society for Mechanical Engineers, New York, NY, 2000.
7. J.C. Newman and I. Raju, "Analyses of Surface Cracks in Finite Plates under Tension and Bending", NASA Technical Report 1578, 1979.

HOMEWORK PROBLEMS

1. In plane stress, write the components of strain that are not zero. Why are only three components of strain independent? For plane strain, similarly write the components of stress that are not zero and why are only three components independent?
2. Based on the assumptions used to derive the crack tip stress fields, what limitations apply in the use of K as a fracture parameter?
3. Using the relationship between G and K, describe an experimental method that can be used to determine K for planar cracked body.
4. How much difference there is in the magnitudes of K for a given value of G for plane stress and plane strain conditions?
5. On your own, given the following Z-function for a crack in a semi-infinite body, derive the crack tip stress fields.
6. $Z = \dfrac{K}{\sqrt{2\pi z}}$
7. Why is the stress intensity parameter, K, suitable for predicting fracture?
8. Derive an expression for the stress intensity parameter for crack subjected to a uniform pressure, p, on the crack surfaces as shown in the figure below:

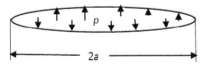

9. Show that the expressions 3.20, 3.21, and 3.22 all reduce to equation 3.18 when $a/W \rightarrow 0$.
10. Derive a relationship between the K-calibration functions F (associated with point loads) and f (associated with a distributed stress, σ, for finite size center crack panel with a width of $2W$, thickness, B, and a crack size, $2a$.
11. The K-solution for single edge notch specimen subject to three-point bend loading is given in Table 3.1 and the K-solution for an edge-crack specimen subjected to pure bending with a moment M, is given in Table 3.1. Find the relationship between $f(a/W)$ and $F(a/W)$ in the two equations for estimating K.
12. A flat plate that is 20 cm wide and 5 cm thick has a semi-elliptical crack and is subjected to a normal stress of 100 MPa. The major axis of the crack ($2c$) is 2 cm and the depth, a, is 0.5 cm. Calculate the value of K at the deepest point and the point on the external surface along the crack front.

APPENDIX 3A

3A.1   CAUCHY–REIMANN EQUATIONS

Let $ReZ(x,y)$ and $ImZ(x,y)$ be the real and imaginary parts, respectively of a complex function $Z(x+iy)=Z(z)=ReZ(x,y)+iImZ(x,y)$ of the complex variable $z=x+iy$. Let ReZ and ImZ also be differentiable at any point in the domain D. This implies that the partial derivatives of ReZ and ImZ exist. Then Z is complex-differentiable at that point if and only if the partial derivatives of ReZ and ImZ satisfy the Cauchy–Riemann equation (3A.1a and b) at that point.

$$\frac{\partial ReZ}{\partial x} = \frac{\partial ImZ}{\partial y} \quad (3A.1a)$$

$$\frac{\partial ReZ}{\partial y} = \frac{\partial ImZ}{\partial x} \quad (3A.1b)$$

Thus, if $Z(z)$ is complex analytic function in a specified domain and other functions are as described below:

$$Z' = \frac{dZ}{dz}; \quad Z = \frac{d\bar{Z}}{dz}; \quad \bar{Z} = \frac{d\bar{\bar{Z}}}{dz} \text{ it also follows that,}$$

$$\bar{\bar{Z}} = \int \bar{Z} dz; \ \bar{Z} = \int Z dz; \ Z = \int Z' dz$$

Applying the Cauchy–Riemann conditions to the real and imaginary parts of Z, we get the following relationships:

$$\frac{\partial Re\bar{Z}}{\partial x} = \frac{\partial Im\bar{Z}}{\partial y} = Re\bar{Z}$$

$$\frac{\partial Im\overline{Z}}{\partial x} = -\frac{\partial Re\overline{Z}}{\partial y} = Im\overline{\overline{Z}}$$

$$\frac{\partial Re\overline{Z}}{\partial x} = \frac{\partial Im\overline{Z}}{\partial y} = Re\overline{Z}$$

$$\frac{\partial Im\overline{Z}}{\partial x} = -\frac{\partial Re\overline{Z}}{\partial y} = ImZ$$

$$\frac{\partial ReZ}{\partial x} = \frac{\partial ImZ}{\partial y} = ReZ'$$

$$\frac{\partial ImZ}{\partial x} = -\frac{\partial ReZ}{\partial y} = ImZ'$$

3A.2 DERIVATION OF THE CRACK TIP DISPLACEMENT FIELDS

We will use the Cauchy–Reimann conditions above to derive equations for the displacement components u and v in front of the crack tip. From the strain-displacement relationships:

$$\varepsilon_x = \frac{\partial u}{\partial x} = \frac{1}{E}\left[\sigma_x - v\sigma_y\right] = \frac{1}{E}\left[(ReZ - yImZ') - v(ReZ + yImZ')\right]$$

$$= \frac{1}{E}\left[(1-v)ReZ - yImZ'(1+v)\right]$$

$u = \int \varepsilon_x dx$, and implementing the Cauchy–Reiman conditions,

$$u = \frac{1}{E}\left[(1-v)Re\overline{Z} - y(1+v)ImZ\right]$$

$$Z = \frac{K}{\sqrt{2\pi z}} = \frac{K}{\sqrt{2\pi r}}\left(\cos\frac{\theta}{2} - i\,\sin\frac{\theta}{2}\right)$$

$$\overline{Z} = \frac{2K}{\sqrt{2\pi}}z^{1/2} = \frac{2K}{\sqrt{2\pi}}r^{1/2}\left(\cos\frac{\theta}{2} + i\,\sin\frac{\theta}{2}\right)$$

$$\text{Thus, } u = \frac{2K(1+v)}{E}\sqrt{\frac{r}{2\pi}}\cos\frac{\theta}{2}\left[\frac{(1-v)}{1+v} + \sin^2\frac{\theta}{2}\right]$$

$$u = \frac{K}{\mu}\sqrt{\frac{r}{2\pi}}\cos\frac{\theta}{2}\left[\frac{(1-v)}{1+v}+\sin^2\frac{\theta}{2}\right] \text{ where } \mu \text{ is the shear modulus} = \frac{E}{2(1+v)}$$

If we substitute $\kappa = \dfrac{3-v}{1+v}$, for plane stress,

$$u = \frac{K}{2\mu}\sqrt{\frac{r}{2\pi}}\cos\frac{\theta}{2}\left[\kappa - 1 + 2\sin^2\frac{\theta}{2}\right] \tag{3A.2}$$

The above equation also applies for plane strain if substitute $\kappa = 3 - 4v$.

The displacement component v can similarly be obtained by the following equation:

$$v = \int \varepsilon_y dy = \frac{1}{E}\left[\int (\sigma_y - v\sigma_x)dy\right] = \frac{1}{E}\left[\int \left(ReZ + yImZ' - v\left(ReZ - yImZ'\right)\right)dy\right]$$

$$\text{or } v = \frac{1}{E}\left[\int \left((1-v)ReZ + (1+v)yImZ'\right)dy\right]$$

From the Cauchy–Reimann equations, we can write the following relationships:

$$\int ReZdy = Im\bar{Z}$$

$$\frac{\partial}{\partial y}(yReZ) = -yImZ' + ReZ$$

$$\int \left(\frac{\partial}{\partial y}(yReZ)dy\right) = -\int (yImZ')dy + \int ReZdy$$

$$\int (yImZ')dy = -yReZ + Im\bar{Z}$$

Substituting these equations into the equation for v gives:

$$v = \frac{1}{E}\left[(1-v)Im\bar{Z} + (1+v)\left(-yReZ + Im\bar{Z}\right)\right]$$

$$= \frac{1}{E}\left[2Im\bar{Z} - (1+v)yReZ\right]$$

For a Mode I crack problem, in the vicinity of the crack, we have:

$$Z = \frac{K}{\sqrt{2\pi z}} = \frac{K}{\sqrt{2\pi}}z^{-1/2} = \frac{K}{\sqrt{2\pi}}r^{-1/2}\left(\cos\frac{\theta}{2} - i\sin\frac{\theta}{2}\right)$$

$$\bar{Z} = \frac{2K}{\sqrt{2\pi}}r^{1/2}\left(\cos\frac{\theta}{2} + i\sin\frac{\theta}{2}\right)$$

Substituting from the above equations into the equation for v gives:

$$v = \frac{K}{E}\sqrt{\frac{r}{2\pi}}\left[4\sin\frac{\theta}{2} - (1+v)(r\sin\theta)\cos\frac{\theta}{2}\right]$$

$$= \frac{K(1+v)}{E}\sqrt{\frac{r}{2\pi}}\sin\frac{\theta}{2}\left[\frac{4}{(1+v)} - 2\left(\cos\frac{\theta}{2}\right)^2\right]$$

$$= \frac{K}{2\mu}\sqrt{\frac{r}{2\pi}}\sin\frac{\theta}{2}\left[\kappa + 1 - 2\left(\cos\frac{\theta}{2}\right)^2\right]$$

where $\kappa = \dfrac{3-v}{1+v}$ for plane stress

$$v = \frac{K}{2\mu}\sqrt{\frac{r}{2\pi}}\sin\frac{\theta}{2}\left[\kappa + 1 - 2\left(\cos\frac{\theta}{2}\right)^2\right] \tag{3A.3}$$

Equations (3A.2) and (3A.3) can be used for plane strain conditions also if we substitute $\kappa = (3-4v)$.

4 Crack Tip Plasticity

Stress intensity parameter, K, is based on linear elasticity theory but we have observed that fracture in engineering materials is accompanied by plastic deformation that is characterized by elastic-plastic stress–strain behavior. Also, elastic analysis predicts that the stresses at the crack tip are infinite which is physically impossible because metals undergo plastic deformation once the stresses exceed the yield strength. George Irwin [1] was one of the early researchers to come to terms with plasticity within a dominantly linear-elastic framework to deal with fracture problems. He introduced the term small-scale-yielding (SSY) to represent situations where fracture occurs in dominantly linear-elastic bodies but under the conditions of local plasticity. He then went on to adjust the stress magnitudes predicted by elastic crack tip field equations to account for plasticity under SSY. Some background knowledge about plasticity under multiaxial conditions is relevant here. Those not familiar with it are encouraged to go to Appendix 4A to review it. Even those familiar with the topic might take a quick look to understand the terminology used in this chapter.

4.1 ESTIMATE OF THE PLASTIC ZONE SIZE

The simplest material behavior to account for plasticity is an elastic-perfectly plastic material in which once the applied stress exceeds the yield strength, stress–strain relationship becomes flat as shown in Figure 4.1. Such stress–strain behavior is labelled as the elastic-perfectly plastic material and does not include strain hardening and therefore can only be considered approximate representation of behavior of more complex engineering materials. To enforce the conditions of SSY, it is important to estimate the size of the crack tip plastic zone. A simple, and an order of magnitude estimate of the crack tip plastic zone size, r_y, was obtained by Irwin [1] as in Figure 4.2. He took the elastic stress distribution and truncated it when σ_y became equal to or greater than σ_{ys} which led to equation (4.1). In his analysis, he assumed plane stress conditions for which the Von Mises yield criterion, equation (4.2), applied to the crack tip field equations in which the normal stresses in the x and y directions are the principal stresses, σ_1 and σ_2 respectively, and are equal in magnitude. For details of the Von Mises criterion, please see Appendix 4A. The third principal stress, σ_3 for plane stress $=0$. This leads to the effective stress at yield, σ_e, to be equal to the uniaxial yield strength, σ_{ys}, as can be seen from equation (4.2).

$$r_y = \frac{1}{2\pi}\left(\frac{K}{\sigma_{ys}}\right)^2 \tag{4.1}$$

$$\sigma_e = \frac{1}{\sqrt{2}}\left[(\sigma_1 - \sigma_2)^2 + (\sigma_2 - \sigma_3)^2 + (\sigma_3 - \sigma_1)^2\right]^{1/2} = \sigma_{ys} \tag{4.2}$$

DOI: 10.1201/9781003292296-4

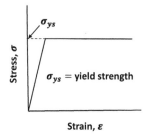

FIGURE 4.1 Schematic of the stress–strain relationship of an elastic-perfectly plastic material.

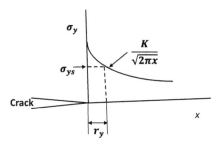

FIGURE 4.2 A schematic representation of Irwin's simple estimation of the crack tip plastic zone size under plane stress conditions.

Under plane strain conditions at the crack tip, $\tau_{xy}=0$, thus, σ_x, σ_y are the principal stresses σ_1 and σ_2, respectively and are equal, and $\sigma_z=\sigma_3=2v\sigma_x$. From the Von Mises' criterion, equation (4.2), we can derive equation (4.3).

$$\sigma_e = \frac{1}{\sqrt{2}}\left[\left(\sigma_y - 2v\sigma_y\right)^2 + \left(2v\sigma_y - \sigma_y\right)^2\right]^{1/2} = \sigma_{ys} \quad \text{or} \quad \sigma_y = \frac{\sigma_{ys}}{\left(1-2v\right)} \quad (4.3)$$

For $v = \dfrac{1}{3}$, yielding occurs when $\sigma_y=3\sigma_{ys}$. Thus, according to this simple estimate, the plane strain plastic zone size must be 1/9th of the plane stress plastic zone size for the same value of applied K.

A more accurate estimate of the plastic zone size can be made by recognizing that the elastic stress distribution is modified by plastic deformation and the new stress distribution must have a flat region within a plastic zone that is greater than r_y, followed by an elastic stress distribution that is also different from the original elastic stress distribution. This modified stress distribution is labelled as the elastic-plastic stress distribution and is schematically shown in Figure 4.3.

The new plastic zone size is labelled as r_p and we will now estimate its size. For force equilibrium to hold, the areas under the elastic and elastic-plastic stress

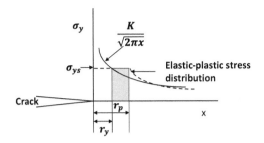

FIGURE 4.3   A more accurate determination of the plastic zone size based on an elastic-plastic stress distribution.

distributions must be equal. For large, wide bodies that meet the requirements of SSY, the areas under the two stress distributions for $x > r_p$ must be the same. This then implies that the filled rectangular area must equal the unfilled area shown in Figure 4.3. Thus,

$$\int_0^{r_y} \frac{K}{\sqrt{2\pi x}} dx - \sigma_{ys} r_y = \sigma_{ys} \left(r_p - r_y \right)$$

$$\left[\frac{2K}{\sqrt{2\pi}} x^{1/2} \right]_0^{r_y} = r_p \sigma_{ys}$$

$$r_p = \frac{1}{\pi} \left(\frac{K}{\sigma_{ys}} \right)^2 = 2r_y \quad \text{for plane stress} \tag{4.4a}$$

$$r_p = \frac{1}{\pi} \left(\frac{K}{3\sigma_{ys}} \right)^2 \quad \text{for plane strain} \tag{4.4b}$$

The stress state near the crack tip is never quite plane strain even for thick sections because there is a free surface at the crack tip which promotes plane stress conditions. Three-dimensional finite element stress analyses have shown that a better estimate of the plane strain plastic zone size is given by equation (4.5):

$$r_p = \frac{1}{3\pi} \left(\frac{K}{\sigma_{ys}} \right)^2 \tag{4.5}$$

Regardless of the details of these derivations leading to different constants that depend on the underlying assumptions, the most significant observation is that the crack tip plastic zone size under SSY conditions is uniquely related to the stress intensity parameter, K. This further increases its appeal as a fracture parameter for cracked bodies.

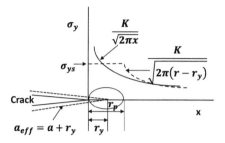

FIGURE 4.4 Schematic representation of the elastic and elastic-plastic crack tip stress distributions and the concept of an "effective crack size," a_{eff}.

4.2 PLASTICITY MODIFIED CRACK TIP STRESS FIELD FOR SSY

Irwin hypothesized that plastic zone is a soft region that is unable to *fully* transmit the elastic stresses across the plastic region and is essentially analogous to a crack except that the crack is so soft that it is unable to transmit *any* stresses across its surfaces. The plastic zone, on the other hand, can still transmit stresses that are equal to the yield strength. Therefore, a plastic zone can be considered as an extended crack that is larger than the physical crack size by an amount equal to a fraction of the plastic zone size. He then noted that due to plasticity, the elastic stress distribution can be shifted to the right by a distance equal to r_y so equations (4.6), are expected to approximately capture the plasticity modified stress in the y-direction. This is illustrated schematically in Figure 4.4.

$$\sigma_y = \sigma_{ys} \quad \text{for} \quad r \le 2r_y \tag{4.6a}$$

$$\text{and} \quad \sigma_y = \frac{K}{\sqrt{2\pi\left(r - r_y\right)}} \quad \text{for} \quad r > 2r_y \tag{4.6b}$$

Let us examine the elastic-plastic stress distribution. If we assume the effective crack size, a_{eff} as $a + r_y$, and relocate the origin to the new crack tip and reference the value of K, as K_{eff} to the larger crack size, the correct elastic stress distribution beyond the plastic zone is then obtained. Thus, $K_{eff} = K(a_{eff})$ which is also the plasticity modified value of K that will be higher than the K from purely elastic considerations. K_{eff} can be estimated iteratively by first calculating the elastic K and then estimating r_y and using the new value of K_{eff} and recalculating r_y and continuing it until the value of K_{eff} converges to essentially the same value. However, for the center crack panel, of infinite width a closed form solution for estimating K_{eff} can be derived as shown in equations (4.7) and (4.8).

$$K = \sigma\sqrt{\pi a}$$

$$K_{eff} = \sigma\sqrt{\pi\left(a_{eff}\right)} = \sigma\sqrt{\pi\left(a + r_y\right)} = \sigma\sqrt{\pi\left(a + \frac{1}{2\pi}\left(\frac{K_{eff}}{\sigma_{ys}}\right)^2\right)} \tag{4.7}$$

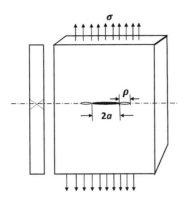

FIGURE 4.5　Schematic of Dugdale's rendering of the crack tip plastic zone in a thin semi-infinite plate.

$$K_{eff}^2 = \sigma^2 \pi \left(a + \frac{1}{2\pi} \left(\frac{K_{eff}}{\sigma_{ys}} \right)^2 \right)$$

$$K_{eff} = \frac{K}{\sqrt{\left(1 - \frac{1}{2} \left(\frac{\sigma}{\sigma_{ys}} \right)^2 \right)}} \tag{4.8}$$

In practice, the plasticity correction may be negligible; however, for high values of K and in situations that demand high accuracy, it may be necessary to use the correction.

Dugdale (1960) [2] proposed another method to account for stress re-distribution at the crack tip due to plastic deformation. He noticed that the plastic zone in thin plates containing cracks were shaped like strips (Figure 4.5). This is because slip lines in thin strips emanate from the crack tip at 45° angles and intersect the surface and stop growing in the y-direction. This derivation applies to thin sheets only because it is primarily for plane stress conditions. This alternate expression for plastic zone size is used when analyzing thin sheets used in aircraft structures and is therefore important.

Like Irwin, Dugdale also realized that infinite stresses at the crack tip was a mathematical anomaly and had to be eliminated based on physics of the problem. Irwin's derivation accounted for it by capping the stresses and Dugdale did that by first extending the crack by a distance equal to the plastic zone, ρ, at each end and then by applying a compressive stress equal to σ_{ys} over the distance from the real crack tip to ρ as shown in Figure 4.6.

The K for a wedge load applied on a crack is taken from Chapter 3:

$$K = \frac{P}{\sqrt{\pi a}} \sqrt{\frac{a+b}{a-b}}$$

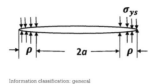

FIGURE 4.6   Closing the length of the crack over the distance of the plastic zone.

If K_{cl} is the value of K due to the closure stresses, it is estimated as follows:

$$K = K_{cl} = \frac{\sigma_{ys}}{\sqrt{\pi(a+\rho)}}\left[\int_a^{a+\rho}\sqrt{\frac{a+\rho+x}{a+\rho-x}}\,dx + \int_{-a}^{-(a+\rho)}\sqrt{\frac{a+\rho+x}{a+\rho-x}}\,dx\right]$$

$$K_{cl} = \frac{\sigma_{ys}}{\sqrt{\pi(a+\rho)}}\left[\int_a^{a+\rho}\sqrt{\frac{a+\rho+x}{a+\rho-x}}\,dx + \int_a^{(a+\rho)}\sqrt{\frac{a+\rho-x}{a+\rho+x}}\,dx\right]$$

$$K_{cl} = \frac{\sigma_{ys}}{\sqrt{\pi(a+\rho)}}\left[\int_a^{a+\rho}\frac{a+\rho+x+a+\rho-x}{\sqrt{(a+\rho)^2-x^2}}\,dx\right]$$

$$K_{cl} = \frac{2\sigma_{ys}}{\pi}\sqrt{\pi(a+\rho)}\int_a^{a+\rho}\frac{dx}{\sqrt{(a+\rho)^2-x^2}}$$

$$K_{cl} = \frac{2\sigma_{ys}}{\pi}\sqrt{\pi(a+\rho)}\cos^{-1}\left(\frac{a}{a+\rho}\right) \tag{4.9}$$

To cancel the singularity, Dugdale equated the K due to the external stress σ and the K_{cl}. Thus,

$$\sigma\sqrt{\pi(a+\rho)} = \frac{2\sigma_{ys}}{\pi}\sqrt{\pi(a+\rho)}\cos^{-1}\left(\frac{a}{a+\rho}\right)$$

$$\frac{\rho+a}{a} = \sec\left(\frac{\pi}{2}\frac{\sigma}{\sigma_{ys}}\right)$$

$$\rho = a\left[\sec\left(\frac{\pi\sigma}{2\sigma_{ys}}\right)-1\right] \tag{4.10}$$

Also, $\dfrac{a}{\rho+a} = \cos\left(\dfrac{\pi}{2}\dfrac{\sigma}{\sigma_{ys}}\right) \approx 1 - \dfrac{1}{2!}\left(\dfrac{\pi}{2}\dfrac{\sigma}{\sigma_{ys}}\right)^2 + \dfrac{1}{4!}\left(\dfrac{\pi}{2}\dfrac{\sigma}{\sigma_{ys}}\right)^4 - \cdots$

For $\sigma/\sigma_{ys} < 1$, we can neglect terms after the second term and show that,

$$\rho = \frac{\pi}{8}\left[\frac{\sigma\sqrt{\pi(a+\rho)}}{\sigma_{ys}}\right]^2 = \frac{\pi}{8}\left(\frac{K_{eff}}{\sigma_{ys}}\right)^2 = 0.392\left(\frac{K_{eff}}{\sigma_{ys}}\right)^2 \qquad (4.11)$$

Irwin's estimate was $0.318\left(\dfrac{K_{eff}}{\sigma_{ys}}\right)^2$ thus making the two estimates of plastic zone sizes quite similar. It is always reassuring when we get identical or similar results from two different approaches that are both based on good Physics.

4.3 PLASTIC ZONE SHAPE

Next, we will estimate the shape of the crack tip plastic zone first under the plane stress conditions and subsequently under plane strain conditions. Von Mises' yield criterion was presented in equation (4.2). From the Mohr's circle, or the stress transformation equations (2A.25) in Appendix 2A.5, we know that the principal stresses can be obtained from:

$$\sigma_1, \sigma_2 = \frac{\sigma_x + \sigma_y}{2} \pm \sqrt{\left(\frac{\sigma_x - \sigma_y}{2}\right)^2 + \tau_{xy}^2} \quad \text{and} \quad \sigma_3 = 0 \qquad (4.12)$$

Substituting from the values of the various stress components from the field equations we get,

$$\sigma_1, \sigma_2 = \frac{K\cos\dfrac{\theta}{2}}{\sqrt{2\pi r}}\left[1 \pm \sqrt{\left(\sin\frac{\theta}{2}\sin\frac{3\theta}{2}\right)^2 + \left(\sin\frac{\theta}{2}\cos\frac{3\theta}{2}\right)^2}\right] = \frac{K\cos\dfrac{\theta}{2}}{\sqrt{2\pi r}}\left[1 \pm \sin\frac{\theta}{2}\right]$$

We can then show that

$$\sigma_e = \frac{K}{\sqrt{2\pi r}}\cos\frac{\theta}{2}\left[1 + 3\sin^2\frac{\theta}{2}\right]^{1/2}$$

At the plastic zone boundary, $\sigma_e = \sigma_{ys}$; $r = r_y$

$$r_y(\theta) = \frac{1}{2\pi}\left(\frac{K}{\sigma_{ys}}\right)^2\cos^2\frac{\theta}{2}\left[1 + 3\sin^2\frac{\theta}{2}\right] \qquad (4.13)$$

For plane strain, $\sigma_3 = \dfrac{2vK}{\sqrt{2\pi r}}\cos\dfrac{\theta}{2}$

Substituting into the yield criterion and solving for r_y, we get

$$r_y(\theta) = \frac{1}{4\pi}\left(\frac{K}{\sigma_{ys}}\right)^2\left[(1-2v)^2(1+\cos\theta) + \frac{3}{2}\sin^2\theta\right] \qquad (4.14)$$

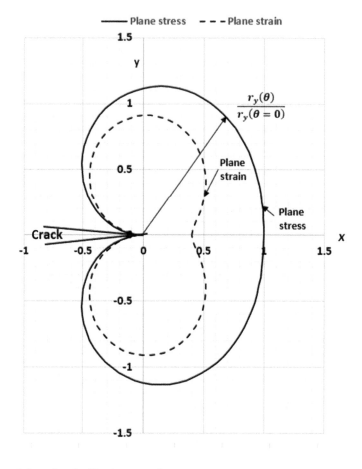

FIGURE 4.7   Boundary of the crack tip plastic zone illustrating its shape under plane stress and plane strain conditions using the Von Mises' yield criterion.

Figure 4.7 shows the shape of the plastic zone for plane stress and plane strain using equations (4.13) and (4.14). The plastic zone scales with the value of $(K/\sigma_{ys})^2$ in a self-similar manner, provided the state-of -stress remains either plane stress or plane strain. In other words, as the load (or K) increases, the cracked body must maintain plane stress or plane strain conditions. If a body starts as being under plane strain and then as K value is increased it tends toward plane stress, the expansion of the plastic zone will no longer be self-similar. The shape of the plastic zone shown in Figure 4.7 will be different if instead of Von Mises' criterion for yielding, we had chosen the Tresca criterion of yielding. Even though the shape of the plastic zone is different but the dependence of its size on K along the x-axis is identical to the one from the Von Mises' criterion. Thus, the plastic zone size and shape scale up and down with square of the value of K according to both the Von Mises and Tresca yield criteria.

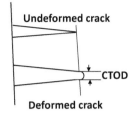

Information classification: general

FIGURE 4.8 Definition of crack tip opening displacement and blunting of the crack from plasticity.

4.4 CRACK TIP OPENING DISPLACEMENT (CTOD)

It is of interest to estimate the CTOD as a function of K in cracked bodies. CTOD is defined as the displacement in the y-direction between the crack faces at the original crack tip as shown in Figure 4.8. The original crack tip is shown to blunt from plastic deformation.

$$v = \frac{K}{2\mu}\sqrt{\frac{r}{2\pi}}\sin\frac{\theta}{2}\left[\kappa + 1 - 2\cos^2\frac{\theta}{2}\right]$$

CTOD $=2v$ for $r=r_y$ and $\theta=\pi$ and $K=K_{eff}$. Thus,

$$\text{CTOD} = 2v = \frac{K_{eff}}{\mu}\sqrt{\frac{r_y}{2\pi}}\left[\kappa + 1\right] \text{ where } \kappa = \frac{3-v}{1+v}$$

$$\text{or} \quad \text{CTOD} = \frac{8}{2\pi}\frac{K_{eff}^2}{E\sigma_{ys}} \sim \frac{K_{eff}^2}{E\sigma_{ys}} \tag{4.15}$$

4.5 SUMMARY

In this chapter, we have estimated the size and shape of the crack tip plastic zone under plane stress and plane strain conditions under SSY conditions. We have shown that the plastic zone size scales linearly with value of $(K/\sigma_{ys})^2$ and expands with increasing value of K in a self-similar manner if the state-of-stress remains either as plane stress or plane strain. If the state of stress switches from one to the other, self-similarity is not preserved. We have derived equations that allow for estimating the K value that is modified for plastic deformation at the crack tip under SSY condition. The relationship between CTOD and K was also derived and it is shown that CTOD is uniquely determined by K. Thus, fracture criteria based on K and CTOD are equivalent.

REFERENCES

1. G.R. Irwin, "Plastic Zone Near a Crack and Fracture Toughness", Proceedings of the 7th Sagamore Conference, 1960.
2. D.S. Dugdale, "Yielding of Steel Sheets Containing Slits", *Journal of Mechanics and Physics of Solids*, Vol. 8, 1960, pp. 100–108.
3. W. Johnson and P.B. Mellor, *Engineering Plasticity*, Van Nostrand Reinhold, New York, NY, 1973.
4. R. Hill, *Plasticity*, Clarendon Press, Oxford, England, 1950.

HOMEWORK PROBLEMS

1. Explain the differences between the plastic zone size estimates, r_y and r_p. Derive a relationship between the two.
2. An aircraft grade Al-alloy panel has a crack 2 cm in length in the middle of the panel. The yield strength of the alloy is 400 MPa. If the panel is loaded to a stress of 200 MPa normal to the crack plane, calculate the value of K without the plasticity correction and with plasticity correction. Estimate the error that occurs in K estimation if plasticity is neglected?
3. What will the plasticity correction be if in problem 2 the material is a high strength steel with a yield strength of 1,000 MPa?
4. Show that the plane strain plastic zone size is one-fourth of the plane stress plastic zone size if the Poisson's ratio for the material is 0.25. Give a physical reason for why plane strain plastic zone is much smaller than the plane stress plastic zone size.
5. Explain what is meant by "effective crack size." Why is this correction necessary?
6. What is the $r_y(\theta)$ for $\theta = 45°$ and K at $50\,\text{MPa}\sqrt{m}$ in a high strength steel with a yield strength of 1,000 MPa *under plane stress*? Repeat the calculation for a material with a yield strength is 500 MPa.
7. What does the term SSY imply and why is it important to ensure that SSY conditions exist for stress intensity parameter, K, to be a viable crack tip parameter.
8. The actual plastic zone size for plane strain is said to be given by,

$$r_p = \frac{1}{3\pi}\left(\frac{K}{\sigma_{ys}}\right)^2$$

The above estimate is larger than the estimated value for pure plane strain conditions, why?
9. Explain with the help of a diagram, what is meant by crack tip opening displacement, CTOD?
10. Compare the value of elastically calculated K to the plasticity corrected value of K for a finite size center crack panel as the load is increased. The expression for K is given below. Assume that $W = 10\,\text{cm}$, $a = 1\,\text{cm}$, and the applied stress is initially 200 MPa and is steadily increased to 600 MPa in increments of 100 MPa. The yield strength of the material is given as 500 MPa.

$$K = \sigma\sqrt{\pi a}\left[\sec\left(\frac{\pi a}{2W}\right)\right]^{1/2}\left[1 - 0.025\left(\frac{a}{W}\right)^2 + 0.06\left(\frac{a}{W}\right)^4\right]$$

11. Calculate the plastic zone shape using the Tresca yield criterion and compare it the one calculated using Von Mises' criterion shown in Figure 4.7.

APPENDIX 4A: PLASTIC YIELDING UNDER UNIAXIAL AND MULTIAXIAL CONDITIONS

4A.1 UNIAXIAL STRESS–STRAIN CURVE

As also briefly described in Appendix 2A, in a uniaxial state of stress, when the applied stress, σ, exceeds the yield strength of the material, σ_{ys}, permanent deformation (or plastic deformation) occurs. Often, it is not possible to precisely define the critical stress at which plastic deformation commences, therefore, the operational definition of yield strength is used where the stress required to cause a plastic strain of 0.2%. Beyond the yield strength, the material continues to deform plastically until instability is reached. The stress at which instability occurs, σ_u, is the ultimate tensile strength and the corresponding strain, ε_u, is called the maximum uniform strain. The latter term results from the observation that up to ε_u the strain is distributed uniformly in the specimen and beyond $\varepsilon = \varepsilon_u$, Figure 2A.4, it is concentrated in the region where a neck develops and eventually fracture occurs. The stress–strain relationship can be described by a relationship called the Ramberg–Osgood relationship:

$$\varepsilon = \frac{\sigma}{E} + \alpha\left(\frac{\sigma}{\sigma_{ys}}\right)^m \tag{4A.1}$$

where α and m are material constants derived from regression of the stress–strain curve and $\varepsilon_o = \sigma_{ys}/E$. Both m and α are dimensionless. m is also inverse of the strain hardening exponent which is frequently reported in the literature along with data from tensile tests. Elastic, perfectly-plastic materials are those which have a flat stress–strain curve when $\sigma = \sigma_{ys}$.

4A.2 VON MISES YIELD CRITERION FOR MULTIAXIAL LOADING

Stress states in the crack tip region are more complex than in a uniaxial tensile test. Therefore, it is essential to choose a yield criterion for multiaxial stress states. The one used commonly in fracture mechanics is the Von Mises criterion. For a detailed discussion on other yield criteria, the reader should consult other books [3,4]. The assumptions which accompany the use of the Von Mises criterion are:

- The yield strengths in tension and compression are the same.
- The volume during plastic deformation is conserved, thus the Poisson's ratio is 0.5.
- The mean normal stress (or the hydrostatic stress), $\sigma_m = \frac{1}{3}(\sigma_1 + \sigma_2 + \sigma_3)$, does not participate in the plastic deformation process.

The last assumption implies that yielding must occur when the function f of $(\sigma_1 - \sigma_2)$, $(\sigma_2 - \sigma_3)$ and $(\sigma_3 - \sigma_1)$ is a constant:

$$f\left[(\sigma_1 - \sigma_2),(\sigma_2 - \sigma_3),(\sigma_3 - \sigma_1)\right] = C \tag{4A.2}$$

Thus, if a stress state defined by σ_1, σ_2, σ_3, causes yielding, an equivalent stress state, $\sigma_1' = \sigma_1 - \sigma_m$, $\sigma_2' = \sigma_2 - \sigma_m$, $\sigma_3' = \sigma_3 - \sigma_m$ will also cause yielding because the two stress states differ only by the mean normal stress which does not participate in yielding. Mises postulated that yielding occurs when the root mean square of the three maximum values of the shear stresses becomes equal to or exceeds a certain constant value:

$$\left[\frac{(\sigma_1 - \sigma_2)^2 + (\sigma_2 - \sigma_3)^2 + (\sigma_3 - \sigma_1)^2}{3}\right]^{1/2} = C_1$$

$$\text{or} \quad (\sigma_1 - \sigma_2)^2 + (\sigma_2 - \sigma_3)^2 + (\sigma_2 - \sigma_3)^2 = C_2 \tag{4A.3}$$

Applying the above criterion to the tensile test yields:

$$C_2 = 2\sigma_{ys}^2 \tag{4A.4}$$

In the more general form, the criterion can be written as:

$$\left(\sigma_x - \sigma_y\right)^2 + \left(\sigma_y - \sigma_z\right)^2 + \left(\sigma_z - \sigma_x\right)^2 + 6\left(\tau_{xy}^2 + \tau_{yz}^2 + \tau_{zx}^2\right) = 2\sigma_{ys}^2 \tag{4A.5}$$

Equations (4A.3) and (4A.5) lead to the definition of effective (or equivalent) stress, σ_e, as being:

$$\sigma_e = \frac{1}{\sqrt{2}}\left[(\sigma_1 - \sigma_2)^2 + (\sigma_2 - \sigma_3)^2 + (\sigma_3 - \sigma_1)^2\right]^{1/2} \tag{4A.6}$$

For uniaxial tensile loading, it is seen that $\sigma_e = \sigma_1$. Plastic deformation during multi-axial loading occurs when $\sigma_e = \sigma_{ys}$. Also, under plane stress conditions at the crack tip, if $\sigma_1 = \sigma_2$, and $\sigma_3 = 0$, σ_e is also $= \sigma_{ys}$

4A.3 TRESCA YIELD CRITERION

The Tresca criterion states that plastic deformation in a body subjected multiaxial stresses occurs when the largest of $(\sigma_1 - \sigma_2)$, $(\sigma_2 - \sigma_3)$, and $(\sigma_3 - \sigma_1)$ exceeds σ_{ys}.

5 Fracture Toughness and its Measurement

In the pre-fracture mechanics era (prior to 1960s), the impact energy absorbed during fracture was measured primarily by a simple test called the Charpy impact test, first proposed by Georges Charpy in 1905. The experimental set-up used for conducting this test is shown in Figure 5.1. It consists of a hammer of a standard weight that is dropped from a fixed height to strike a specimen of standard size containing a sharp and precisely machined notch. The difference in the potential energy of the hammer at the start of the swing and at the end of the swing is the Charpy energy absorbed by the specimen during fracture. Several identical specimens are tested to measure the energy absorbed as a function of temperature to characterize the ductile to brittle transition behavior of steels as schematically shown in Figure 5.2. The energy absorbed in ferritic steels as a function of temperature can be divided in three distinct regimes (a) the lower-shelf behavior in which the energy absorbed is consistently low, (b) the transition region in which the toughness rises sharply with temperature, and (c) the high toughness upper-shelf region characterized by a plateau. The temperature at which the fracture behavior is a mix of 50% brittle and 50% ductile is called the fracture appearance transition temperature (FATT) and is an important parameter in design and in choosing materials and heat treatments specified for given applications. The service temperatures of components made from the material must always be greater than FATT and preferably in the upper shelf region.

As an example, the actual Charpy energy data for a weld from a A302B steel in aged and unaged conditions is shown in Figure 5.3 [1]. The data clearly show the ability of the test to measure changes in the fracture resistance of the steel as a function of the aging treatment. In this example, the aging treatment increases the FATT and restricts the useful temperature range of the weldment. Thus, a heat treatment that lowers FATT enhances the apparent fracture toughness of the material and conversely, increasing FATT is an indication of embrittlement in the material.

Another observation from Figure 5.3 is that the impact energy absorbed remain the same in lower-shelf and upper-shelf regions with temperature. The aging treatment only shifts the FATT to a higher value. These are useful practical observations in the toughness behavior that have guided material selection for over a 100 years.

Despite developments in fracture mechanics, Charpy test is still used today as a quality control test by material producers because it is inexpensive and can be performed very quickly. The results of these tests along with results from other tests such as material chemistry and tensile tests are used to ensure material quality. However, Charpy data is unable to provide quantitative measure of fracture toughness for estimating allowable stresses in components and/or selecting nondestructive inspection techniques leading to accept/reject criteria for components, consequently not fulfilling important needs of the material selection and the design process. Fracture

DOI: 10.1201/9781003292296-5

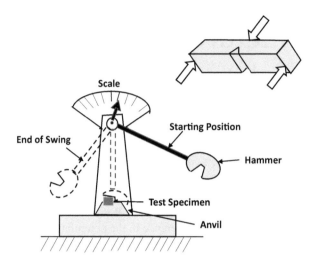

FIGURE 5.1   The test specimen and a set up for the Charpy impact test.

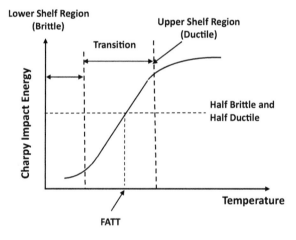

FIGURE 5.2   Schematic of Charpy impact energy as a function of temperature for ferritic steels.

toughness tests based on fracture mechanics provide this important complementary information.

In the previous chapters, we have built a rigorous analytical framework and have identified stress intensity parameter, K, as a candidate parameter for predicting fracture in engineering materials and structures. In this chapter, we will explore how K is used to measure the ability of engineering materials to resist fracture, and the onset of fracture instability. Fracture toughness is a term used to represent the material's intrinsic resistance to fracture and is a material property that can be used in quantitative assessments of the risk of fracture. This chapter is about describing methods available for reliably measuring fracture toughness of materials.

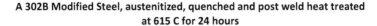

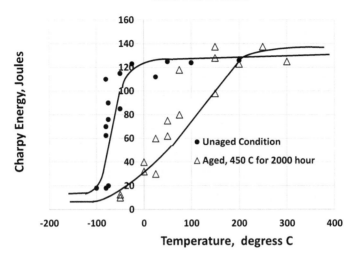

FIGURE 5.3   Charpy energy data as a function of test temperature.

As one can imagine that the fracture toughness of brittle materials such as concrete, ceramics, and glasses are expected to be very low and therefore these materials
are primarily used in applications involving only compressive stresses. The critical K for fracture in these materials is $0.5 - 2\,\mathrm{MPa}\sqrt{m}$. Among metals, there are a
wide variety of alloys with fracture toughness values ranging from low to very high.
Examples of low toughness metals include high strength aluminum alloys, cast irons,
and ultra-high strength steels with tensile strength levels in the range of $2\,\mathrm{GPa}$. By
comparison, even these low toughness metallic alloys have fracture toughness values
in the range of $20 - 40\,\mathrm{MPa}\sqrt{m}$ that are an order of magnitude higher than those
found in ceramics, glasses, and concrete. Examples of very high fracture toughness
materials (critical K for fracture of $200 - 300\,\mathrm{MPa}\sqrt{m}$) include copper and its alloys
and austenitic stainless steels that have fracture toughness values that are an order
of magnitude higher than the low toughness metallic alloys. There are also several
engineering alloys whose toughness values lie between the extremes of low toughness and the very high toughness materials. Thus, quantitatively measuring fracture
toughness of engineering materials is important in material selection and in design
of components made from those materials.

5.1   SIMILITUDE AND THE STRESS INTENSITY PARAMETER, K

In Chapters 3 and 4, we established the mechanics underpinnings of the stress intensity parameter, K and its limitations. We argued that K characterizes fracture and
crack growth *under small-scale yielding* (SSY) conditions because:

1. *K uniquely* characterizes the crack tip stress, strain, and displacement fields
   in the vicinity of the crack tip,

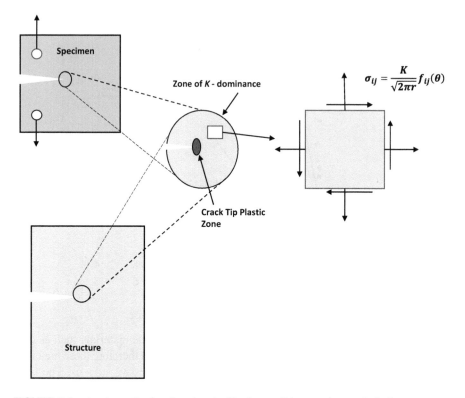

FIGURE 5.4   A schematic showing the similitude conditions at the crack tip in test spec-imens and structures of different geometry and sizes provided SSY conditions are main-tained (T.L. Anderson, Fundamentals of Fracture Mechanics, CRC Press, reproduced with permission.).

2. Through its relationship with *G* for both plane stress and plane strain, it *uniquely* characterizes the energy available for crack extension, and

3. It *uniquely* determines the size and shape of the crack tip plastic zone under plane stress or plane strain conditions under SSY. In other words, for the same applied *K* and stress state, the crack tip plastic zone shape and size are identical for different crack geometries and loading configurations if plane stress or plane strain conditions are maintained.

In Figure 5.4, similitude is shown to exist between the crack tip region in a labora-tory specimen shown in the upper left-hand corner of the figure and the crack tip region of a hypothetical structure shown at the bottom of the figure [2]. Even though the geometries and loading configurations of the specimen and the structure are dif-ferent, the crack tips in both cases see identical environments if the values of *K* are the same, SSY conditions are maintained, and the body remains either in plane stress or in the plane strain conditions. Thus, laboratory specimens of a variety of sizes and shapes may be used to measure fracture toughness of a material and the measured value of fracture toughness may be used to predict the fracture behavior

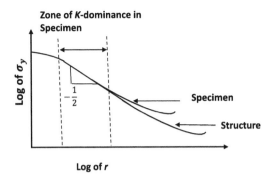

FIGURE 5.5 Crack tip stress distribution considering plastic deformation at the crack tip but also showing the matching stresses in the elastic region of the test specimen and the structure where K dominates the stresses.

of structures of completely different shape and size but made from the same material with identical chemical composition and heat-treatment. For example, if the fracture toughness in 7075-T6[1] aluminum alloy is $30\,\mathrm{MPa}\sqrt{m}$, then fracture in a component of this material is also expected to occur at $30\,\mathrm{MPa}\sqrt{m}$. Considerable effort has gone into development of test methods that are described next for accurately measuring the value of fracture toughness.

The existence of similitude and its limits in the crack tip regions of differently shaped and loaded cracked bodies subjected to the same value of K is explored next. In Figure 5.5, the normal stress in the y-direction in front of crack tip is schematically shown for the test specimen and for the structural component. In the first region plasticity occurs, the size and shape of which, is fully determined by K if SSY prevails. Thus, the stresses in this region are identical in test specimens and in the structure if the K values are the same. In the next region, stresses are elastic and are also governed by the K-dominated crack tip stress fields so, they are the same in the test specimen and the structure loaded to equal values of K. In the third region, the stress distributions are characteristic of the geometry and loading configuration and may be different in the specimen and the structure. However, if the first two regions are sufficiently large so the fracture processes occur entirely within those regions, the fracture behavior of the specimen and the structure is expected to be characterized by K. This is the justification for use of K for predicting fracture in components. By fracture processes we mean the nucleation and growth of voids in the case of ductile fracture and the development of cleavage conditions in the case of brittle fracture.

5.2 FRACTURE TOUGHNESS AS A FUNCTION OF PLATE THICKNESS

In the previous chapter, we considered how crack tip plasticity depends on whether we choose plane stress or plane strain conditions. Plane stress conditions occur in thin sheets because free surfaces cannot support stresses. Thus, on free surfaces, the normal stress and two components of shear stress are zero, and since in thin bodies,

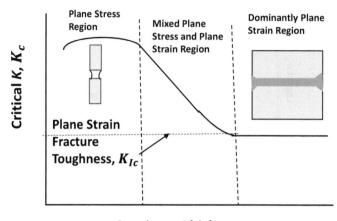

Specimen Thickness

FIGURE 5.6 A schematic showing how fracture toughness is expected to vary between constant values for plane stress and plane strain, and through the transition in the mixed mode region. Schematics of plastic deformation patterns in a plane parallel to the back face of the specimen but near the crack tip region are shown in the inset of the figure. On the surface of the specimen, the state-of-stress is always plane stress, but in the interior, it is plane strain for thick bodies.

the distance between the two surfaces is small, the normal stress across the thickness must also be approximately zero leading to the conditions of plane stress throughout the thickness of the specimen. Plane strain, on the other hand, develops in thick planar bodies. If one considers several plates with variety of thicknesses ranging from say 1 mm to several tens of mm but with identical planar dimensions containing edge cracks, we can expect dominantly plane stress conditions to exist at the crack tips of thin plates and dominantly plane strain conditions to exist at the tips of cracks in thick plates. We use the term dominantly plane strain because the conditions at the two free surfaces even in thick bodies are under plane stress but in the bulk of the body, conditions are closer to that of plane strain. Fracture toughness values measured in thin plate materials that develop much larger plastic zones can be expected to be higher than those for thick plates that are under plane strain conditions. This is illustrated schematically in Figure 5.6.

In Figure 5.6, we have shown the fracture toughness as being constant in the plane stress and plane strain regions and as being thickness dependent in the region in-between plane stress and plane strain. The plane stress fracture toughness is an idealized concept and has not been experimentally demonstrated because the thickness range over which it occurs is limited. The plane strain fracture toughness region, on the other hand, occurs for all thicknesses greater than a certain value. Also, since the plane strain fracture toughness represents the lower bound of the material's toughness, there is more interest in measuring that value. This value is designated as K_{Ic} and is a material property that does not depend on specimen geometry and size, and it represents the K value at which fracture is expected to occur in thick sections that meet the requirement of being under dominantly plane strain condition.

5.3   DUCTILE AND BRITTLE FRACTURE
AND THE LEFM APPROACH

We will now explore the types of fractures that can be addressed by linear elastic fracture mechanics (LEFM). Because of the underlying assumptions, not all fractures can be addressed by LEFM. At the macro-level, ductile and brittle fractures are characterized by respectively high and low amounts of energy absorbed in the form of plastic deformation during fracture. There are underlying mechanisms that operate at the atomic and micro-meter length-scales (dimensions of a grain) that determine whether the fracture will be brittle or ductile as discussed previously in Chapter 2 but are discussed here in more depth as part of LEFM.

Cleavage fracture proceeds along definite crystallographic planes, Figure 2.2b, such as along {100} planes in BCC metals that have the lowest planar atomic density among primary planes and consequently require breaking the fewest atomic bonds per unit crack area. The direction of fracture plane is expected to vary as it moves from one grain to the next following the orientation of the low atomic density planes among adjoining grains. Cleavage fracture is brittle by nature and if it is the dominant mechanism of fracture, LEFM is the appropriate analytical framework for addressing it. This will be discussed further in the next section.

Ductile fracture occurs by nucleation, growth and coalescence of micro-voids that form on impurities present in metals known as inclusions. All engineering materials contain these inclusions that are potential sites for facilitating fracture. Figure 2.3 shows how ductile fractures appear at high magnifications of about $5,000x$ under an electron microscope. The plastic deformation required to grow the voids prior to coalescence between adjoining voids absorbs energy and contributes to ductile behavior. Thus, ductile fractures require that voids that nucleate at inclusions or at large incoherent precipitates, be able to grow by plastic deformation. Unstable fracture is then preceded by stable crack growth. LEFM framework may have limitations in dealing with such fractures because they could potentially be accompanied by large scale plasticity and no longer occur under SSY conditions. However, if the specimens are large and SSY conditions are met, LEFM can be used to characterize the stable crack growth resistance.

When adjoining voids are located very close to each other and do not have much space to grow, or if the material between the voids is brittle because of coherent precipitates[2] that raise strength but decrease ductility, fracture can be brittle despite being the micro-void growth and coalescence type. An example of such brittle fracture is seen in high-strength aluminum alloys where voids that form on the larger incoherent particles are unable to grow significantly before coalescing with the neighboring voids. LEFM can be used to describe such fractures if SSY conditions are maintained during the test. Such fractures occur suddenly without any significant stable crack growth, much like cleavage fractures. High strength, precipitation hardened aluminum alloys used extensively in aircraft structures are examples of such materials.

Conditions of plane strain in the crack tip region constrain plastic deformation and cause materials to behave in a brittle fashion. Fracture toughness of ferritic steels that have body-centered cubic (BCC) crystal structure depends on temperature much like the Charpy energy as shown in Figure 5.2. The fracture toughness differences between the "lower shelf" and the "upper shelf" regions in ferritic steels can be as large as a factor of five. Other materials such as aluminum alloys, stainless steels, and Nickel based alloys that have a face-centered cubic (FCC) crystal structures do not go through a ductile-to-brittle transition. Only BCC steels and hexagonal closed packed materials undergo this transition.

To prevent fracture during service, materials are chosen such that the service temperature is in the upper transition or the upper shelf regions. If this cannot be assured, extreme care must be taken to avoid sudden high loads on components when they are cold. For example, there is the high fracture risk in an engine block in cold climates when a car is started from cold condition and is quickly accelerated, particularly if the engine block is made from cast steel. If the engine block is made from aluminum, the risk is lower because aluminum does not undergo a ductile to brittle transition. Never-the-less, to avoid high thermal stresses, it is always good practice to warm up the engine by letting it run on idle until the temperature gradients stabilize, prior to subjecting it to high speeds.

5.4 MEASUREMENT OF FRACTURE TOUGHNESS

This discussion is divided into measuring the plane strain fracture toughness, K_{Ic}, for thick bodies with cracks and the fracture toughness of thin panels in which the fracture toughness may be dependent on the thickness of the specimen, as in Figure 5.6.

5.4.1 MEASUREMENT OF PLANE STRAIN FRACTURE TOUGHNESS, K_{Ic}

Choice of test specimens: The specimen geometries typically used for fracture toughness testing are shown in Figure 5.7. Several other geometries may be used, depending on the product form from which specimens are machined. For example, if the specimen is being taken from an aircraft panel, the middle crack tension specimen[3] might be most suitable specimen type and if the product form is cylindrical in shape, the disk-shaped compact type specimen might be best. The American Society for Testing Materials (ASTM) standardizes [3] the geometries of fracture specimens as part of the test specifications. The key dimension of the specimen in each case that must be specified by the user is the width, W. The thickness, B is 0.5 W for plane strain fracture toughness testing. The reason for this will be explained later. ASTM Standard E399 [3], in addition to specimen geometry, specifies the test procedure and data analysis procedures for measuring Plane Strain Fracture Toughness, K_{Ic}.

Once the specimen type is chosen and machined, the specimen is pre-cracked using fatigue loading to produce a sharp and a reproducible crack ahead of the machined notch. Then the load is gradually applied under the conditions of constant load-line displacement rate until the specimen fractures. The force-displacement diagram is recorded during loading so the point of fracture can be precisely identified.

The K-solutions for the various geometries are listed in the form given by equation (5.1):

$$K = \frac{P}{BW^{\frac{1}{2}}}F\left(\frac{a}{W}\right) \qquad (5.1)$$

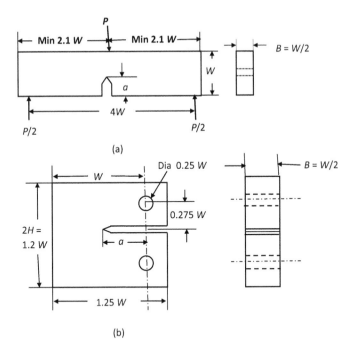

(a)

(b)

FIGURE 5.7 Geometries of some select specimen types commonly used in measuring fracture toughness using the ASTM Standard E399. (a) Single edge crack bend, SEC(B) specimen standard compact type C(T) specimen.

$F(a/W)$ values for C(T) specimens is given in equation (5.2) and in equation (5.3) for single edge crack-bend, SEC(B), specimen.

$$F\left(\frac{a}{W}\right) = \frac{2 + a/W}{(1 - a/W)^{3/2}}$$

$$\times \left[0.886 + 4.64(a/W) - 13.32\left(\frac{a}{W}\right)^2 + 14.72\left(\frac{a}{W}\right)^3 - 5.6(a/W)^4 \right] \quad (5.2)$$

$$F(a/W) = 3\sqrt{\frac{a}{W}} \left[\frac{1.99 - (a/W)(1 - a/W)\left[2.15 - 3.93\left(\frac{a}{W}\right) + 2.7(a/W)^2\right]}{2(1 + 2(a/W))(1 - a/W)^{3/2}} \right] \quad (5.3)$$

The force, P, versus displacement, V, diagrams during fracture testing are measured and recorded. The displacement can be measured along the load-line, or it can be measured at the crack-mouth. The diagrams may show initial nonlinearity in the force-displacement behavior due to settlement in the seating of the displacement gage, but that disappears once the load levels are increased. All diagrams show a linear force-displacement region in which the inverse of the stiffness is the elastic

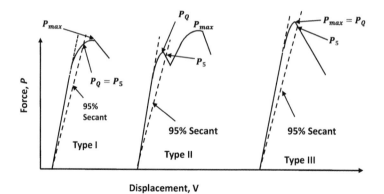

FIGURE 5.8　Types of force-displacement diagrams recorded during fracture toughness testing.

compliance of the specimen, and it depends on the initial crack size after fatigue pre-cracking. Different types of force-displacement behavior are observed that vary with the material type when the force levels approach the fracture point.

The force-displacement diagrams fall in one of the following categories designated as Type I, Type II, or Type III shown in Figure 5.8. In Type I behavior, the force-displacement diagram becomes nonlinear due to plasticity or due to stable crack growth and fracture occurs at a force equal to P_{max}. In Type II behavior, events called "pop-in" occur that are caused by small amounts of localized crack extension along the crack front, resulting in a sudden force drop; subsequently, the fracture stabilizes, and the force levels rise again until fast fracture occurs at P_{max}. The Type III force-displacement diagram stays essentially linear until fracture suddenly occurs at a force of P_{max}.

Next, we must choose a force level that we will call P_Q, to be used as the candidate force level for calculating the critical value of the stress intensity parameter. To determine P_Q, we draw a line with a slope that is 5% less than the slope of the linear elastic region of the force-displacement diagram as shown in Figure 5.8. The intersection of this 5% secant offset line and the force-displacement diagram is noted as P_5.

Type III behavior is the simplest for choosing P_Q from the force-displacement behavior which remains linear elastic almost all the way to fracture. In this case, P_{max} lies to the left of P_5. This implies that fracture occurred before any substantial plastic deformation or stable crack growth occurred in the specimen and the fracture is brittle and sudden. This is an ideal scenario for measuring fracture toughness using fracture mechanics, provided it meets some qualification requirements discussed later. The fracture force chosen as P_Q is the maximum force and is used for calculating the fracture toughness, K_Q, the candidate value of K_{Ic}.

In Type I force-displacement diagram, P_5 is less than P_{max}. For this case, $P_Q = P_5$. The nonlinearity in the force-displacement diagram is due to either stable crack growth or due to increasing plasticity with increasing K value, or due to both. Thus, P_Q is chosen at a point where we limit the amount of nonlinearity in choosing its value.

For Type II force-displacement behavior, if the point of pop-in is to the left of P_5, then P_Q is the load at pop-in. If not, $P_Q = P_5$. Again, P_5 sets the limits of acceptable nonlinearity whether its source is crack tip plasticity or stable crack growth.

We next measure the crack size at fracture on the surface of the fractured specimen and calculate the value of K_Q using equation (5.1). For K_Q to qualify as K_{Ic}, the conditions specified in equations (5.4) and (5.5), must be met.

$$B, a, (W - a) \geq 2.5 \left(\frac{K_Q}{\sigma_{ys}} \right)^2 \qquad (5.4)$$

$$\text{and} \quad \frac{P_{max}}{P_Q} \leq 1.1 \qquad (5.5)$$

If either condition is not met, $K_Q \neq K_{Ic}$ and it signifies that a larger specimen size is needed to measure a valid K_{Ic}. We can also see why B and crack size, a, are 0.5 W. The optimum specimen is when a, $(W-a)$ and B are all the same which implies that B and the crack size, a, must be as close to 0.5 W as possible. Thus, if one dimension satisfies equation (5.4), the others will also satisfy that equation.

The purpose of the requirement in equation (5.4) is to ensure that the specimen is thick enough to be in dominantly plane strain condition and the planar dimensions are large enough to be considered as a dominantly linear elastic body or under SSY. The factor 2.5 ensures that the thickness and the other length dimensions of the specimen are about 25 times of the plane strain plastic zone size. The requirement in equation (5.5) further ensures that the fracture occurs under dominantly elastic, or under the SSY conditions. If plasticity is no longer small scale, the force-displacement curve will become nonlinear at lower values of force and P_Q will be smaller in comparison to P_{max} and will not meet the condition in equation (5.5).

Example Problem 5.1

Figure 5.9 shows the force-displacement diagram of a compact type specimen tested according to ASTM standard E399 to measure K_{Ic} of a high strength steel used in construction of bridges. The specimen dimensions are given as $W = 10$ cm, $B = 5$ cm, $a = 5$ cm. Determine the K_{Ic} and provide justification for whether it is a valid measurement. Assume that the yield strength of the material is 1,000 MPa.

$$K_Q = \frac{P_Q}{BW^{1/2}} \frac{\left(2 + \dfrac{a}{W}\right)}{\left(1 - \dfrac{a}{W}\right)^{3/2}} \left[0.886 + 4.64(a/W) - 13.32\left(\frac{a}{W}\right)^2 + 14.72\left(\frac{a}{W}\right)^3 - 5.6(a/W)^4 \right]$$

$$B = 5 \text{ cm}, \quad W = 10 \text{ cm}, \quad a = 5 \text{ cm}, \quad \text{and} \quad \sigma_{ys} = 1{,}000 \text{ MPa}$$

$$K_Q = \frac{0.156 \times 9.66}{0.05 \times (0.01)^{1/2}} = 95.31 \text{ MPa}\sqrt{m}$$

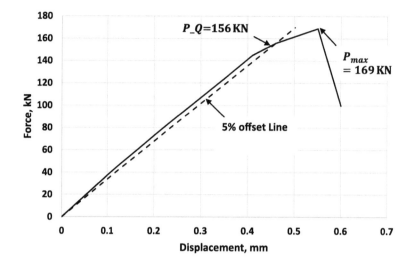

FIGURE 5.9   The force-displacement diagram for Example Problem 5.1.

$$\text{Thus, } 2.5\left(\frac{K_Q}{\sigma_{ys}}\right)^2 = 2.5\left(\frac{95.31}{1,000}\right)^2 = 0.0227 \text{ m} = 2.27 \text{ cm}$$

$$B, (W-a), a \text{ are } > 2.5\left(\frac{K}{\sigma_{ys}}\right)^2$$

Also, $\dfrac{P_{max}}{P_Q} = 1.0833$ which is less than 1.1

Since K_Q meets all the requirements to qualify as K_{Ic}, the $K_{Ic} = 95.31\,\text{MPa}\sqrt{m}$.

5.4.2 FRACTURE TOUGHNESS OF THIN PANELS

Fracture toughness of ductile materials that sustain stable crack extension prior to fracture, especially in the case of thin panels, is not represented by K_{Ic}. In these materials, as stable crack growth progresses, the resistance to fracture rises so the K level must be raised to continue crack extension. Fracture behavior is thus characterized by an R-curve of the type we discussed in Chapter 2. An example of such an R-curve is schematically shown in Figure 5.10. ASTM Standard E561 [4] addresses the method to obtain R-curves and is briefly described next. This method is used for determining toughness of aircraft panels among other applications.

If we superimpose applied K-trends with crack size for a constant applied external stress of different magnitudes ranging from σ_1 to σ_4 as in Figure 5.10, we can predict how much stable crack growth is expected to occur prior to fast-fracture. For stable crack growth to occur,

$$K > K_R \tag{5.6}$$

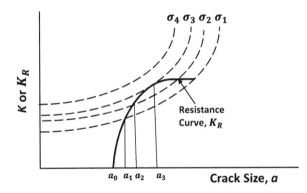

FIGURE 5.10 A schematic diagram showing the K versus crack size curves for different applied stress levels superimposed on the K-resistance curve (also known as the R-curve) to demonstrate the concept of stable crack growth and instability in ductile materials, especially in thin panels under plane stress conditions.

and for crack growth to become unstable, in addition to the above condition,

$$\left(\frac{\partial K}{\partial a}\right)_{\sigma} \geq \frac{dK_R}{da} \tag{5.7}$$

Under constant displacement rate conditions, instability in ductile materials occurs only after sustaining some stable crack growth. The tests recommended in E561 are thus conducted under constant displacement rate conditions so the R-curve can be characterized for significant amounts of crack growth. This phenomenon is very similar to one discussed in Chapter 2 where G was used as the crack extension force for driving stable crack growth. Since K and G are uniquely related to each other, this behavior is entirely expected.

The ASTM E561 standard recommends using primarily C(T) specimens and center crack panels also referred to as the middle crack tension M(T) specimen for generating stable crack growth data. In this case, W determines the planar dimensions of the specimen and the specimen thickness, B, is treated as an independent dimension and is selected to match the thickness of the component because the R-curve, also referred to as the K_R-curve, is sensitive to thickness and applies only to cracked bodies of the same thickness.

Determining the K_R-curve requires that we measure crack size at several points on the force-displacement curve. This is done using the compliance relationship between elastic compliance and the crack size using the slopes of the force-displacement lines at various points in the test as shown Figure 5.11. The crack size at any point can be obtained from the measured compliance value at that load level. In this construction, the various points at defined intervals in the force-displacement diagram are connected to the origin and inverse of the slope is equal to the compliance. The compliance and crack sizes are related by equation (5.8).

$$C_i = \frac{V_i}{P_i} = \frac{1}{BE} g\left(a/W\right) \tag{5.8}$$

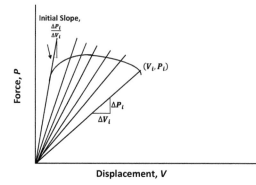

FIGURE 5.11 Schematic of a force-displacement diagram with superimposed compliance plots at various points along the diagram during a *K*-resistance test.

The function $g(a/W)$ depends on the specimen geometry and the measurement location of the displacement vis-à-vis, the position of the displacement gage in the sample. These functions have been derived by Saxena and Hudak [5] for C(T) and M(T) specimens and are described in Appendix 5A. These relationships are used in several ASTM standards for measurement of fracture toughness and crack growth testing. When we connect the various points on the force-displacement diagram such as in Figure 5.11, to the origin and compute the crack length from the compliance, we are essentially determining the effective crack size that includes the physical crack size plus r_y. This is because we have not separated the contribution due to plastic deformation from the total displacement.

The *K*-resistance curve is a plot of K_R with the amount of physical crack growth, Δa. Thus, we must subtract the value of r_y from the effective crack length determined from the compliance relationship to determine the physical crack growth. This procedure is illustrated in Example Problem 5.2.

The K_R-resistance is a measure of the resistance to stable crack growth under plane stress or mixed mode between plane stress and plane strain, but the planar dimensions must meet the conditions of SSY. Thus, size criteria have been established to exclude data that lie outside the bounds of dominantly elastic conditions. The following conditions have been stipulated by ASTM-E561 to ensure SSY conditions in the specimens tested [4].

For C(T) specimens, the qualified K_R versus Δa data points must meet the condition that the remaining ligament must be greater than four plane stress plastic zone sizes.

$$(W - a) \geq \frac{4}{\pi}\left(\frac{K_R}{\sigma_{ys}}\right)^2 \tag{5.9}$$

For M(T) specimens it is stipulated that the net section stress in the uncracked ligament must be less than the yield strength. This condition is expressed as follows:

$$R_V = \left(P/(2B(W - a)/\sigma_{ys}\right) < 1 \tag{5.10}$$

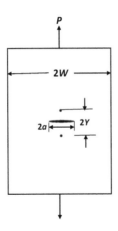

FIGURE 5.12 M(T) specimen used for characterizing K_R-curve for an Al alloy panel in Example Problem 5.2.

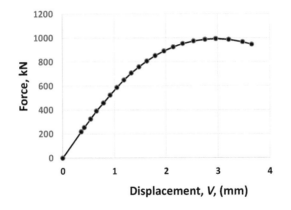

FIGURE 5.13 The force-displacement diagram for the specimen in Example Problem 5.2.

Only data that meet the conditions specified in equations (5.9) and (5.10) for C(T) and M(T) specimens respectively, can be qualified as having met the SSY requirement. The equations for estimating K and for compliance as a function of crack size and the inverse equations for estimating crack size from compliance for C(T) and M(T) specimens are given in Appendix 5A.

Example Problem 5.2

A long and wide middle crack tension, M(T), specimen of the type shown in Figure 5.12 was used to characterize the K_R-curve of an aluminum alloy. The following information is provided about the material and the specimen. $W = 760$ mm, $B = 6.5$ mm, Initial half crack length after fatigue pre-cracking, $a_{or} = 125$ mm, elastic modulus, $E = 70$ GPa and $\sigma_{ys} = 500$ MPa. Half gage length of the displacement gage, $Y_{0r} = 14.1$ mm. The force-displacement points measured are shown in Figure 5.13 and are listed in Table 5.1.

TABLE 5.1

Parameters Needed to Derive the *K*-Resistance Curve from the Measured Force-Displacement Behavior of an M(T) Panel from an Aluminum Alloy

V (mm)	P (kilo N)	BE (V/P)	a_{eff}/W	a_{eff} (mm)	Δa_{eff} (mm)	K_{eff} ($MPa\sqrt{m}$)	r_y (mm)	a (mm)	Rv	Δa (mm)
0	0			125	0.00	0	0.00	125.00	0.00	0.00
0.36	218.7	0.510	0.327	124.2	−0.77	29.6	0.56	123.67	0.05	−1.33
0.42	254	0.513	0.328	124.6	−0.39	34.5	0.76	123.85	0.06	−1.15
0.54	324	0.519	0.330	125.5	0.54	44.2	1.24	124.30	0.08	−0.70
0.66	392.5	0.519	0.330	1.5	0.50	53.5	1.82	123.68	0.09	−1.32
0.79	459.3	0.534	0.336	127.9	2.86	63.4	2.56	125.30	0.11	0.30
0.92	524.7	0.544	0.340	129.2	4.21	72.9	3.39	125.82	0.13	0.82
1.05	588	0.554	0.344	130.8	5.76	82.4	4.32	126.44	0.14	1.44
1.19	650	0.567	0.349	132.6	7.64	91.9	5.38	127.26	0.16	2.26
1.33	708	0.584	0.355	134.9	9.95	101.3	6.53	128.41	0.17	3.41
1.48	759	0.605	0.363	138.0	12.98	110.2	7.74	130.24	0.19	5.24
1.63	807	0.628	0.371	141.1	16.05	119.0	9.02	132.04	0.20	7.04
1.80	851.6	0.655	0.381	144.7	19.66	127.8	10.40	134.26	0.21	9.26
1.96	890.1	0.686	0.391	148.7	23.66	136.2	11.81	136.85	0.22	11.85
2.14	923.8	0.720	0.403	153.1	28.08	144.4	13.27	139.81	0.23	14.81
2.33	952.6	0.759	0.415	157.8	32.79	152.2	14.76	143.03	0.24	18.03
2.54	974.2	0.809	0.431	163.8	38.75	160.1	16.34	147.42	0.24	22.42
2.76	987.5	0.867	0.449	170.4	45.44	167.5	17.87	152.57	0.25	27.57
2.96	993.5	0.927	0.466	176.9	51.93	173.7	19.22	157.70	0.25	32.70
3.21	986.7	1.011	0.488	185.6	60.57	179.7	20.56	165.01	0.26	40.01
3.47	966.2	1.117	0.515	195.9	70.88	184.7	21.72	174.16	0.25	49.16
3.65	945.2	1.200	0.535	203.4	78.38	187.2	22.32	181.06	0.25	56.06

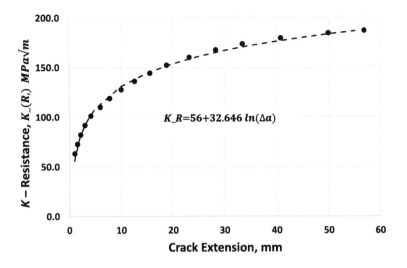

FIGURE 5.14 The K-resistance curve for the aluminum alloy in Example Problem 5.2.

Selected points on the force-displacement diagram in Fig. 5.13 that are spaced evenly are connected to the origin. The inverse slope of these points is related to the effective crack length at various points during the test.

The equations for estimating K and the compliance relationships with crack size for M(T) specimens are given in Appendix 5A.2, equations 5A.4–5A.7. Table 5.1 provides the summary of calculations of various parameters to construct the K_R versus Δa or the crack growth resistance curve. The various columns in Table 5.1 consist of:

- the dimensionless compliance of the specimen, (BEV/P), at various points along the force-displacement diagram,
- the predicted effective crack size of the specimen, (a_{eff}/W) from equation (5A.7),
- K_{eff} from equation (5A.4),
- the value of Δa_{eff}
- the physical crack extension, $\Delta a = \Delta a_{eff} - r_y$,
- the ratio of the net section stress and the yield strength, R_V, from equation (5.10),

From the values listed in Table 5.1, it is seen that R_V is always less than 1 and $(W-a) \geq \dfrac{4}{\pi}\left(\dfrac{K}{\sigma_{ys}}\right)^2$. The value of $\dfrac{4}{\pi}\left(\dfrac{K}{\sigma_{ys}}\right)^2$ can be calculated for the last point as 178 mm, while the $W-a$ at that point is 579 mm. Thus, all data gathered are valid points in the crack growth resistance curve that is plotted in Figure 5.14.

The crack growth resistance curve for this alloy is given by the following equation obtained from regression analysis.

$$K_R = 56 + 32.646 \ln(\Delta a)$$

where K_R is in MPa$\sqrt{m}$ and Δa in mm.

5.5   CORRELATIONS BETWEEN CHARPY ENERGY AND FRACTURE TOUGHNESS

Charpy energy data for ferritic steels is available for large number of materials so a natural question arises if this data can be used to estimate fracture toughness of these materials. A single correlation between Charpy energy and fracture toughness throughout the temperature regime does not exist. So, the correlations described here are broken down between the various regimes of the Charpy impact data and are derived empirically by inspecting data from several ferritic materials to establish trends. The correlations between fracture toughness and Charpy energy in the upper transition region are quite complex and depend on several factors. The readers are referred to reference [6] for a detailed discussion that is outside the scope of an intro-ductory text such as this one.

5.5.1   CHARPY ENERGY VERSUS FRACTURE TOUGHNESS CORRELATION FOR LOWER-SHELF AND LOWER TRANSITION REGION

This correlation covers the lower-shelf region and the region below the transition temperature within the transition region of the Charpy energy versus temperature curve where at least 50% of the fracture during a Charpy test is brittle. The equation below provides lower bound estimates of the fracture toughness:

$$K_{Ic} = \left[\left(12\sqrt{C_v} - 20 \right) \left(\frac{25}{B} \right)^{1/4} + 20 \right] \qquad (5.11)$$

where C_v = Charpy energy in Joules, B is the specimen thickness in mm, and K_{Ic} is in MPa$\sqrt{m}$. The lowest value of Charpy energy for steels is about 2.8 joules that gives rise to a minimum value of $K_{Ic} = 20\,$MPa$\sqrt{m}$ for a specimen that is 25 mm thick.

5.5.2   CHARPY ENERGY VERSUS FRACTURE TOUGHNESS CORRELATION IN THE UPPER-SHELF REGION

The empirical relationship between Charpy energy and fracture toughness in the upper-shelf region are given as follows:

$$\left(\frac{K_{Ic}}{\sigma_{ys}} \right)^2 = 0.52 \left(\frac{C_v}{\sigma_{ys}} - 0.02 \right) \qquad (5.12)$$

where K_{Ic} is in MPa$\sqrt{m}$ C_v in Joules and σ_{ys} is in MPa.

Equations (5.11) and (5.12) may be used to get conservative estimates of the frac-ture toughness of ferritic steels in the lower-shelf and the lower transition region and in the upper-shelf region, respectively.

5.6 SUMMARY

This chapter described quantitative measures of fracture toughness for brittle and ductile materials. We first discussed the historical Charpy impact test for measuring energy absorbed during brittle and ductile fracture and the phenomenon of brittle to ductile transition behavior in steels. While useful as a quality assurance test, Charpy test does not provide a quantitative measure of the conditions for fracture in structural components. Thus, the fracture mechanics approach for the measurement of fracture toughness is necessary. In brittle materials, fracture toughness is represented by a single parameter known as the plane strain fracture toughness, K_{Ic}. In ductile materials, unstable fracture is preceded by stable crack growth and the instability point depends on the geometry, size and loading conditions, and is not a material constant. But stable crack growth as a function of K is a material property. Therefore, the emphasis in ductile materials is on characterizing the stable crack growth behavior of the material represented by the K_R-resistance curve.

A method for measuring fracture toughness under plane strain conditions, K_{Ic}, was described with reference to the ASTM standard E399. The qualification criteria for establishing the validity of K_{Ic} were stated along with the rationale for the various requirements. Next, the test method used for establishing stable crack growth resistance curve using thin panels was discussed as described in ASTM Standard E561. In these tests in-plane, SSY conditions are maintained but the specimens are not sufficiently thick to generate plane strain conditions. Thus, the K_R-resistance curve is dependent on the thickness of the specimen used and therefore the chosen thickness of the specimens must be the same as that of the component.

Example problems are used to illustrate the use of these test methods. Correlations between the Charpy impact data and the K_{Ic} were also briefly described.

REFERENCES

1. R. Nanstad, M.A. Sokolov, and D.E. McCabe, "Applicability of Fracture Toughness Master Curve to Irradiated Highly Embrittled Steel and Intergranular Fracture", *Journal of ASTM International*, Vol. 5, 2008, pp. 1–14.
2. T.L. Anderson, *Fundamentals of Fracture Mechanics*, 2nd Edition, CRC Press, Boca Raton, FL, 1995.
3. Standard Test Method for Linear-Elastic Plane-Strain Fracture Toughness, K_{Ic}, of Metallic Materials, ASTM Standard E399, ASTM Book of Standards, Vol. 3, 2016.
4. Standard Test Method for Measurement of R-Curve, ASTM Standard E561, ASTM Book of Standards, Vol. 3, 2016.
5. A. Saxena and S.J. Hudak, Jr., "Review and Extension of Compliance Relationships for Common Crack Growth Specimens", *International Journal of Fracture*, Vol. 14, 1978, pp. 453–469.
6. C. Bannister, *Determination of Fracture Toughness from Charpy Impact Energy: Procedure and Validation*, British Steel Report, Moorgate, United Kingdom, 1998.

HOMEWORK PROBLEMS

1. A C(T) specimen from a high strength steel is tested for measuring the plane strain fracture toughness, K_{Ic}. The thickness, $B = 10$ cm, $W = 20$ cm, $a = 10$ cm. The force-displacement diagram obtained from the test is a Type 3 diagram (see Figure 5.8). The maximum force during the test is 465 kilo-N and P_5 is 440 kilo-N. Calculate the K_{Ic} of the steel and show whether the test meets the ASTM 399 size requirements for a valid test.

2. What are the sources of nonlinearity in a load-displacement diagram produced during a fracture toughness test?

3. What is the purpose of fatigue pre-cracking prior to fracture toughness testing?

4. A car engine block made from a steel casting is being evaluated for use in very cold climates. What would you recommend for reducing the risk of brittle fracture in the engine block? Explain your answer.

5. Calculate the size of the smallest compact specimen that will be needed to measure K_{Ic} in 304 stainless steel if the yield strength of stainless steel is 250 MPa and the estimated K_{Ic} is $300 \, \text{MPa}\sqrt{m}$. What load capacity machine will be needed to conduct this test? Comment on its practicality?

6. What is the primary difference between ASTM standards E399 and E561 referenced in the discussion here?

7. Show that the factor of 2.5 in equation (5.4) ensures that the length dimensions of the specimen, B, $W - a$, and a, are all approximately 25 times of the plane strain plastic zone size.

8. A 200 mm wide and 6.25 mm thick C(T) specimen is used to generate the K_R-curve of a structural steel. The yield strength of the material is 650 MPa, the elastic modulus, $E = 200$ GPa, and the initial crack size, $a_0 = 90$ mm. The displacement is measured at the mouth of the C(T) specimen on the front-face of the specimen. The force-displacement data are given in Table 5.2 and shown graphically in Figure 5.15.

TABLE 5.2
The Force-Displacement Data Obtained during a Fracture Resistance Test on a C(T) Specimen

P (kilo-N)	V (mm)	P (kilo-N)	V (mm)	P (kilo-N)	V (mm)
0	0	24.29387	0.940986	25.07267	1.267024
5	0.182861	24.77153	0.983138	24.7655	1.315587
10	0.365722	25.07698	1.019478	24.39866	1.363622
15	0.548583	25.27333	1.05233	23.98251	1.411513
20.10952	0.73545	25.45309	1.111712	23.52487	1.459573
22.75324	0.847975	25.44381	1.166037	23.0319	1.508069
23.47635	0.886739	25.3058	1.217419	22.50858	1.55724

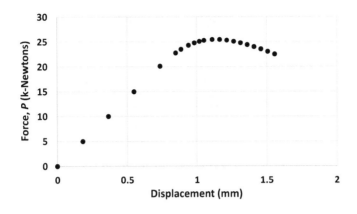

FIGURE 5.15 Force-displacement diagram for a C(T) specimen in Homework Problem 8.

APPENDIX 5A: COMPLIANCE RELATIONSHIPS FOR C(T) AND M(T) SPECIMENS

Elastic compliance measurements play an important role in experimental fracture mechanics. In the days prior to the use of finite element analysis for determining stress intensity parameters, compliance measurements were also used for determining stress intensity parameter, K. Compliance measurements are routinely used in measuring crack sizes during fracture toughness, fatigue crack growth, stress corrosion cracking, and corrosion-fatigue testing. Therefore, it is worthwhile to describe compliance relationships in commonly used fracture mechanics specimens, the compact type, C(T), and middle crack tension, M(T), specimens.

5A.1 COMPLIANCE RELATIONSHIPS FOR C(T) SPECIMEN

We start by recognizing that elastic compliance, C, for any cracked body is given in the functional form as in equation (5.8)

$$C_i = \frac{V_i}{P_i} = \frac{1}{BE} g\left(a/W\right)$$

We note that the dimensionless quantity, CBE, is dependent only on a/W provided the planar geometry of the body remains the same and all other planar dimensions of the body are proportioned to W. Figure 5A.1 shows a plot of CBE as a function of a/W for a C(T) specimen and includes experimental and numerical data which agree well with each other. If the point of displacement measurement is other than the load-line, the compliance relationships are given by:

$$\frac{BEV_x}{P} = \frac{\dfrac{BEV_0}{P}\left|\dfrac{x_0}{W} - \dfrac{x}{W}\right|}{\left[\dfrac{x_0}{W} + 0.25\right]} \tag{5A.1}$$

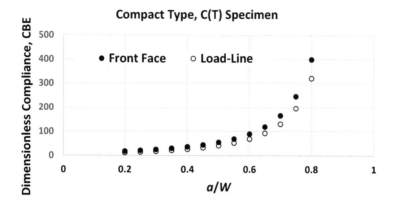

FIGURE 5A.1   The relationship between nondimensional compliance (CBE) and crack size for compact-type specimens [5].

where $\dfrac{BEV_0}{P}$ is the compliance value when the displacements are measured on the front-face of the specimen using machined knife-edges to attach the clip gage and is given by equation (5A.2).

$$\frac{BEV_0}{P} = \left(1 + \frac{0.25}{a/W}\right)\left(\frac{1+a/W}{1-a/W}\right)^2$$

$$\left[\left(1.6137 + 12.678\,(a/W) - 14.231\right)(a/W)^2 + 16.61\,(a/W)^3\right.$$

$$\left. + 35.05\,(a/W)^4 - 14.494\left(\frac{a}{W}\right)^5\right] \tag{5A.2}$$

where $x=$ distance between the loading line and the point of displacement measurement. x is negative when we move toward the front-face of the specimen from the load-line and positive as we move toward the crack tip. x_0 is the location of the point of rotation and is given by:

$$\frac{x_0}{W} = 0.009953 + 3.02437\left(\frac{a}{W}\right) - 7.95768\left(\frac{a}{W}\right)^2$$

$$+13.546\left(\frac{a}{W}\right)^3 - 10.627\left(\frac{a}{W}\right)^4 + 3.113\,(a/W)^4 \tag{5A.3}$$

The x/W value for the data in Figure 5A.1 is -0.25.

Frequently, it is necessary to express crack size as a function of compliance. This inverse relationship is expressed by [5]:

$$a/W = C_0 + C_1 U_x + C_2 U_x^2 + C_3 U_x^3 + C_4 U_x^4 + C_5 U_x^5 \tag{5A.4}$$

TABLE 5A.1
Constants in Equation (5A.3) for Different Measurement Locations [5]

x/W	C_0	C_1	C_2	C_3	C_4	C_5
0	1.0002	−4.0632	11.242	−106.04	464.33	−650.68
0.25	1.001	−4.6695	18.46	−236.82	1214.9	−2143.6

$$\text{where} \quad U_x = \frac{1}{(C_x BE)^{1/2} + 1}$$

where C_x=compliance corresponding to the measurement location x. The constants C_0, C_1, C_2 are given for various x values in reference [5] and are listed for select values in Table 5A.1. These equations provide considerable flexibility in choosing the measurement location of the displacement while making compliance measurements.

5A.2 Compliance and K-Relationships for M(T) Specimens

A schematic of the M(T) specimen is shown in Figure 5.12 defining all the planar dimensions. The displacement in this case is measured across two points located on the sides of the crack faces separated by a distance$=2Y_0$. The displacement across those points $2Y$ apart is labelled as $2V$.

$$K = \frac{P}{2BW}\sqrt{\pi a \cdot \sec\frac{\pi a}{2W}} \tag{5A.5}$$

$$(BEC) = \frac{2BEv}{P} = \left[\frac{Y_0}{W}\sqrt{\frac{\pi a/2W}{\sin(\pi a/2W)}} \right]$$

$$\left\{ \frac{2W}{\pi Y_0}\cosh^{-1}\left(\frac{\cosh\left(\dfrac{\pi Y}{2W}\right)}{\cos\left(\dfrac{\pi a}{2W}\right)}\right) - \frac{1+v}{\sqrt{1+\left(\dfrac{\sin\left(\dfrac{\pi a}{2W}\right)}{\sinh\left(\dfrac{\pi Y}{2W}\right)}\right)^2}} + v \right\} \tag{5A.6}$$

In equation (5A.6), $C = \dfrac{2V}{P}$ where V=half of the displacement across the measurement points for an $\dfrac{Y_0}{W} = 0.0371$. Note, that use equation (5A.6) it is necessary to consider the full displacement ($2V$) across the gage length of $2Y_0$.

$$U = U\left(\frac{Y_0}{W} = 0.0371\right) = \frac{1}{1 + \sqrt{BEC}} \qquad (5A.7)$$

$$\frac{a}{W} = -0.3862 + 15.981U - 62.456U^2 + 99.987U^3 + 74.735U^4 - 21.731U^5 \quad (5A.8)$$

NOTES

1 7075-T6 aluminum alloy is a designation used to describe a class of high strength aluminum alloy used as a structural material in the aerospace industry because of its high strength, low density and good corrosion resistance.

2 The term coherent precipitates refers to GP (Guinier–Preston) zones that form in precipitation hardened alloys where the typical particles are of the size less than 10 nm and have atomic planes that are coherent with the surrounding matrix. The same precipitates when they become larger have atomic planes that are incoherent with the matrix.

3 In the fracture mechanics literature, the center crack tension, CC(T) specimen is also referred to as middle crack tension specimen M(T).

6 Fatigue Crack Growth

6.1 INTRODUCTION

Fatigue failures occur when a structural component is subjected to periodically fluctuating stresses referred to as cyclic loading. The amplitudes of cyclic stress at which fatigue cracks can form and subsequently grow are significantly below the yield strength of the material. Shafts in rotating machinery and rail-road wheels are subjected to many loading/unloading cycles during service and are, therefore, prime examples of components where preventing fatigue failures is one of the most important concerns during design and material selection. If the cyclic load amplitudes are above a certain threshold known as the endurance limit of the material, microscopic cracks form as fatigue cycles accumulate preferentially at points of stress concentration such as at notches and machining marks. Even in well-polished, smooth surfaces, fatigue cracks can first form on the surface and then propagate under continued cyclic loading. Surfaces are more vulnerable to fatigue crack formation in comparison to the interior because surface atoms are at a higher energy state due to unsatisfied chemical bonds and are consequently preferred sites for dislocation activity [1]. A detailed discussion of dislocation motion due to cyclic plasticity is not within the scope of this book that focusses on engineering aspects of fatigue crack growth. Readers are referred to other books specifically about fatigue [2] for that information.

Fatigue is involved in a large fraction of mechanical failures encountered in engineering components. Fatigue cracks, as mentioned before, form due to local dislocation activity that occurs even when the stresses are below the yield strength of materials. Cracks that form with accumulation of fatigue cycles eventually reach a critical size and fracture occurs suddenly under normal loading conditions, and without warning. As an example, a picture of fatigue failure in a crank shaft arm is shown in Figure 6.1. The crack in the crank shaft arm first formed on the surface in the dark region seen in the picture and it grew at an increasingly faster rate as the crack size increased under the influence of cyclic loading encountered during service. The region of fatigue crack growth appears as the dark region in the picture because of the machine oil that seeps between the crack faces by capillary action and covers the fracture surfaces. The shiny area is the region of fast fracture that occurred during the last fatigue cycle prior to fracture and remains relatively uncontaminated by machine oil.

In this chapter, we will explore the characteristics of fatigue crack growth by formulating a theory for predicting the rates at which cracks grow under cyclic loading. If cracks are either present in components or they are presumed to be present because the nondestructive methods used to detect crack-like defects have resolution limits, the entire design life of the component is spent in growing the cracks from its initial size to a critical size at fracture. In the case of the crank shaft, fracture occurred when crack size reached the boundary between the shiny and the dark regions on the fracture surface.

DOI: 10.1201/9781003292296-6

FIGURE 6.1   Fatigue fracture in an aluminum crank arm. The dark area is where the fatigue cracks formed and grew, and the shiny area is the region of brittle fracture that occurred suddenly and without warning. http://en.wikipedia.org/wiki/Fatigue_(material).

6.2   FATIGUE CRACK GROWTH (OR PROPAGATION) RATES

6.2.1   DEFINITIONS

Some definitions and pertinent terms related to fatigue are described in this section. Figure 6.2a shows cyclic load applied to a cracked body that translates into stress intensity parameter, K, as function of time, as shown in Figure 6.2b. Some parameters that are characteristic of the loading cycle are defined next.

$$\Delta P = P_{max} - P_{min} = \text{Cyclic Load}$$

Where, P_{max} and P_{min} are the maximum and minimum loads, respectively during the fatigue cycle.

$$\text{The mean or average load} = P_{\text{mean}} = \frac{P_{max} + P_{min}}{2}$$

$$P_a = \frac{\Delta P}{2} = \frac{P_{max} - P_{min}}{2} = \text{Cyclic Load Amplitude}$$

$$\Delta K = K_{max} - K_{min} = \text{Cyclic Stress Intensity Parameter}$$

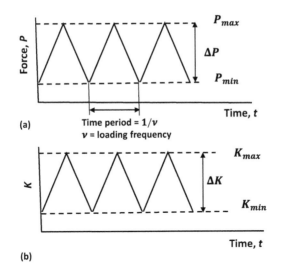

FIGURE 6.2 (a) Cyclic load applied to a cracked body and (b) the corresponding value of the K as a function of time.

where K_{max} and K_{min} are the maximum and minimum stress intensity factors during the cycle.

$$\text{Loading Frequency} = v = \frac{1}{\text{time period}}$$

$$\text{For } P_{min} \geq 0, \quad \text{Load ratio,} \quad R = \frac{P_{min}}{P_{max}} = \frac{K_{min}}{K_{max}}$$

$$\text{For } P_{min} < 0, \quad \text{Load ratio,} \quad R = \frac{P_{min}}{P_{max}}$$

Note that for negative values of the minimum force, P_{min}, the load ratio is defined only in terms of the ratio of minimum and maximum forces (or loads). K has no meaning for negative loads because cracks do not open under compressive loads and stresses can be transmitted across the crack faces, distinct from when the cracks are open under tensile loading.

Let's conduct the following thought experiment. If we take multiple cracked bodies that are identical in all ways and subject them to different levels of cyclic loads, $\Delta P_1 > \Delta P_2 > \Delta P_3 \ldots$ we can expect the relationships between crack size, a, and fatigue cycles, N, to be as schematically shown in Figure 6.3 for the various load ranges. The crack progresses at a faster rate as ΔP increases and for a constant ΔP as the crack size increases. The slope of the a versus N curve at any point is the fatigue crack growth rate per cycle, da/dN and can be functionally represented by equation (6.1).

$$\frac{da}{dN} = f(\Delta P, a, \text{geometry}) \tag{6.1}$$

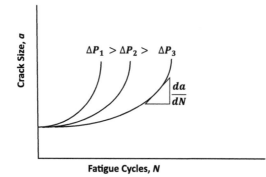

FIGURE 6.3 Expected crack size as function of fatigue cycles, N, behavior for various cyclic load levels.

Paul Paris in 1961 [3] was the first to express the relationship in equation (6.1) in terms of the cyclic stress intensity parameter, ΔK. He recognized that when the load ratio, R and loading frequency are kept constant, the variables ΔP, a, and geometry can be combined into a single parameter, ΔK. Paris [3] and coworkers experimentally explored the relationship between da/dN and ΔK by plotting the results from several fatigue crack growth tests on a log-log plot and discovered that all the data from various tests that were conducted at different load ranges lay on a single straight line on a log-log plot represented a power-law, equation (6.2).

$$\frac{da}{dN} = c\left(\Delta K\right)^{n} \tag{6.2}$$

Equation (6.2), shown schematically in Figure 6.4, is known as the Paris law. The constants n and c are constants that are obtained from regression of experimental data points relating da/dN to ΔK for a given material. These are constants that are characteristic of the material much like fracture toughness, K_{Ic}, values. The values of n for several structural materials range between 2 and 6. These constants can be influenced by the loading frequency (ν), temperature (T), and the load ratio, R. The use of ΔK for correlating fatigue crack growth rates also comes with the limitation that the cracked body is under dominantly elastic conditions.

A more general form of the Paris law can be written as in equation (6.3).

$$\frac{da}{dN} = f_{1}\left(\Delta K, R, \nu, H, T\right) \tag{6.3}$$

where H is a parameter dependent on the history of loading, and T is temperature. During constant load amplitude (CLA) testing, R, H, T, and ν are held constant and the Paris law, as originally proposed, holds. The history effects originate due to, among other factors, from the residual plasticity in the wake[1] of the growing crack. The material that was once ahead of the crack tip and experienced cyclic plastic deformation is located behind the crack tip after passage of the crack tip and carries with it residual plastic deformation as shown Figure 6.5. The plasticity in the wake of the crack results in the deformed crack faces that come in contact during

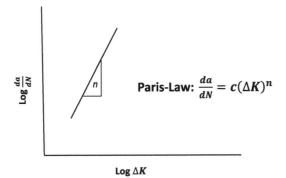

FIGURE 6.4 Graphical representation of the Paris law that is typically valid for crack growth rates in the range of $2.5 \times 10^{-5} \leq \dfrac{da}{dN} \leq 2.5 \times 10^{-2}$ mm/cycle.

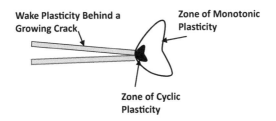

FIGURE 6.5 Crack closure resulting from plasticity in the wake of the crack.

unloading before the load reaches its minimum for the cycle and shield the crack tip. This phenomenon was first observed by Elber [4,5] and is known as crack closure. Elber proposed that the portion of the fatigue cycle below the closure load does not participate in driving the fatigue crack since the crack remains closed below the closure load. The residual plastic deformation in the wake, and thus the crack closure load, is affected by the loading history. For example, a higher than usual K_{max} during a fatigue cycle will result in a larger plastic zone ahead of the crack tip and that will subsequently result in a higher crack closure load due to higher levels of residual plasticity as the region once ahead of the crack tip transitions into the wake region. During constant force amplitude loading, the effects of crack closure remain constant, therefore, H can be treated as a constant. However, if the force amplitude were to vary, H will also vary and will result in a crack growth rate that is a function of the load history. This will be considered in more detail later in the chapter while discussing variable amplitude loading.

6.2.2 Mechanisms of Fatigue Crack Growth

The mechanisms by which fatigue cracks grow in engineering materials are also of considerable interest because it can provide approaches for improving fatigue crack

(a)   (b)

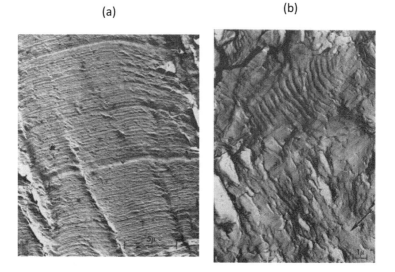

FIGURE 6.6   Fatigue striations observed in an electron microscope (a) in a 7178 Al Alloy, and (b) in 4340 steel [6] (Pictures reproduced with permission from ASTM, Conshohocken, PA).

growth resistance of materials. Examination of fatigue fracture surfaces under an electron microscope at magnifications in the range of 5,000–10,000x is commonly employed to study fatigue crack growth. An example of such an image is shown in Figure 6.6a for a 2024 T3 aluminum alloy and in Figure 6.6b for a high strength low alloy steel [6] that show clear marks that are related to growth of cracks, called fatigue striations. The local direction of crack growth of the crack is also evident from the striations.

Fatigue striations associated with fatigue crack growth are observed frequently but not in all cases. Their appearance on the fracture surface is dependent on the material, the stress intensity parameter amplitudes, and environment. Aluminum alloys and titanium alloys that rapidly form tenacious oxides on freshly created fracture surfaces, show striations such as in Figure 6.6a. On the other hand, striations or striation like features are less obvious in steels, Figure 6.6b. Figure 6.7a and b show the correlation between striation spacings and measured fatigue crack growth rates in an Al alloy and in 1,018 steel [7]. In the Al alloy, the correlation between da/dN and striation spacing, S, is excellent at da/dN values greater than approximately 1 µm/cycle and begins to diverge at smaller crack growth rates. In the carbon steel, the excellent correlation between da/dN and S continues to 0.01 µm. Striation spacings can be a good indicator of how fast the crack was growing in service, when used in failure analyses.

Figure 6.8 shows a process first proposed by Laird [8] that explains the formation of striations during loading/unloading as the crack extends in size. The first three diagrams, a–c, show how the process evolves during loading part of the cycle from no applied force to the maximum level while diagrams d and e are process steps

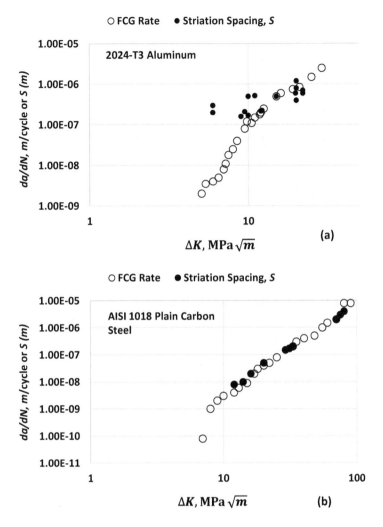

FIGURE 6.7 Comparison between measured FCG rates and striation spacings in (a) 2024-T3 aluminum alloy and (b) AISI 1018 steel [7].

during unloading. The step f represents the beginning of re-enactment of the loading/unloading sequence after the crack has advanced. Plastic deformation is expected to occur at $\pm 45°$ from the crack plane where the shear stresses are the highest, followed by crack blunting and stretching. During unloading, the dislocation movement direction reverses and forms the wings on the sides of the extended crack on both surfaces that appear as striations in the photomicrographs of the fracture surface.

6.2.3 FATIGUE CRACK GROWTH LIFE ESTIMATION

Fatigue crack growth life from an initial flaw size, a_i, can be calculated as illustrated in the following example using a semi-infinite center cracked plate loaded with a

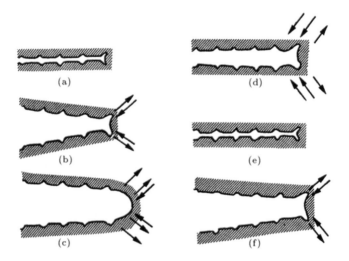

FIGURE 6.8 A schematic showing Laird's theory of fatigue striation formation [8] (The figure has been reproduced with permission of ASTM, Conshohocken, PA).

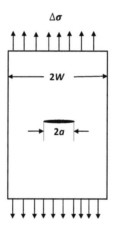

FIGURE 6.9 A center-crack panel subjected to remote uniform cyclic stress, $\Delta\sigma$.

fatigue stress of $\Delta\sigma$, Figure 6.9. The cyclic life of the panel, N_f, is obtained in the following manner starting from equation (6.2):

$$\int_0^{N_f} dN = \int_{a_i}^{a_f} \frac{da}{c\Delta K^n}$$

$$N_f = \int_{a_i}^{a_f} \frac{da}{c\left[\left(\Delta\sigma\sqrt{\pi a}\right)\right]^n}$$

$$N_f = c^{-1}\Delta\sigma^{-n}\pi^{-\frac{n}{2}}\left[a^{1-\frac{n}{2}}\right]_{a_i}^{a_f} \cdot \frac{1}{\left[1-\dfrac{n}{2}\right]} = \frac{a_f^{1-\frac{n}{2}} - a_i^{1-\frac{n}{2}}}{c(\Delta\sigma)^n \pi^{\frac{n}{2}}\left(1-\dfrac{n}{2}\right)} \tag{6.4}$$

The final crack size, a_f, is estimated from the fracture toughness of the material, K_{Ic}, and the maximum stress during the cycle, σ_{max} as follows:

$$a_f = \frac{1}{\pi}\left(\frac{K_{Ic}}{\sigma_{max}}\right)^2 \tag{6.5}$$

Equation (6.4) provides a relationship that documents the role of fracture toughness, applied cyclic stress, the load ratio, R, and the two material parameters, c, and n, in determining the useful life of the structural panel as illustrated in the following Example Problem.

Example Problem 6.1

An aircraft fuselage structure is made from a high strength aluminum alloy with a yield strength of 350 MPa, a fracture toughness, K_{Ic}, of 35 MPa$\sqrt{m}$, and the values of $n=5$, $c = 5\times10^{-12}$ m$\left(\text{MPa}\sqrt{m}\right)^{-5}$. The panel has a width (W) of 50 mm and is 5 mm thick and can leave behind edge cracks from the metal forming process that can potentially be as large as 1 mm deep. This flaw dimension represents the smallest crack size that can be detected with 100% reliability during nondestructive testing that is performed on the panel prior to assembly. Thus, in the damage tolerant approach, we assume that a flaw of that size exists, but the structure should not fail during service. The panel is expected to experience a cyclic stress of 120 MPa with a frequency of 1 cycle/day. Estimate the guaranteed design life of the panel in number of cycles.

Solution

The K expression for this geometry is given by (from Table 3.1):

$$K = \sigma\sqrt{\pi a}\sqrt{\frac{2W}{\pi a}\tan\left(\frac{\pi a}{2W}\right)}\left\{\frac{0.752+2.02(a/W)+0.37\left(1-\sin\dfrac{\pi a}{2W}\right)^3}{\cos\left(\dfrac{\pi a}{2W}\right)}\right\}$$

where $W=50$ mm (0.05 m). The predicted fatigue crack growth rates during service are given by substituting the values of c and n in equation (6.2). The value of K as function of crack size is calculated in excel and given in Table 6.1.

From Table 6.1, we see that the K_{Ic} value is reached at a crack size of approximately 0.0125 m so that is the end of life of the component. It corresponds to a fatigue life of 4,702 cycles, which is the calculated design life. At 1 cycle/day, it is approximately 12.8 years. The component must be inspected prior to reaching that number of cycles. Also, it is common to apply a factor of safety on the calculated design life. The factor of safety varies with the criticality of the component and the

TABLE 6.1

Crack Size as Function of Fatigue Cycles in Example Problem 6.1

a (m)	$\Delta K \left(MPa\sqrt{m} \right)$	da/dN (m/cycle)	Cycles (N)
0.001	7.59	1.26E−07	0
0.0015	9.34	3.55E−07	2,269
0.002	10.84	7.48E−07	3,226
0.0025	12.19	1.34E−06	3,725
0.003	13.43	2.19E−06	4,018
0.0035	14.61	3.33E−06	4,205
0.004	15.74	4.83E−06	4,331
0.0045	16.84	6.76E−06	4,420
0.005	17.91	9.20E−06	4,485
0.0055	18.96	1.22E−05	4,533
0.006	20.00	1.60E−05	4,570
0.0065	21.05	2.06E−05	4,598
0.007	22.09	2.63E−05	4,620
0.0075	23.14	3.32E−05	4,637
0.008	24.19	4.15E−05	4,651
0.0085	25.26	5.15E−05	4,662
0.009	26.35	6.35E−05	4,671
0.0095	27.45	7.80E−05	4,679
0.01	28.58	9.53E−05	4,685
0.0105	29.72	1.16E−04	4,690
0.011	30.89	1.41E−04	4,694
0.0115	32.09	1.70E−04	4,697
0.012	33.32	2.05E−04	4,700
0.0125	34.58	2.47E−04	4,702

consequences of failure. If the design life is inadequate, the component must be redesigned to reduce the stresses.

One more consideration remains before we finalize the calculation of design life. For fracture to occur at $K_{max} = 35\,MPa\sqrt{m}$, we must also ensure that the thickness of the panel and the remaining ligament meet the following equations from Chapter 5:

$$B, a, (W-a) \geq 2.5 \left(\frac{K_{max}}{\sigma_{ys}} \right)^2 = 2.5 \left(\frac{35}{350} \right)^2 = 0.025 \text{ m}$$

From Table 6.1, we see that, when $K_{max} = 35\,MPa\sqrt{m}$, $a = 0.0125$ m, $W-a = 0.0375$ m and $B = 0.005$ m. Thus, the condition for a and $W-a$ are met but the thickness is less than the requirement to ensure plane strain conditions. In this case, it is likely that the critical K for fracture is $\geq K_{Ic}$. The cyclic life calculated is therefore conservative because the crack can grow to a size larger than 0.0125 m prior to failure. Such conservatism is always desirable in structural life assessments.

Example Problem 6.2

A spring-loaded tension arm that activates a relay switch is 25 mm wide and is 8 mm thick and is forged from American Iron and Steel Institute (AISI) 1018 carbon steel. It is designed to support a force of 30 kilo-Newtons (kN) for up to 200,000 cycles during service. The fatigue crack growth properties of this steel are represented by:

$$\frac{da}{dN} = 6 \times 10^{-11} (\Delta K)^{2.2}$$

where da/dN is in m/cycle and ΔK in MPa$\sqrt{m}$. The plane strain fracture toughness, K_{Ic}, of the material is 50 MPa$\sqrt{m}$. The arm is visually inspected after forging and final machining to ensure that there are no defects that are larger than 5 mm deep on either side of the arm along its length. An arm failed in approximately 10,000 cycles in service causing an expensive unplanned outage in the plant, so the Plant Manager has asked the manufacturer for compensation. The manufacturer of the arm suspects that the spring used by the customer was stiffer than specified and was therefore applying a higher force on the arm than used in design calculations. He has engaged a fracture mechanics expert to investigate his suspicion that the service load applied to the arm was higher than 30 kN.

Solution

The fracture surface is brought into the laboratory and is examined under the electron microscope, and it is found that there is indeed a manufacturing crack that is 4 mm deep and that this crack grew during service to 6.5 mm under service loading when the failure occurred. The fracture surface reveals striations that were spaced 0.228 μm at a crack depth of 4.5 mm and the spacing increased to 0.306 μm at a crack depth of 6.0 mm.

The K-expression for this configuration is the same as in Example Problem 6.1. Next, we determine values of ΔK, da/dN, and the accumulated fatigue cycles, N, between crack sizes of 4–6.5 mm at intervals of 0.5 mm for values of $\Delta \sigma$ of 150, 200, and 300 MPa, as presented in the Table 6.2. Note that the design stress is $\frac{30,000}{0.025 \times 0.008} = 150 \times 10^6$ Pa $= 150$ MPa. At this level of cyclic stress, the K_{max} for a crack size of 6.5 mm is only 25.17 MPa$\sqrt{m}$, that is not sufficient to cause fracture. The predicted accumulated fatigue cycles are 46,000. We repeat the calculation for a stress level of 200 MPa and find that the K level at a crack size of 6.5 mm is still insufficient to cause fracture and the number of predicted accumulated cycles is 24,000. Next, we raise the stress level to 300 MPa and find that the K_{max} at a crack size of 6.5 mm is approximately 50 MPa$\sqrt{m}$, and the life is about 10,000 cycles, so that seems to be the likely stress level encountered during service and deserves to be further investigated. The crack growth rates at crack sizes of 4.5 and 6.0 mm were found to be 2.3×10^{-7} and 3.1×10^{-7} m/cycle, respectively. These compare well with the striation spacings measured on the fracture surfaces at those points providing further proof that the service loads were higher than the loads that the component was designed to withstand. Thus, it can be concluded that the service stress was two times that of the planned design stress resulting from a stiffer spring used during installation. As further proof, the stiffness of the spring can be

TABLE 6.2

The Values of K, da/dN, and Accumulated Cycles, N as a Function of Crack Size in the Tension Arm Subjected to Various Stress Ranges

	$\Delta\sigma = 150\,MPa$			$\Delta\sigma = 200\,MPa$			$\Delta\sigma = 300\,MPa$		
a		da/dN	N		da/dN			da/dN	
(mm)	$K\,MPa\sqrt{m}$	(m/cycle)	(Cycles)	$K\,MPa\sqrt{m}$	(m/cycle)	N	$K\,MPa\sqrt{m}$	(m/cycle)	N
4	20.10	4.4E−08	0.0E+00	26.80	8.3E−08	0.0E+00	40.21	2.0E−07	0
4.5	21.21	5.0E−08	1.1E+04	28.28	8.3E−08	6.0E+03	42.41	2.3E−07	2,462
5	22.26	1.0E+01	3.0E+04	29.68	1.0E−07	1.1E+04	44.52	2.5E−07	4,652
5.5	24.24	6.1E−08	3.9E+04	31.02	1.1E−07	1.6E+04	46.53	2.8E−07	6,620
6	24.24	6.7E−08	4.6E+04	32.32	1.3E−07	2.1E+04	48.47	3.1E−07	8,405
6.5	25.17	7.2E−08	4.6E+04	33.56	1.4E−07	2.4E+04	50.35	3.3E−07	10,037

measured in the laboratory to provide more evidence. This analysis puts the blame on to the user for the failure of the tension arm and absolves the manufacturer of any blame. The initial flaw size of 4 mm was smaller than the largest flaw size that could be missed during visual nondestructive inspections.

6.3 THE EFFECT OF LOAD RATIO, TEMPERATURE, AND FREQUENCY ON FATIGUE CRACK GROWTH RATE IN THE PARIS REGIME

Having established the relationship between da/dN and ΔK and its importance in determining the rates at which cracks grow under fatigue loading, it is necessary to explore how this relationship changes with other secondary variables such as loading frequency, load ratio and temperature because these can vary in service loading of components. We will see that these are less important variables compared to ΔK in determining the da/dN behavior in the Paris regime but important enough to deserve systematic consideration. Their contributions vary depending on the material and environmental factors.

Figure 6.10 schematically shows how load ratio affects the fatigue crack growth rate relationship with ΔK. We notice that as the load ratio increases (or the K_{max} increases while ΔK remains constant), the fatigue crack growth rates also increase. In benign or innocuous environments, it is widely believed based on experimental data that loading frequency does not affect fatigue crack growth rates. But in corrosive environments or under high temperature creep conditions, frequency of loading does become important. These effects will be considered in more detail in later chapters. Temperature, in the sub-creep range ($T < 0.4\ T_M$ where, T_M is the melting point in degrees Kelvin) is also a secondary variable and its effects on fatigue crack growth rates are small but should not be assumed to be negligible without experimental proof. Thus, if the service temperature in which the component operates is higher than ambient or much lower than ambient, tests should be performed at those temperatures and the constants c and n should be obtained for the service temperature conditions.

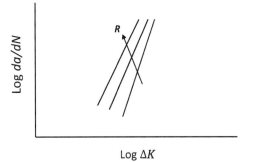

FIGURE 6.10   A schematic of the effects of load ratio, R on the FCG rate behavior in the Paris regime.

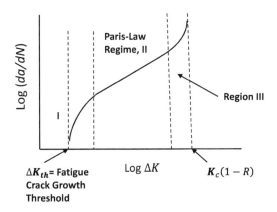

FIGURE 6.11   Wide range FCGR behavior of structural metals. FCGR, fatigue crack growth rate.

6.4   WIDE RANGE FATIGUE CRACK GROWTH BEHAVIOR

If we consider fatigue crack growth rates over several decades ranging from 10^{-7} to 10^{-1} mm/cycle, the relationship with ΔK is not linear anymore on a log-log scale and is shown schematically in Figure 6.11. The behavior can be divided into three regions. Region I is characterized by a steep decrease in the crack growth rates with a decrease in ΔK such that a threshold value ΔK_{th} is reached below which the crack growth rate becomes negligible for practical purposes. ΔK_{th} is a material property and is used in design of components that are subjected to high frequency loading such as shafts in rotating equipment or in turbine blades vibrating at their natural frequencies due to aerodynamic forces. Region II is the Paris regime described earlier where the da/dN and ΔK have a power-law relationship. Region III occurs when the K_{max} approaches the fracture toughness of the material, and the crack growth rates rise rapidly with increase in ΔK. Recall that $K_{max} = \dfrac{\Delta K}{1-R}$, therefore, the final point on the da/dN versus ΔK curve is given by $K_c(1-R)$ where K_c is the fracture toughness or the critical K at which fast fracture occurs. Note, that we have not used K_{Ic} here because K_{Ic} is reserved for plane strain fracture toughness that only applies to bodies of sufficient thickness and in-plane dimensions to ensure not only dominantly linear elastic, but also plane strain conditions.

A three-component model was developed to represent all three regions of the $da/dN - \Delta K$ behavior by a single equation as follows [9]:

$$\frac{1}{da/dN} = \frac{1}{A_1 \Delta K^{n_1}} + \frac{1}{A_2}\left(\frac{1}{(\Delta K)^{n_2}} - \frac{1}{\left(K_c(1-R)\right)^{n_2}}\right) \tag{6.6}$$

In Region II, the above equation reduces to the form identical to the Paris law, equation (6.2). The influence of load ratio, R, also depends on the crack growth rate regime as shown in Figure 6.12. The importance of load ratio is the highest in Regions I and III and the lowest in Region II. In Region II, the fatigue crack growth behavior is more sensitive to R in aluminum and titanium alloys than in steels. On the other hand, the near threshold fatigue crack growth rates are sensitive to R in all materials. The constants in equation (6.6) for 2219-T851 Al alloy are given in Table 6.3

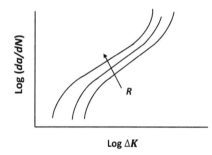

FIGURE 6.12   Wide range fatigue crack growth behavior at different load ratios, R, illustrating how R has a larger influence in the near threshold region and the fast crack growth region.

TABLE 6.3

Constants in the Three-Component Model Representing the Wide-Range Fatigue Crack Growth Rate Behavior in 2219-T851 Aluminum Alloy at $0.1 \le R \le 0.8$ [9]

Load Ratio, R	n_1	n_2	A_1, m/ $(MPa)^m$	A_2, m/ $(MPa)^m$	K_c, $MPa\sqrt{m}$
0.1	12.5	3.7	5.35×10^{-16}	2.5×10^{-11}	36.6
0.3	12.5	3.3	9.96×10^{-15}	6.83×10^{-11}	36.6
0.5	12.5	3.3	2.76×10^{-12}	9.01×10^{-11}	38.8
0.8	12.5	3.3	1.24×10^{-12}	1.98×10^{-10}	38.8

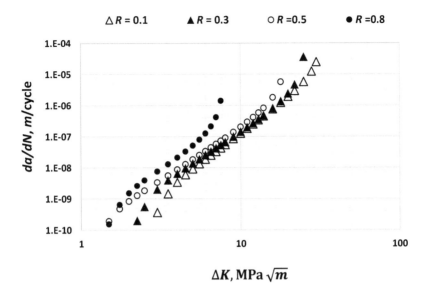

FIGURE 6.13 FCGR trends for aluminum alloy 2219-T851 over a wide range of crack growth rates and for $0.1 \leq R \leq 0.8$ [9] as represented by the constants in the three-component model. FCGR, fatigue crack growth rate.

for several load ratios. The trends described by equation (6.6) using constants in Table 6.3 for this material are plotted in Figure 6.13.

An equation, that is valid over a wide range of fatigue crack growth behavior of the type shown in equation (6.6) is required for estimating lives of members designed for long life where the initial ΔK levels lie outside of the Paris regime. This is illustrated in the following example.

Example Problem 6.3

A 200 mm wide and 12.5 mm thick plate made from 2219-T851 aluminum alloy is subjected to a maximum cyclic stress of 160 MPa on the front-face that linearly decreases to 80 MPa on the back face. These stresses decrease to levels of 80 and 40 MPa on the front and back faces, respectively at the minimum load during the fatigue cycle. The plate is inspected for surface flaws using ultrasonics that can reliably detect flaws that are 1 mm deep at their maximum depth and 10 mm long on the surface. Calculate the number of cycles and sketch the profile of the flaw as it evolves from its original size and grows to a depth of 5 mm.

Solution

The stress profile of the plate at maximum and minimum stress is shown in Figure 6.14. We assume that a crack of the size that can escape detection is present on the surface at the front face of the plate where the stresses are highest.

The K-expressions for semi-elliptical surface flaws in plates subjected to tension and bending were given by equation (3.31). All parameters in the above equation

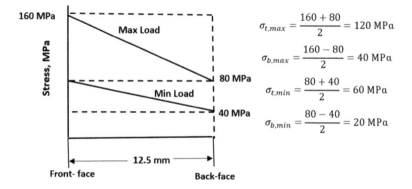

FIGURE 6.14 Maximum and minimum cyclic stresses applied on the plate containing a semi-elliptical flaw.

were defined in Chapter 3 and are detailed there. The values of σ_t and σ_b at the maximum and minimum load are shown in Figure 6.14 at the maximum and minimum stress on the component.

$$K = \left(\sigma_t + H\sigma_b\right)\sqrt{\frac{\pi a}{Q}} f\left(\frac{a}{t}, \frac{a}{c}, \frac{c}{W}, \phi\right)$$

The load ratio, R, for this loading is 0.5 obtained from the ratio of the minimum to maximum values of K in the plate during a cycle. This is calculated as follows:

$$R = \frac{K_{min}}{K_{max}} = \frac{\sigma_{t,min} + H\sigma_{b,min}}{\sigma_{t,max} + H\sigma_{b,max}} = \frac{60 + 20H}{120 + 40H} = 0.5$$

Thus, from Table 6.3, the constants, A_1, A_2, n_1, n_2, and K_c describing the fatigue crack growth rate are chosen accordingly for $R=0.5$.

$$\frac{da}{dN} = \left[\frac{1}{2.76 \times 10^{-12} \Delta K^{12.5}} + \frac{1}{9.01 \times 10^{-11}} \left(\frac{1}{\Delta K^{3.3}} - \frac{1}{\left(38.8 \times (1-.5)\right)^{3.3}}\right)\right]^{-1}$$

The ΔK corresponding to the deepest point of the crack, A, and the point of intersection with the front-face of the plate, C, can be estimated from these equations using excel. These values are then used to calculate the respective fatigue crack growth rates at those points from the da/dN (or dc/dN) versus ΔK relationship. The number of fatigue cycles, ΔN to advance the crack depth, a, by an incremental amount such as 0.25 mm along the depth is calculated. The corresponding growth in the c direction is then calculated for the same number of fatigue cycles. The new a and c values are updated in the equations and the calculations are repeated. The results are shown in the Table 6.4. These calculations can be used to plot the crack shapes as a function of fatigue cycles corresponding to crack depths ranging from 1 to 5 mm as shown in Figure 6.15. To estimate the crack profile, we assume that the shape of the crack remains a semi-ellipse with the major axis$=2c$ and the minor axis$=2a$. We note from Table 6.4 and from the figure that the crack shape or the a/c values evolve continuously as fatigue cycles accumulate.

TABLE 6.4
Evolution of the Crack Size and Shape with Fatigue Cycles

a (mm)	c (mm)	ΔK_A, MPa$\sqrt{m}$	da/dN (mm/cycle)	ΔK_C, MPa$\sqrt{m}$	dc/dN (mm/cycle)	ΔN (Cycles)	N (Cycles)
1	5	9.35	1.5E−05	1.26	4.1E−08	16,936	0
1.25	5.001	10.17	2.0E−05	1.64	3.4E−07	12,797	16,936
1.5	5.005	10.82	2.4E−05	2.01	8.6E−07	10,413	29,734
1.75	5.014	11.33	2.8E−05	2.38	1.6E−06	8,924	40,147
2	5.028	11.73	3.1E−05	2.74	2.5E−06	7,942	49,071
2.25	5.048	12.04	3.4E−05	3.10	3.8E−06	7,273	57,013
2.5	5.075	12.28	3.7E−05	3.44	5.3E−06	6,806	64,287
2.75	5.112	12.46	3.9E−05	3.77	7.2E−06	6,473	71,093
3	5.158	12.60	4.0E−05	4.10	9.5E−06	6,229	77,565
3.25	5.218	12.72	4.1E−05	4.41	1.2E−05	6,043	83,794
3.5	5.291	12.81	4.2E−05	4.71	1.5E−05	5,894	89,838
3.75	5.380	12.90	4.3E−05	5.01	1.9E−05	5,762	95,731
4	5.488	12.98	4.4E−05	5.30	2.2E−05	5,634	101,493
4.25	5.614	13.07	4.5E−05	5.59	2.7E−05	5,500	107,127
4.5	5.761	13.18	4.7E−05	5.87	3.2E−05	5,353	112,628
4.75	5.931	13.30	4.8E−05	6.16	3.7E−05	5,188	117,981
5	6.123	13.44	5.0E−05	6.45	4.3E−05	5,003	123,169
5.25	6.340	13.61	5.2E−05	6.74	5.0E−05	4,799	128,172
5.5	6.582	13.80	5.5E−05	7.03	5.8E−05	4,577	132,971
5.75	6.849	14.01	5.8E−05	7.34	6.7E−05	4,342	137,548
6	7.141	14.25	6.10E−05	7.65	7.8E−05	4,097	141,890

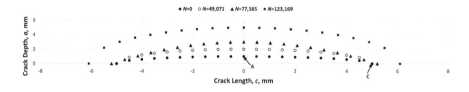

FIGURE 6.15 The predicted shape of the semi-elliptical surface crack with fatigue cycles.

6.5   CRACK TIP PLASTICITY DURING CYCLIC LOADING

To gain further insight into the fatigue crack growth process, it is of interest to examine the plastic deformation that occurs ahead of the crack tip under cyclic loading. The aspects of crack tip plasticity encountered during cyclic loading in the form of a monotonic plastic zone, a zone of reversed cyclic plasticity, and crack closure are discussed in more detail in this section. This will give us additional insights into history effects such as due to different load ratios, cyclic overloads, and during variable amplitude fatigue.

6.5.1   Cyclic Plastic Zone

Since the plastic deformation near the crack tip is surrounded by a much larger elastic region, the material elements closest the crack tip experience reversed plastic deformation during unloading. The region that experiences reversed plasticity is called the cyclic plastic zone and to better explain it we divide the crack tip area into four regions as shown schematically in Figure 6.16, and then examine the stress-strain behavior in these regions during a fatigue cycle.

The region marked "1" in Figure 6.16 is in the elastic region of the body where during loading/unloading, stresses and strains increase and decrease proportionally and linearly with the applied force as shown in Figure 6.17a. In this region, the cyclic stress and strain ranges are expected to scale proportionally with the applied value of ΔK as per the crack tip field equation (3.15). In Region "2," Figure 6.17b, which is within the plastic zone corresponding to K_{max}, the material experiences compressive stresses when the external load is removed but the magnitude of the compressive stress is not sufficient to cause reversed plastic deformation. Past the first fatigue cycle, the material in that region loads and unloads elastically. In other words, this region does not experience plastic strains beyond the first cycle.

Region "3," Figure 6.18a, is at the boundary of the cyclic plastic zone where plastic deformation occurs both during loading and unloading. The range of stress $\Delta\sigma$ in this case equals to two times the cyclic yield strength, $2\sigma_{ys}^c$, of the material. Cyclic yield strength is defined as the stress amplitude, $\Delta\sigma/2$, at which 0.2% plastic deformation occurs during both loading and unloading. If a material is nonhardening or softening, the value of σ_{ys}^c is the same as σ_{ys}, but for materials that either cyclically harden or soften, it will be higher than σ_{ys} for hardening type materials and lower than σ_{ys} for softening materials. The locus of points or boundary along which the stress amplitude is σ_{ys}^c is the cyclic plastic zone. In Region "4," Figure 6.18b, as the

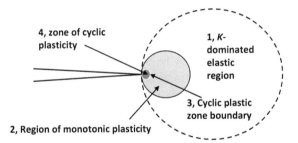

FIGURE 6.16   Deformation regions ahead of the crack tip during cyclic loading.

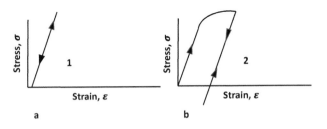

FIGURE 6.17   The figure (a) shows pure elastic behavior in region 1 during loading and unloading and (b) that applies to region 2 in Figure 6.14, shows plastic deformation during the first cycle and elastic behavior subsequently.

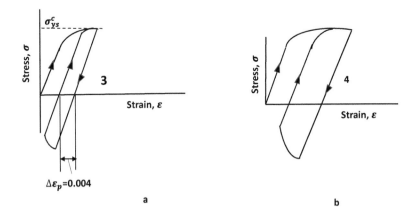

FIGURE 6.18   The figure (a) shows the stress-strain behavior in region 3 at the boundary of the cyclic plastic zone during loading and unloading, and the figure (b) shows higher amounts of cyclic plastic deformation experienced by the material as it approaches the crack tip that is deeper into the cyclic plastic zone.

crack tip is approached, the amount of cyclic plasticity during loading and unloading is even more pronounced. The stress amplitude, $\Delta\sigma/2$, in this region exceeds σ_{ys}^c and the plastic strain amplitude is greater than 0.2%.

As a first order estimate, σ_{ys}^c may be taken as being equal to σ_{ys} for monotonic loading. In practice, it is expected to be (about 10%–20%) higher or lower than σ_{ys} for cyclically hardening and cyclically softening materials, respectively. For select materials, that are known as cyclically stable, the monotonic and cyclic yield strength is the same.

The cyclic plastic zone size, r_p^c, is then given by substituting ΔK for K and $2\sigma_{ys}^c$ for σ_{ys} in equation (4.4a):

$$r_p^c = \frac{1}{\pi}\left(\frac{\Delta K}{2\sigma_{ys}^c}\right)^2 \tag{6.7}$$

If we assume that $\sigma_{ys} = \sigma_{ys}^c$, we get the result that the cyclic plastic zone size is approximately 1/4th the size of the monotonic plastic zone if the loading cycle consists of cycling between K_{max} and 0.

$$r_p \approx 4r_p^c \tag{6.8}$$

At higher load ratios, the difference in sizes between the monotonic plastic zone associated with K_{max} and the cyclic plastic zone associated with ΔK will be larger than estimated by equation (6.8).

6.5.2 Crack Closure during Cyclic Loading

Crack closure was briefly introduced in Section 6.2. In this section, its significance in determining fatigue crack growth rates over a wide range of growth rates is explored. Elber [4,5] was the first to notice the physical signs of crack closure by conducting experiments in which he mounted strain gages above and below the crack plane and recorded the variation in measured strain during a cycle as the fatigue crack passed by that point. He noticed that strain gages in the region behind the crack tip recorded compressive stresses as the applied force approached zero. In numerous later studies that followed, it was observed that the force-displacement relationship during loading-unloading showed a change in stiffness of the cracked body as the force approached zero, as schematically shown in Figure 6.19. Elber proposed that the residual plastic deformation on the crack faces near the crack tip causes the crack faces to come in contact prior to the force reaching zero. Thus, the crack appears to begin to close at a force, $P_{op} > 0$ during unloading. During loading, a force equal to P_{op} must be applied to fully open the crack. Crack closure caused by plastic deformation was named plasticity induced crack closure.

A clearer picture of crack opening and closing was obtained by Pippan [11] who examined the opening and closing of cracks in a scanning electron microscope during the various portions of the loading/unloading cycle in a specimen containing a naturally grown fatigue crack subjected to cyclic plastic deformation as shown

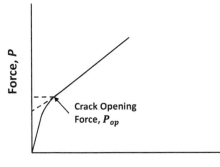

Displacement, V

FIGURE 6.19   Force-displacement diagram showing the change in compliance (or stiffness) as the force approaches zero. The diagram can be used to determining the crack opening force.

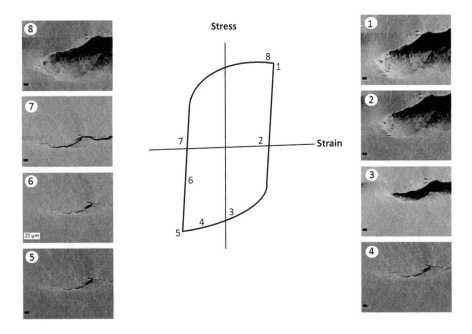

FIGURE 6.20   Photomicrographs obtained from a scanning electron microscope of a fatigue crack during different stages of loading/unloading in a specimen subjected to cyclic plastic deformation. (Reprinted from Pippan and Hohenwarter [11] with permission).

in Figure 6.20. Loading under reversed plasticity conditions enhances the physical appearance of crack closing and opening. Going from Point, 1, at the highest stress/strain value to minimum stress/strain value, the crack is seen to begin closing at Point 4 ahead of the minimum stress/strain, Point 5. During the loading portion of the fatigue cycle, the crack does not open until Point 6, stress higher than the minimum level. Experiments such as this and ones reported by Elber four decades prior, show physical evidence of crack closure.

Elber proposed that the K level up to the crack opening load, K_{op}, does not participate in growing fatigue cracks and therefore should not be considered in estimating the ΔK value. Hence, the ΔK level that drives the crack should be adjusted according to the following equation:

$$\Delta K_{eff} = K_{max} - K_{op} \qquad (6.9)$$

It was experimentally shown that when fatigue crack growth rates obtained under constant load amplitude (CLA) at several load ratios in aluminum alloys were correlated with ΔK_{eff}, they collapsed into a single trend. This result was indeed significant, and it motivated several research studies on the crack closure phenomenon during the subsequent years. There was also the potential for characterizing fatigue crack growth rates under variable amplitude loading using ΔK_{eff} as another motivation to study crack closure. However, consistent, and accurate measurements of crack closure (or opening) levels were found to be difficult and estimating their magnitudes in components was even more challenging. This has kept this approach from being widely used in applications outside of aerospace structures where extensive studies, such as by Newman [12] have been conducted to understand, measure, and compute crack closure levels in aluminum alloys used in the industry. Besides plasticity, crack closure levels are also affected by geometry, roughness of the fracture surfaces that relates to the underlying microstructure of the material, and by oxide debris produced by fretting due to frequent contact between fracture surfaces. This further complicates the estimation of ΔK_{eff}. For these reasons, engineers have continued to use ΔK for characterizing fatigue crack growth rates and the test methods for measuring fatigue crack growth rates are also based on ΔK, leaving load ratio and test frequency as controlled external variables.

6.6   FATIGUE CYCLES INVOLVING COMPRESSIVE LOADING

Frequently, cyclic loading in structural components include negative or compressive loads and it is therefore necessary to consider the effect of negative loads on the fatigue crack growth rates. Positive loads, all the way from zero to maximum load result in Mode I crack opening. On the other hand, compressive loads do not open cracks and theoretically can transmit compressive stresses across the crack surfaces and therefore there is no crack tip stress singularity as is the case when cracks are open. However, compressive stresses can crush the plastically deformed region behind the crack tip and affect the crack closure load and therefore have influence on the fatigue crack growth behavior, so it deserves careful consideration.

Since a negative K is a misnomer, the lowest value of K is zero, and ΔK for cycles where the minimum load is compressive is equal to K_{max}. Further, the R value is specified by the ratio of the minimum to maximum load (P_{min}/P_{max}) or the minimum to maximum stress ($\sigma_{min}/\sigma_{max}$) and can have negative values. Thus, for $R \geq 0$, $\Delta K = K_{max} - K_{min}$ but for $R < 0$, $\Delta K = K_{max}$.

6.7   MODELS FOR REPRESENTING LOAD RATIO EFFECTS ON FATIGUE CRACK GROWTH RATES

Unifying the effects of load ratio on the wide-range fatigue crack growth behavior from the threshold levels to high crack growth rates going all the way to critical K for fracture, is important in engineering applications. In this section, models based on two parameters, ΔK and K_{max} are discussed. Recall from definitions in Section 6.1, that,

$$K_{max} = \frac{\Delta K}{1-R}$$

If fatigue crack growth rates are written as being a function of ΔK and K_{max} instead of ΔK and R as in equation (6.3), we can rewrite the equation for constant frequency as:

$$\frac{da}{dN} = f(\Delta K, K_{max}) \qquad (6.10)$$

The effects due to history at different load ratios are embedded in the functional form of a single parameter that includes both ΔK and K_{max} (or R). Walker [13] proposed the first such parameter as follows:

$$\frac{da}{dN} = f\left(K_{max}^{\gamma}(\Delta K)^{1-\gamma}\right) = f\left(\frac{\Delta K}{(1-R)^{\gamma}}\right) \qquad (6.11)$$

where γ is a regression constant. This parameter was successful in unifying the data from different load ratios, but only in the Paris regime. Vasudevan and Sadananda [14] proposed the plot shown in Figure 6.21 to represent the unified effects of ΔK and K_{max}. Their hypothesis is that at high ΔK levels, the da/dN is controlled by K_{max} and at lower ΔK levels it is dominantly a function of ΔK. The curves are L-shaped because the transition from ΔK to K_{max} happens rather abruptly and there is only a small region where both parameters are important. At high K_{max} levels, ductile or brittle fracture and environmental effects come into play and therefore K_{max} becomes dominant. The governing equation for their approach is as follows:

$$\frac{da}{dN} = f\left(\Delta K^{\beta_1} K_{max}^{\beta_2}\right) \qquad (6.12)$$

where β_1 and β_2 are regression constants. This equation explicitly includes the role of ΔK and K_{max} in controlling the fatigue crack growth behavior. Kujawski [15] proposed a parameter, ΔK_d, that is a subset of the more general parameter in equation (6.12) and assumes $\beta_1 = \beta_2 = 0.5$. Thus,

$$\Delta K_d = \sqrt{\frac{K_{max} \Delta K}{2}} = \frac{\Delta K}{\sqrt{2(1-R)}} \qquad (6.13)$$

In the above equation at $R=0.5$, by definition, ΔK_d and ΔK are equal. In Figure 6.22, we replot the data for aluminum alloy 2219-T851 shown in Figure 6.13 in the

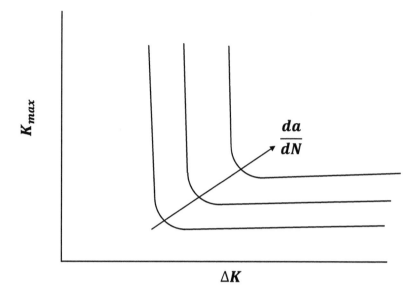

FIGURE 6.21 Schematic representation of fatigue crack growth rates in the K_{max} and ΔK space showing an L-type behavior after Sadananda and Vasudevan [12].

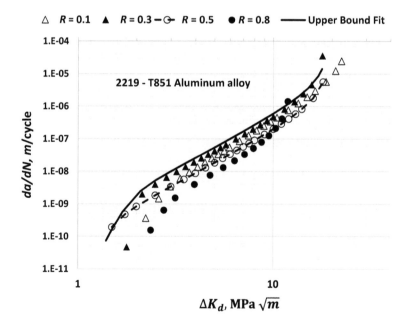

FIGURE 6.22 Fatigue crack growth behavior of 2219-T851 aluminum alloy at several load ratios ranging from 0.1 to 0.8 in the $da/dN - \Delta K_d$ space to normalize the effects of R. The solid line represents the upper-bound trend exhibited by the data.

$\Delta K_d - da/dN$ plot. The fatigue crack growth rates at various values of R seem to come together in a narrower band with no systematic variation with R. The variation in the da/dN versus ΔK can be arguably attributed to data scatter. Equation (6.6) can be rewritten in the $da/dN - \Delta K_d$ space as follows:

$$\frac{1}{da/dN} = \frac{1}{A_{1d}\Delta K_d{}^{n_1}} + \frac{1}{A_{2d}} \left[\frac{1}{\Delta K_d{}^{n_2}} - \frac{1}{\left(K_c\sqrt{\frac{1-R}{2}}\right)^{n_2}} \right] \tag{6.14}$$

The constants $A_{1d}, A_{2d}, n_1, n_2,$ and K_c are values of $A_1, A_2, n_1, n_2,$ and K_c in equation (6.6) at $R=0.5$. For other values of R, one can replace ΔK_d in equation (6.6) by $\Delta K / \sqrt{2(1-R)}$ leading to the following relationships:

$$A_1 = \frac{A_{1d}}{\left(2(1-R)\right)^{n_1/2}} \quad \text{and} \quad A_2 = \frac{A_{2d}}{\left(2(1-R)\right)^{n_2/2}}$$

While $n_1, n_2,$ and K_c remain the same as their counter parts in the da/dN and ΔK space, note that at $R=0.5$ where, $\Delta K_d = \Delta K$, all constants in equations (6.6) and (6.14) are identical. The predicted trend from the $R=0.5$ condition is plotted in Figure 6.22 using dotted line to demonstrate that the model can be used to represent the trends in the data.

For design purposes, it is useful to represent all the data by an upper bound equation that can provide conservative yet reasonable estimates of fatigue crack growth rates at all values of $R \geq 0$. The upper bound trend in terms of ΔK_d utilizing the three-component model is described by the following equation.

$$\frac{1}{da/dN} = \left[\frac{1}{1.21 \times 10^{-12}\,\Delta K_d{}^{12.5}} + \frac{1}{2.66 \times 10^{-10}} \left(\frac{1}{\Delta K_d{}^{3.3}} - \frac{1}{\left(K_c\sqrt{\frac{(1-R)}{2}}\right)^{3.3}} \right) \right] \tag{6.15}$$

Equation (6.15) was derived by first choosing the values of $n_1, n_2,$ and the value of K_c from equation (6.6). Then the A_{1d} and A_{2d} values were adjusted to represent the upper bound trend as shown by the solid line in Figure 6.22. The following example shows how this equation can be used to obtain upper bound fatigue crack growth rate behavior at any value of R that is >0.

Example Problem 6.4

For structural integrity assessment of the panel in Example 6.3 for a different service load history, you are asked to provide the fatigue crack growth behavior for

$R = 0.1$. Using equation (6.15), estimate the constants suitable for use at that load ratio and show that the model is a conservative representation of the fatigue crack growth behavior at $R = 0.1$.

Solution

Substituting for ΔK_d in equation (6.15), we get:

$$\frac{1}{da/dN} = \left[\frac{1}{1.21 \times 10^{-12} \left(\dfrac{\Delta K}{\sqrt{2(1-R)}} \right)^{12.5}} \right.$$

$$\left. + \frac{1}{2.66 \times 10^{-10}} \left(\frac{1}{\left(\dfrac{\Delta K}{\sqrt{2(1-R)}} \right)^{3.3}} - \frac{1}{\left(K_c \sqrt{\dfrac{1-R}{2}} \right)^{3.3}} \right) \right]$$

For $R = 0.1$, the above equation reduces to

$$\frac{1}{da/dN} = \left[\frac{1}{3.07 \times 10^{-14} (\Delta K)^{12.5}} + \frac{1}{1.0 \times 10^{-10}} \left(\frac{1}{(\Delta K)^{3.3}} - \frac{1}{(26)^{3.3}} \right) \right]$$

The da/dN versus ΔK trend from the above equation is plotted Figure 6.23 along with the actual trend. The upper bound trend is clearly a conservative representation of the data.

6.8 FATIGUE CRACK GROWTH MEASUREMENTS (ASTM STANDARD E647)

Scientists and engineers have measured the resistance to fatigue crack formation and fatigue crack growth for decades and consequently, several advances have occurred in test techniques and equipment for ensuring that the measured properties relate to the material behavior and are free, as much as possible, from random scatter caused by variability in test procedures and lack of proper control in the test equipment. American Society for Testing and Materials (ASTM) has produced a test standard that was first published in 1978 and is known as E-647 Standard Test Method for Measurement of Fatigue Crack Growth Rates [10]. The standard addresses precision in equipment performance for applying the fatigue loads, devices for measuring crack extension as a function of cycles, test specimen geometries to be used and procedures for analysis of the data.

Figure 6.24a shows a set-up for fatigue crack growth testing of a compact type specimen at high temperature including the furnace that surrounds the specimen for

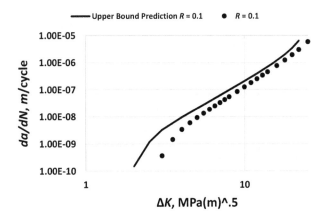

FIGURE 6.23   Comparison between the upper-bound fatigue crack growth rate behavior with the mean trend at $R = 0.1$.

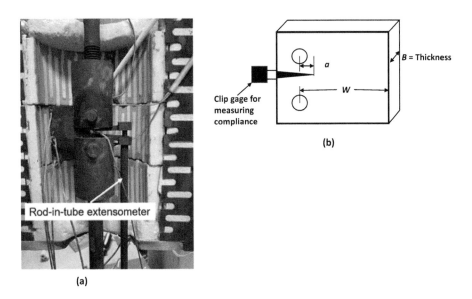

FIGURE 6.24   (a) A set up for fatigue crack growth testing at high temperatures using a compact type specimen and a spring-loaded rod and tube type extensometer for measurement of load-line displacements and (b) schematic of a C(T) specimen and a clip-on displacement gage for measuring displacements on the front-face.

heating it to the test temperature. In this set up, the displacement gage consists of a rod and tube assembly to transfer the load-line displacement outside the furnace, so the displacement gage is not subjected to heat from the furnace. Figure 6.24b shows a schematic of a compact type (C(T)) specimen and the associated clip-on gage that measures the displacement along the front-face of the specimen. Such a set-up is used in fatigue crack growth rate (FCGR) testing under ambient conditions. The clip gages in the two set-ups are used to measure compliance of the specimen as a function of fatigue cycles which in turn can be used to convert to crack length as a function of cycles. Crack lengths can also be measured using an optical traveling telescope at a magnification of 10–20x which requires specimen surface preparation and special lighting to enhance the optical image of the crack tip. Electrical resistance measurement is another method of measuring crack extension as a function of cycles. All these techniques, if used properly, yield similar results and are therefore part of the standard. In the discussion below, we describe the compliance technique in more detail.

The relationship between compliance and crack size is given by the relationship [16] shown in equation (6.16).

$$\frac{a}{W} = C_0 + C_1 U + C_2 U^2 + C_3 U^3 + C_4 U^4 + C_5 U^5 \tag{6.16}$$

$$\text{where,} \quad U = \frac{1}{\left(\dfrac{BEV}{P}\right)^{1/2} + 1} \tag{6.17}$$

The constants C_i depend on the point at which the displacements are measured. In other words, they will depend on whether the measurements are being made along the load-line or on the front-face of the specimen or any other location along the crack faces. The constants for when displacements are measured along the load-line and the front-face, the most popular locations are given in Table 6.5. These fitting parameters provide accurate compliance values with high levels of accuracy (0.5% or better). The expression for calculating ΔK is given as follows for the C(T) specimen from equation (5.2) given in Chapter 5:

$$\Delta K = \frac{\Delta P}{BW^{1/2}} F(a/W)$$

$$= \frac{\Delta P}{BW^{1/2}} \left[\frac{2 + a/W}{(1 - a/W)^{3/2}} \right]$$

$$\left[0.886 + 4.64(a/W) - 13.32\left(\frac{a}{W}\right)^2 + 14.72\left(\frac{a}{W}\right)^3 - 5.60\left(\frac{a}{W}\right)^4 \right]$$

TABLE 6.5

Constants in Equation (6.16) When Displacements Are Measured on the Load-Line and the Front-Face of the Compact Specimen

Measurement Location	C_0	C_1	C_2	C_3	C_4	C_5
Load-line	1.0002	−4.0632	11.242	−106.04	464.33	−650.68
Front-face	1.0010	−4.6695	18.460	−236.82	1,214.9	−2,143.6

Prior to testing, all specimens are fatigue pre-cracked to produce sharp reproducible cracks from the machined samples, much like those for fracture toughness testing. These cracks are then grown by fatigue loading using constant amplitude loading cycles while the size of the crack is periodically measured, and the count of the associated cycles is maintained using a digital cycle counter. Typically, the crack size is measured in pre-selected increments (0.25 mm is commonly used) and is related to the applied fatigue cycles, N, so that the crack growth rates can be calculated at any crack size along the measured a versus N trend as shown in Figure 6.25a. The slope of this curve at any crack size, a_i, is (da/dN) and the ΔK at the same crack size is the corresponding $(\Delta K)_i$. $(da/dN)_i$, and $(\Delta K)_i$ constitute a data pair that are plotted on the log (da/dN) versus log ΔK plot as shown in Figure 6.25b. Several such data pairs can be obtained from a single test conducted at CLA and constant value of R. Depending on the range of crack growth rates at which data are desired, the load amplitudes for subsequent tests can be chosen/adjusted. By keeping the values of P_{max} and P_{min} the same throughout the test, the load ratio is also held constant. $(da/dN)_i$ and $(\Delta K)_i$ are calculated as follows:

$$\left(\frac{da}{dN}\right)_i = \frac{a_i - a_{i-1}}{N_i - N_{i-1}}$$

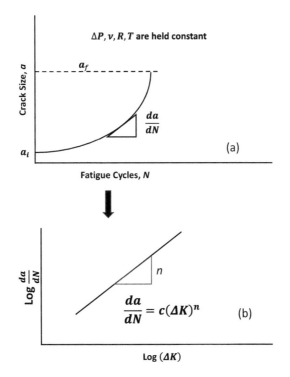

FIGURE 6.25   (a) Estimation of crack growth rates and the corresponding ΔK levels from the FCGR data, and (b) plot of da/dN versus ΔK trend on a log-log plot to accommodate fatigue crack growth rates over several orders of magnitude.

$$\Delta K_i = \frac{\Delta P}{BW^{1/2}} F\left(\frac{a_{i-1} + a_i}{2W}\right)$$

The validity requirements are applied to each data point to qualify the data to ensure dominantly linear elastic conditions. In this case, the requirement determined by equation (6.18) must be met according to ASTM E647:

$$W - a \geq \frac{4}{\pi}\left(\frac{K_{max}}{\sigma_{ys}}\right)^2 \tag{6.18}$$

If this condition is not satisfied, the data are not valid by the ASTM standard because small-scale-yielding cannot be ensured.

An alternate method for measuring fatigue crack growth rates: The CLA test has a disadvantage that the ΔK range over which the data are generated is dependent on the geometry and size of the specimen and the applied load range. There is no ability to control the increment in ΔK as the crack size increases. The ΔK-gradient is uneven with its value being low initially and increasing with crack growth and becoming very steep as the specimen approaches end of life.

A more desirable ΔK-gradient would be an approximately linear increase (or decrease) in ΔK with crack size so the distribution in the crack growth rate data would be more uniform. Such a test is considerably more efficient in collecting data because it distributes the *da/dN* and ΔK points evenly across the crack growth interval. These tests can be conducted with either a positive ΔK-gradient or a negative ΔK-gradient. In the former, the ΔK rises from an initially chosen minimum value to the highest value at the end of the test and in the latter, it is the opposite. This is achieved by continuously measuring the crack size as the test progresses and then utilizing the current crack size to adjust the force amplitude to yield the targeted ΔK value. To minimize the effects of history, especially while using a decreasing ΔK-gradient, the percentage change in plastic zone size (r_p) for a given increment in crack size is held constant. This is accomplished by following equations (6.19) and (6.20) first proposed by Saxena et al. [17].

$$\frac{1}{r_p}\frac{dr_p}{da} = \pi\left(\frac{\sigma_{ys}}{K_{max}}\right)^2 \cdot \frac{2}{\pi}\left(\frac{K_{max}}{\sigma_{ys}^2}\right)\frac{dK_{max}}{da} = C_1 \tag{6.19}$$

$$\text{or} \quad \frac{1}{K_{max}}\frac{dK_{max}}{da} = C$$

$$\text{For constant } R, \quad \frac{1}{\Delta K}\frac{d(\Delta K)}{da} = \frac{1}{K_{max}}\frac{dK_{max}}{da} = C$$

$$\text{Thus,} \quad \frac{d(\Delta K)}{da} = C(\Delta K) \text{ or } \int_{\Delta K_0}^{\Delta K}\frac{d(\Delta K)}{\Delta K} = \int_{a_0}^{a} Cda$$

That leads to the following relationships:

$$\Delta K = \Delta K_0 \exp\left(C(a - a_0)\right) \tag{6.20a}$$

$$K_{max} = K_{max,0} \exp\left(C(a - a_0)\right) \tag{6.20b}$$

$$K_{min} = K_{min,0} \exp\left(C(a - a_0)\right) \tag{6.20c}$$

where ΔK_0 is the initial value of ΔK and a_0 is the initial crack size. By choosing an appropriate value of C, we can select the rate of increase (or decrease in the case of ΔK decreasing test) of ΔK. A negative C value is chosen for a decreasing ΔK test and a positive C value is chosen for an increasing ΔK test. A negative C value is often used for determining crack growth rates at low ΔK values in the near threshold region of fatigue crack growth. ASTM E647 recommends using $|C| \leq 0.08 \ \text{mm}^{-1}$ during decreasing ΔK test based on the experimental studies by Saxena et al. [17]. Such variation of ΔK with crack size is schematically shown in Figure 6.26 for decreasing and increasing ΔK for a fixed value of C when compared to the variation in ΔK for a CLA test. These tests are often more efficient because we can begin the test at a ΔK where we are sure that the crack growth rates are in the range where the cracks will grow in a short timeframe and then decrease ΔK gradually until the targeted low crack growth rates are achieved. In an increasing ΔK test, if we incorrectly guess the ΔK level to grow the crack, we might either miss the targeted low crack growth rates all together if we guessed high, or if we guess low, spend too much time at a ΔK level where the cracks will not grow. This guess work is removed when using decreasing ΔK-controlled testing to determine near-threshold crack growth rates.

A second reason in favor for using decreasing ΔK testing is the more efficient utilization of the specimen, especially for low-strength materials such as 304 stainless steel that has a yield strength of 250 MPa and a fracture toughness value that

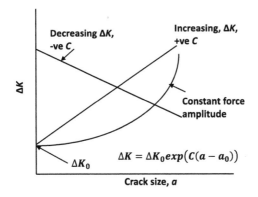

FIGURE 6.26 Variation of ΔK with crack size in FCGR tests conducted under decreasing ΔK and increasing ΔK with constant values of C compared to the variation in ΔK for tests conducted under constant force amplitude.

is on the order of $300\,\mathrm{MPa}\sqrt{m}$. The ΔK_{max} for which fatigue crack growth rate data can be developed may even be higher than $100\,\mathrm{MPa}\sqrt{m}$. From equation (6.18), that determines the validity of the data, we can see that a large ligament of $(4/\pi)$ $(100/250)^2 = 0.2\,\mathrm{m}$ is needed. The ligament at the start of test is the highest so it makes better sense to start the test at the high ΔK and decrease its value as the ligament gets smaller with crack growth.

The corresponding cyclic loads needed to achieve the desired ΔK-gradient can be estimated as follows:

$$\Delta K = \frac{\Delta P}{BW^{1/2}} F(a/W) = \frac{\Delta P_0}{BW^{1/2}} F\left(\frac{a_0}{W}\right) \exp(C(\Delta a))$$

$$\Delta P = \Delta P_0 \frac{F\left(\dfrac{a_0}{W}\right)}{F\left(\dfrac{a}{W}\right)} \exp(C\Delta a) \tag{6.21}$$

The ΔP values according to equation (6.21) will ensure a more planned ΔK-gradient. This requires, as mentioned before, continuous monitoring of crack size and using it as a feedback signal to change the value of ΔP continuously as the crack grows. Both the compliance technique and the electrical potential technique are used for monitoring crack size and then used as a feedback signal to change the load amplitude.

The crack size, a, versus fatigue cycles, N, are recorded at small increments of crack length of about 0.25 mm and are processed to obtain the da/dN versus ΔK trends. Then the data can be processed in a manner like the constant force amplitude test to produce the da/dN versus ΔK relationship. Following ASTM standard E647 in generating fatigue crack growth data is good practice because one is assured that the data is of high quality. Regulatory bodies writing design codes/rules require that the data used for design life calculations be generated using ASTM standards.

Example Problem 6.5

You are asked to develop da/dN versus ΔK data at R values of 0.1 and 0.5 in support of design of pressure vessels that will be subjected to 10 pressure cycles/day. Based on applicable design codes you must utilize an ASTM Grade 372 Grade J class 70 steel that has a yield and ultimate strength of 700 and 900 MPa, respectively. The K_{Ic} is estimated at $80\,\mathrm{MPa}\sqrt{m}$. Some vessels will be cycled between p_{max} and $0.5p_{max}$, where p_{max} is the maximum operating pressure, while other vessels will be cycled from p_{max} to $0.1p_{max}$. Based on stress calculations and on nondestructive test (nondestructive evaluation [NDE]) techniques, it is determined that to estimate fatigue lives, we will need fatigue crack growth rate data for the initial K_{max} of $10\,\mathrm{MPa}\sqrt{m}$. The life of the cylinder should conservatively be considered as over when the K_{max} reaches 0.5 K_{Ic}. Describe the fatigue crack growth tests that you will conduct to support the design. The material available from the prototype vessel is suitable for machining single edge crack specimens of the size shown in Figure 6.27. Assume that the initial crack size in the specimen after fatigue precracking is 2 mm.

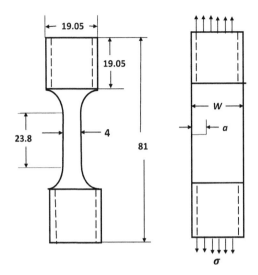

FIGURE 6.27 Schematic of a SEN(T) specimen that can be extracted from an A372-Grade J Class 70 steel pressure vessel for fatigue crack growth rate testing.

Solution

We have a choice of conducting CLA tests to obtain the required data, or we could conduct tests in which the ΔK-gradient is controlled as per equation (6.20). Let's consider both types of tests and compare their advantages/disadvantages before choosing. We are asked to produce data in which the K_{max} value ranges from 10 to 40 MPa$\sqrt{m}$ (half of K_{Ic}). The ΔK corresponding to the given K_{max} values are estimated by the following equation:

$$K_{max} = \frac{\Delta K}{1 - R}$$

Thus, for the deep pressure cycle with an $R=0.1$, ΔK will range from 9 to 36 MPa$\sqrt{m}$ and for the shallow pressure cycle, it will range from 5 to 20 MPa$\sqrt{m}$.

The next step is to determine the minimum remaining ligament size based on equation (6.18) required for generating valid fatigue crack growth data by the ASTM Standard E647 for K_{max} ranging from 10 to 40 MPa$\sqrt{m}$. The K-calibration expression for SEN(T) specimen was previously listed in Table 3.1d and repeated below.

$$\frac{K}{P} BW^{1/2} = F\left(\frac{a}{W}\right) = \sqrt{2\tan\left(\frac{\pi a}{2W}\right)} \left\{ \frac{0.752 + 2.02(a/W) + 0.37\left(1 - \sin\dfrac{\pi a}{2W}\right)^3}{\cos\left(\dfrac{\pi a}{2W}\right)} \right\}$$

The initial $K_{max} = 10$ MPa$\sqrt{m}$, $W=0.01905$ m, $a=0.002$ m, and $B=0.004$ m. Using the above equation, we can estimate the initial value of P_{max} as follows:

$$P_{max} = \frac{K_{max} B\sqrt{W}}{F(a/W)} = \frac{(10)(0.004)\sqrt{0.01905}}{F(0.002/0.01905)} = 0.008 \text{ MN} = 8.0 \text{ kN}$$

Using 8.0 kN as the initial value of P_{max}, we can estimate the K_{max} as a function of crack size in a CLA test as shown in Figure 6.29. The K_{max} reaches a value of approximately $40\,MPa\sqrt{m}$ at a crack size of 8.25 mm at which point we can terminate the test. Thus, the test will consist of growing the crack from 2 to 8.25 mm to obtain the required data. We should check to see if the ligmant size at the end of the test meets the requirement of equation (6.18) to qualify as being valid data.

$$W - a \geq \frac{4}{\pi}\left(\frac{40}{700}\right)^2 = 0.00416 \ m = 4.16 \ mm.$$ The actual remaining ligament will

be $(19.05 - 8.25) = 10.8\,mm$. Therefore, the requirement is met.

For the ΔK-controlled test, we use the size of the minimum required uncracked ligament of 4.16 mm to calculate the K_{max}-gradient, C, as follows:

$$40 = 10\exp\big(C(14.89 - 2)\big)$$

$$C = \frac{1}{(14.89 - 2)}ln\frac{40}{10} = 0.107 \ mm^{-1}$$

$$\text{and } K_{max} = 10\exp\big(0.107(a - 2)\big)$$

Note, that to be consistent with units, we vave chosen C and a in mm. In Figure 6.28, the K_{max} as a function of crack size, a, is also plotted for K-controlled test that will yield the required data. The maximum force as a function of crack size for the two tests are plotted in Figure 6.29.

Either of the two tests will yield the required data for both load ratios by choosing $K_{min} = RK_{max}$. In the CLA test, the range from a K_{max} of 10 to $40\,MPa\sqrt{m}$ is covered within a crack growth from 2 to 8.25 mm while for a controlled K test, it is covered within a crack growth from 2 to 14.89 mm. We will get more da/dN versus ΔK data points in the K-controlled tests than in the CLA tests. More data is always desireable and leads to more confidence in the results. On the other hand, the K-controlled test is harder to carry out because it requires continuous measurement of crack size and adjustment of the maximum and minimum force according to the trend shown in Figure 6.29.

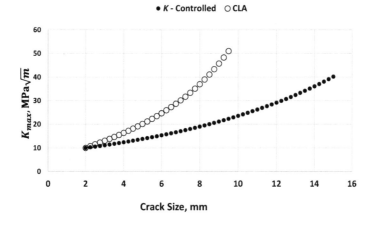

FIGURE 6.28 K_{max} as a function of crack size for CLA and K-gradient controlled tests.

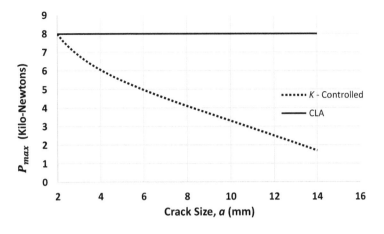

FIGURE 6.29   The applied load (or force) range as a function of crack size for CLA and
K-gradient controlled tests.

6.9   BEHAVIOR OF SMALL OR SHORT CRACKS

For safe design or remaining life calculations, the initial flaw size in components
is assumed to be the smallest size that can be reliably detected by the chosen NDE
technique. The resolution of NDE techniques has been continuously improving
with technology, so the initial crack size assumed in design life calculations is also
expected to decrease to keep pace with these advancements. Small or short cracks
have been shown to grow at rates faster than those predicted by ΔK and are therefore
a concern. Several factors have been suggested to account for the nonunique da/dN
versus ΔK behavior of small or short cracks. The first question, however, is what is
the distinction between "small" or "short" cracks?

Small of short cracks can be divided in three categories. The term "short
crack" is used when only the crack depth is small while the length dimension is
relatively large. The term "small crack" is used when all pertinent dimensions
of the crack are small. The third category consists of "microstructurally small
cracks," in which the crack size is on the order of the microstructural periodicity
such as grain size.

A schematic of small versus large crack growth rate behavior at low load ratios is
illustrated in Figure 6.30. Initially, the small cracks propagate at rates faster than the
large cracks at nominally the same ΔK. As crack size increases, the growth rates of
these cracks decelerate to a minimum, accelerate, and eventually become equal to
the large crack propagation rates, as shown by the curve ABC in Figure 6.30. This
small crack effect is more pronounced in the near-threshold region than in Region
II of the fatigue crack growth behavior. Further, if the applied stress amplitudes
are very low, small cracks, especially those emanating from notches, may propagate
from the notch root at a decreasing rate and then arrest as shown in A′ B′. If we plot
the endurance stress range, $\Delta \sigma_e$, as a function of the starting crack size, a, the plot
will schematically appear like one shown in Figure 6.31 known as the Kitagawa

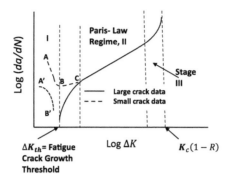

FIGURE 6.30 Schematic of a large and small fatigue crack propagation behavior illustrating the small crack effects.

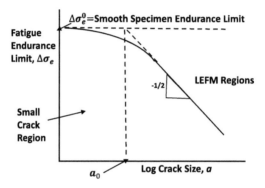

FIGURE 6.31 Plot of fatigue endurance limit as a function of crack size, Kitagawa diagram [17].

diagram [18]. When the crack size is large, one can expect the following relationship between ΔK_{th} and $\Delta \sigma_e$:

$$\Delta K_{th} = 1.12 \Delta \sigma_e \sqrt{\pi a} \qquad (6.22)$$

In equation (6.22), we assume that the crack size correction factor is the surface correction factor 1.12. If ΔK_{th} is a material constant, the relationship between $\log \Delta \sigma_e$ and $\log a$ is linear with a slope of $-1/2$. However, as the crack size decreases, $\Delta \sigma_e$ must approach the smooth specimen endurance stress range of $\Delta \sigma_e^0$ corresponding to zero crack size. Note that $\Delta \sigma_e^0$ is two times the endurance stress amplitude, σ_e^0, for smooth uncracked specimens. A crack size a_0 can be defined below for which the cracks are no longer expected to be uniquely characterized by ΔK:

$$a_0 = \left(\frac{\Delta K_{th}}{1.12 \Delta \sigma_e^0} \right)^2 \qquad (6.23)$$

Equation (6.23) provides a definition of a "short crack."

Example Problem 6.6

The yield strength of plain carbon steel AISI 1018 used in many applications is 300 MPa, the ultimate tensile strength is 500 MPa, and the fatigue endurance limit is 250 MPa. The estimated value of ΔK_{th} is 8 MPa$\sqrt{m}$ at an R value of −1. A shaft made from of this steel is inspected using magnetic particles that can detect surface cracks that are 1 mm deep and 10 mm long in the radial-circumferential plane. The shaft is expected to experience over 10 million cycles of fully reversed loading during service. Estimate the maximum allowable cyclic stress amplitude for the shaft and comment on the use of ΔK for characterizing fatigue crack growth in this case.

Solution

The flaw in this case is a semi-elliptical flaw that is likely to propagate under axial stresses. The ΔK for this configuration is given by:

$$\Delta K \approx 1.12 \sigma_a \sqrt{\frac{\pi a}{Q}}$$

where σ_a is the stress amplitude, and Q is from equation (3.30a) is given by:

$$Q \approx 1 + 1.464 \left(\frac{a}{c}\right)^{1.65}$$

Since the required fatigue life is so large, we want to choose an allowable stress amplitude for which the ΔK is less than ΔK_{th}. We also should evaluate whether the crack will be considered a short crack. Since the crack length is 10 mm, it will not be classified as a small crack. We can write the following equation:

$$\sigma_a = \frac{\Delta K_{th}}{1.12 \times \sqrt{\frac{\pi a}{Q}}} = \frac{8}{1.12 \sqrt{\frac{\pi \times 0.001}{Q}}}$$

The a/c ratio = 1/5 (the surface length = 10 mm = 2c) = 0.2, leading to a value for Q = 1.103. Thus, σ_a = 133.87 MPa. From equation (6.23), a_0 = 0.00082 m = 0.82 mm. Since $a_0 <$ than 1 mm, we can use 133.87 MPa as the maximum allowable stress amplitude in the shaft. If a_0 were to be higher than 1 mm, we would have to make an allowance for short or small crack behavior as described in the following discussion.

6.9.1   LIMITATIONS OF ΔK FOR CHARACTERIZING SMALL FATIGUE CRACK GROWTH BEHAVIOR

There are several conditions which must be met for da/dN to uniquely correlate with ΔK and should be reviewed again in relation to small crack behavior:

- The crack tip stress field must be uniquely characterized by K, in a region which is large compared to size of the crack tip plastic zone.

- The material at the length scale over which fatigue damage develops is iso-tropic and homogeneous.
- The crack grows under dominantly Mode I condition.
- Scales of plasticity associated with the growth of fatigue cracks are small not only in relation to the dimensions of the cracked body but the crack size itself.
- Crack closure levels are independent of crack size.

Meeting the five conditions above are collectively necessary for ensuring that simili-tude exists in the conditions at the tips of cracks in different bodies that are loaded to the same nominal ΔK. Only when all five conditions are met, a unique da/dN versus ΔK relationship can be justified. These limitations are discussed further in this sec-tion in relation to small cracks.

A complete solution for stress ahead of the crack tip that is not just limited to the crack tip region, was derived by Williams [19] and is given by:

$$\sigma_{ij} = \left(\frac{K}{\sqrt{2\pi r}} \right) f_{ij}(\theta) + \sum_{m=0}^{\infty} A_m r^{\frac{m}{2}} g_{ij}^{(m)}(\theta) \tag{6.24}$$

where $f_{ij}(\theta)$ and $g_{ij}^{(m)}(\theta)$ are angular functions. The first term in equation (6.24) is the same as in equations (3.15), the crack tip field equations. The higher order terms, as are evident from the nature of equation (6.24), are only important for higher values of r. In other words, as $r \to 0$, the first singular term dominates and equations (6.24) and (3.15) become the same, regardless of the geometry of the cracked body. The higher order terms do vary with geometry and become significant as r increases. Also, among the higher order terms, the first term in Williams' series does not depend on r. This second term in the series in Equation (6.24) corresponding to m = 0 is a constant and is often referred to as the "T-stress" in the literature.

Figure 6.32 shows a plot of the σ_y stress as a function of distance from the crack tip for two cracks, one of which was 0.25 mm in size and the other 6.25 mm in size, in a 25.0 mm wide, single-edge crack tension, SEC(T), specimen subjected to uniform stress at the ends. The remote stresses in the specimen were adjusted such that the applied K levels in both cases was 22.2 MPa$\sqrt{m}$. Talug and Reifsnider [20] deter-mined several coefficients, A_m, in equation (6.24) by a numerical technique. Thus, full crack tip stress field solutions beyond the first singular term were obtained for the two cracks. The solid line in Figure 6.32 shows the calculation for the full solu-tion for the short crack and the dashed line for the long crack. The Irwin-Westergaard solution, equation (3.15), is also shown for comparison. The long crack full-field solu-tion and the Irwin-Westergaard solutions are indistinguishable while the full-field solution for small cracks is quite distinct from the other two as the distance from the crack tip increases. For this geometry, the region of dominance of the singular field for which the amplitude is characterized by K, is approximately 15% of the crack size. While the distance from the crack tip as a fraction of crack size for which K dominates may vary somewhat with geometry, the important result here is that the

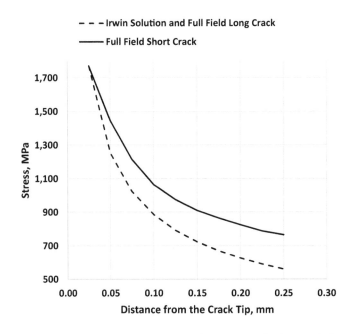

FIGURE 6.32 Comparison of full-field and Irwin solutions for long and short cracks at constant K, as a function of distance r from the crack tip [19].

region of K-dominance is dependent on crack size itself. Thus, the loss of uniqueness of K for characterizing the growth of short cracks becomes evident.

The validity of ΔK as a crack tip parameter is based on the principles of elasticity which also assume that the material is homogeneous and isotropic. When crack sizes are smaller than the average grain diameter, it cannot be assumed that the material is isotropic, or homogeneous. Also, the crack tip experiences considerable nonuniformity in the microstructural features that it encounters. For example, the crack tip stress fields can be influenced significantly by the presence of grain boundaries in the proximity of the crack tip or second phase particles such as inclusions. The back stresses from these microstructural entities can cause considerable variability in the crack growth rates which cannot be addressed within the framework of linear elasticity theory.

The third condition that the crack growth occur in Mode I is also violated frequently when considering the growth of small cracks. For example, small cracks at low ΔK levels often grow along slip bands which form preferentially along certain crystallographic directions. Thus, the crack growth is no longer in pure Mode I and mixed-mode growth becomes a consideration leading to deviation from long crack behavior which is predominantly in Mode I.

The crack closure level can be different for large and small cracks. Since plasticity induced crack closure occurs due to interference in the plastically deformed wake of the crack, differences can occur in the crack closure levels between short and long cracks. Short cracks do not have a plastically deformed wake. Therefore, for the same applied nominal ΔK, the ΔK_{eff} for the short crack can be higher and lead to a higher crack growth rate. Since crack closure is more important in the near-threshold fatigue crack growth region, it follows from this hypothesis that the small crack effect will also be more important in this region, which is consistent with the experimental trends.

If the fatigue crack size at any time is less than 1.0 mm, one must use ΔK with caution for all the reasons discussed above.

6.10 FATIGUE CRACK GROWTH UNDER VARIABLE AMPLITUDE LOADING

Cyclic loading in components is often not constant amplitude, so it is important to understand the effects of variable amplitude loading on fatigue crack growth rates. If the load amplitude is increased and then maintained at the higher level, a linear damage summation model can be used. This model is described by the relationship given in equation (6.25). Let us assume that a cracked body is subjected to a series of m load amplitude blocks $\Delta P_1, \Delta P_2, \Delta P_3, \ldots \Delta P_i \ldots \Delta P_m$ and as a result, the crack grows due to the ith load block from a_{i-1} to a_i starting from an initial crack size a_0, the cycles to failure can be calculated using the linear damage summation rule that assumes no history effects, and is expressed by equation (6.25).

$$N_f = \sum_i^m N_i = \sum_1^m \int_{a_{i-1}}^{a_i} \frac{da}{c(\Delta K_i)^n} \qquad (6.25)$$

The above model does not account for transient effects that occur when the load amplitudes change. If the load amplitude in the ith block is always higher than the previous block of cycles, equation (6.25) is expected to predict the fatigue crack growth rates reasonably accurately. However, if the load amplitude in the ith block is lower than that of the previous block, equation (6.25) is not adequate, there are important load interaction effects that cannot be neglected. These are discussed in the subsequent sections.

6.10.1 EFFECTS OF SINGLE OVERLOADS/UNDERLOADS ON FATIGUE CRACK GROWTH BEHAVIOR

Isolated or occasional overloads can have significant effects on the number of cycles to failure. To illustrate this point, we consider single overloads first. Let us consider a situation in which otherwise CLA loading is interrupted by a single overload as seen in the Figure 6.33a. It is seen that the crack size versus fatigue cycles trend is disrupted by the overload as shown in Figure 6.33b. The crack growth rate following the overload is significantly retarded over N_D cycles, called the delay cycles, and over a crack extension distance of, a_D, following the overload. Beyond that, the original crack growth rate resumes. This retardation in crack growth rate has been attributed to the interactions between the crack tip plastic zone associated with the overload cycle and the plastic zone associated with subsequent cycles. This is illustrated

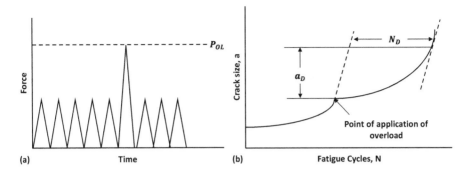

(a)   Time   (b)   Fatigue Cycles, N

FIGURE 6.33   (a) Loading cycles during fatigue interrupted by a single overload cycle and (b) the effect of a single overload on the subsequent fatigue crack growth behavior showing retardation of crack growth rates for N_D cycles over a crack extension of a_D.

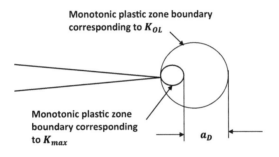

FIGURE 6.34   Plastic zones ahead of the crack tip associated with the regular fatigue cycles and the overload cycle.

schematically in Figure 6.34. A larger plastic zone is formed at the crack tip because of the overload so, the plastic zone associated with the normal value of the load must grow beyond the plastic zone boundary associated with the overload cycle prior to assuming normalcy. Thus, a_D is equal to the difference in the size of the plastic zone due to overload and the one associated with the normal load. The use of overloads to retard the rates at which fatigue cracks grow is selectively used in some industries, including aircraft industry, to enhance fatigue life of components. This must be used with plenty of caution because during the overload cycle itself there is the risk of stable crack growth or fracture.

The simple explanation above for retardation of crack growth rates following a single overload is supported by a couple of other experimental observations. In the first, an under-load is applied instead of an overload and the crack growth rates are observed. Since a compressive load/stress does not create a larger plastic zone at the crack tip, no retardation effects are expected and that is supported by the experimental data. In the second experiment, overload is immediately followed by application of an under-load of equal magnitude. The underload counters the effects of plasticity due to the overload and is expected to annihilate its effects. It has been experimentally shown that the number of delay cycles are very significantly reduced when an overload is followed by an underload of the same magnitude. These observations give credibility to the theory that retardation in the fatigue crack growth rates due to overload are caused by the history effects related to changes in crack tip plasticity due to overloads.

6.10.2   VARIABLE AMPLITUDE LOADING

This section considers crack growth due to a loading spectrum of the type shown in Figure 6.35. The maximum and minimum loads and the amplitude of loading varies frequently during the load spectrum. One method used for predicting fatigue crack growth rates under variable amplitude loading is the rain-flow counting method [21,22]. This method is based on estimating the crack closure load during each loading cycle and using it to estimate ΔK_{eff} to calculate the crack growth rate during that cycle. These calculations are repeated cycle-by-cycle for determining crack extension. The method relies on two assumptions (a) that all overload effects are manifested in crack opening and closing levels and (b) crack opening and closing levels can be accurately estimated. The crack closure load or stress depends on geometry, material, and the loading conditions. Further, crack closure occurs due to plasticity, due to surface roughness, and due to oxide debris from fretting fatigue. These complex phenomena make the calculation of closure load or stress levels challenging, so these calculations in the model are based solely on plasticity induced crack closure and ignores other mechanisms of crack closure. This method has been incorporated in computer codes such as FASTRAN [22] that features "rainflow-on-the-fly" methodology. This approach has been pioneered by James C. Newman [22] and his colleagues at NASA and is motivated by the needs of the aerospace industry. The methodology is based on the following principles/assumptions:

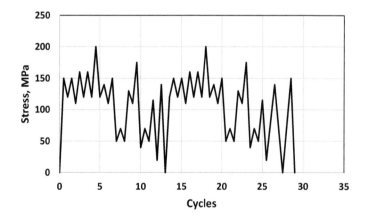

FIGURE 6.35   Stress fluctuation as a function of time during variable amplitude fatigue.

- Damage only occurs during the loading portion of the cycle but unloading may change crack-tip deformation and influence the crack opening stress during the subsequent cycle.
- Stress levels in the loading spectra below the crack opening level do not participate in growing the crack.
- The load history effects are accounted for by the change in the crack opening stress level during the subsequent loading cycles.

ΔK_{eff} is calculated for each loading segment in the "rainflow-on-the-fly" method containing a load reversal from a high to low value and the crack growth rate is estimated by the following equation in which C and n are constants in the Paris fatigue crack growth equation (6.26).

$$\frac{da}{dN} = C\left(\Delta K_{eff}\right)^n \tag{6.26}$$

The implementation of the method is best explained with an example. Let's assume that a segment of the loading spectrum is represented by the loading cycle shown in Figure 6.36. The stress levels at various points of interest in this segment of the loading spectrum are represented by S_i where, i ranges from 1 to 5. We assume that this loading is being applied to a uniformly stressed semi-infinite panel with a center crack of $2a$ for which the effective ΔK can be calculated by the following simple equation.

$$\Delta K_{eff} = (S_i - S_o)\sqrt{\pi a}$$

where S_o is the crack opening stress from the history. For example, the S_o in this load segment is tied to the minimum stress S_1. Since $S_3 > S_o$, it will not affect the crack

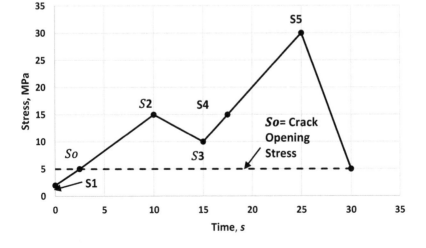

FIGURE 6.36 A segment of a variable amplitude loading spectrum for estimating fatigue crack growth rates using NASA's "rainflow-on-the-fly" method.

opening stress that will remain the same during this entire segment. If S_3 was indeed less than S_o, a new value of crack opening stress would have to be estimated. We estimate the crack growth during the various segments of this loading cycle. Between S_1 and S_2, the crack extension is given by,

$$\Delta a_1 = f(S_2 - S_o)$$

The unloading between S_2 and S_3 does not contribute to crack extension and since $S_3 > S_o$, the crack opening level for the segment between S_3 and S_4 remains S_o.

Between S_3 and S_4, the crack extension is given by,

$$\Delta a_2 = f(S_4 - S_o) - \Delta a_1 + f(S_2 - S_3)$$

The total crack extension is given by,

$$\Delta a = \Delta a_1 + \Delta a_2 = \Delta a_1 + f(S_4 - S_o) - \Delta a_1 + f(S_2 - S_3) = f(S_4 - S_o) + f(S_2 - S_3)$$

Thus, the history effects are captured by adding the additional crack growth due to unloading and loading between stress levels of S_2 and S_3. Since S_5 is not below S_o, no updating of the crack opening stress is necessary during the subsequent loading.

There are limitations of the model that must be acknowledged while applying it to make design decisions. One criticism, perhaps the most important one, is that the uncertainty level is high in both measurement and in analytical estimation of crack opening loads that include consideration of plasticity-induced, surface roughness induced, and oxide debris induced crack closures. Since the model depends strongly on the crack opening load, it leads to uncertainty in the predicted crack growth rates.

The second limitation of this model is based on the ability of crack closure by itself to account for all load interaction effects observed during fatigue crack growth under variable amplitude loading. Crack closure is a result of phenomena that occur behind the crack tip on the fracture surface. If one considers the mechanisms of damage accumulation in front of the crack tip over a wide range of crack growth, a picture as in Figure 6.37 emerges [23]. In the three regions of fatigue crack growth behavior, the mechanisms of damage accumulation are different. In Region I, crack extension occurs primarily through damage accumulation along discrete crystallographic slip planes and therefore the fracture surface appears somewhat featureless compared to Region II where the crack extension occurs by widespread slip leading to striation formation. In Region III, static modes of crack extension begin to appear due to values of K_{max} beginning to approach the fracture toughness of the material. Such variations in fracture mechanisms are unlikely to be fully included in a single parameter such as the crack opening/closure levels. In conclusion, it is reasonable to state that prediction of the effects of variable amplitude loading are still a topic of research and there is no consensus among researchers on a widely applicable method.

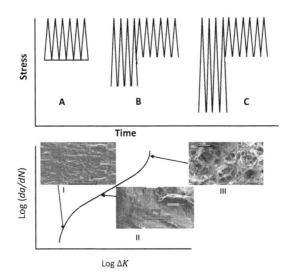

FIGURE 6.37 Fractographs from the surfaces of specimens undergoing crack growth in the three regimes of the *da/dN* versus ΔK behavior showing varying crack growth mechanisms. (Figure adapted from reference [23] with permission).

6.11 SUMMARY

In this chapter, we have learnt how cracks or crack like defects grow under the influence of cyclic or fatigue loading. The fatigue crack growth rates, *da/dN*, over a wide range from 10^{-10} to 10^{-4} m/cycle are uniquely characterized by the cyclic stress intensity parameter, ΔK, for a constant stress (or load) ratio, R, and test frequency, v. The *da/dN* − ΔK relationship changes with R while the effect of loading frequency in innocuous environments is negligible, if any. The threshold for fatigue crack growth, ΔK_{th}, was defined operationally as the ΔK value corresponding to a *da/dN* value of 10^{-10} m/cycle. The relationship between *da/dN* and ΔK is divided in three regions. Region I is the near threshold region characterized by a steep increase in *da/dN* with ΔK. Region II, also known as the Paris regime, is characterized by a power-law relationship between *da/dN* and ΔK. Region III is characterized by accelerating *da/dN* versus ΔK behavior influenced by the fracture toughness of the material. A three-component model was described to represent the *da/dN* versus ΔK over all three regions.

Crack closure models that account for the effects of load ratio, R, on the *da/dN* versus ΔK relationship were described, and it was shown that a parameter, $\Delta K_d = \sqrt{K_{max}\Delta K/2}$ consolidated the effects of load ratio on the *da/dN* versus ΔK behavior quite well for some materials.

Test methods, based on the ASTM Standard E647, for fatigue crack growth testing were discussed using the CLA test and the K-controlled test. Specimen geometry and size selection as well as methods for measuring crack advance as a function of applied fatigue cycles were described.

The chapter concludes with a discussion of fatigue crack growth for small cracks and under variable amplitude fatigue loading. The effects of single overloads and underloads (compressive loads that are significantly larger than the nominal minimum load) were considered and so were the effects of variable amplitude loading. Single overloads are seen to retard the crack growth rates while an underload following an overload seems to negate the effects of overload. The limitations of the model for predicting the effects of variable amplitude loading are discussed.

REFERENCES

1. P. Lukas, M. Klensil, and J. Krejci, "Dislocations and Persistent Slip Bands in Copper Single Crystals Fatigued at Low Stress Amplitudes", *Physica Status Solidi (b)—Basic Solid-State Physics*, Vol. 27, 1968, pp. 545–558.
2. S. Suresh, *Fatigue of Materials*, 2nd Edition, Cambridge University Press, Cambridge, England, 1998.
3. P.C. Paris, M.P. Gomez, and W.E. Anderson, "A Rational Analytic Theory of Fatigue", *The Trends in Engineering*, Vol. 13, 1961, pp. 9–14.
4. W. Elber, "Fatigue Crack Closure under Cyclic Tension", *Engineering Fracture Mechanics*, Vol. 2, 1970, pp. 37–45.
5. W. Elber, "The Significance of Fatigue Crack Closure", in *Damage Tolerance in Aircraft Structures, ASTM STP 486*, 1971, American Society for Testing and Materials, Conshohocken, PA, pp. 230–242.
6. C.D. Beachem and R.M.N. Pelloux, "Electron Fractography—A Tool for Studying Micro mechanisms of Fracturing Processes" in *Fracture Toughness and its Applications, ASTM STP 381*, 1965, American Society for Testing and Materials, Conshohocken, PA, pp. 210–245.
7. H. Cai and A.J. McEvily, "On Striations and Fatigue Crack Growth in 1018 Steel", *Materials Science and Engineering: A*, Vol. 34, 2001, pp. 86–89.
8. C. Laird, "The Influence of Metallurgical Structure on the Mechanisms of Fatigue Crack Propagation", in *Fatigue Crack Propagation, ASTM Special Technical Publication 415*, 1967, American Society for Testing and Materials, Conshohocken, PA, pp. 131–168.
9. A. Saxena, S.J. Hudak, and G.M. Jouris, "A Three Component Model for Representing Wide Range Fatigue Crack Growth Data", *Engineering Fracture Mechanics*, Vol. 12, 1979, pp. 103–115.
10. ASTM Standard E647-15, *Standard Test Method for Measurement of Fatigue Crack Growth Rates, ASTM Book of Standards*, Vol. 3, American Society for Testing and Materials, Conshohocken, PA, 2016.
11. R. Pippan and A. Hohenwarter, "Fatigue Crack Closure: A Review of the Physical Phenomena", *Fatigue and Fracture of Engineering Materials and Structures*, Vol. 40, 2017, pp. 471–495.
12. J.C. Newman, Jr., "A Crack Opening Stress Equation for Fatigue Crack Growth", *International Journal of Fracture*, Vol. 24, 1984, pp. R131–R135.
13. K. Walker, "The Effect of Stress Ratio During Crack Propagation in Fatigue for 2024-T3 and 7075-T6 Aluminum", in *Effects of Environment and Complex Load History for Fatigue Life, ASTM Special Technical Publication, STP 462*, 1970, American Society for Testing and Materials, Conshohocken, PA, pp. 1–14.
14. K. Sadananda and A.K. Vasudevan, "A Unified Frame-work for Fatigue Crack Growth", in G. Lutjering and H. Nowack Editors, *Fatigue '96*, Vol. 1, 1996, Pergamon Press, Berlin, Germany, pp. 375–380.

15. D. Kujawski, "A Two Parameter Fatigue Crack Driving Force and its Application to FCG Analysis", Presentation at the Virtual Symposium on Fatigue, Fracture, and Structural Integrity, FFIA 2021, Nov 9–11, 2021, Indian Structural Integrity Society, and the Indian Institute of Metals.
16. A. Saxena and S.J. Hudak, Jr., "Review and Extension of Compliance Relationships in Common Crack Growth Specimens", *International Journal of Fracture*, Vol. 14, 1978, pp. 453–469.
17. A. Saxena, S.J. Hudak, J.K. Donald, and D. Schmidt, "Computer-Controlled Decreasing Stress Intensity Technique for Low-Rate Fatigue Crack Growth Testing", *Journal of Testing and Evaluation*, Vol. 6, 1978, pp. 167–174.
18. H. Kitagawa and S. Takahashi, "Applicability of Very Small Cracks or Cracks in Early Stage", Proceedings of the Second International Conference on Mechanical Behavior of Materials", ASM International, Metals Park, 1976, pp. 627–631.
19. M.L. Williams, "On the Stress Intensity Distribution at the Base of a Stationary Crack", *Journal of Applied Mechanics*, Vol. 24, 1957, pp. 109–114.
20. A. Talug and K. Reifsnider, "Analysis and Investigations of Small Flaws", in ASTM STP 637, 1977, pp. 81–96. American Society for Testing and Materials, Philadelphia, PA
21. Y. Murakami, "The Rain Flow Method in Fatigue", Proceedings of the International Symposium on Fatigue Damage Measurement and Evaluation under Complex Loading, Fukuoka, Japan, July 25–26, 1991.
22. J.C. Newman, Jr., "Fatigue and Crack Growth under Constant and Variable Amplitude Loading in 9310 Steel Using 'Rainflow-On-the Fly'" Methodology, *Metals*, Vol. 11, 2021, pp. 807.
23. R. Sunder, "Unraveling the Science of Variable Amplitude Loading", *Journal of ASTM International*, Vol. 9, 2012, pp. 103940.

HOMEWORK PROBLEMS

1. Why is ΔK an appropriate parameter for characterizing fatigue crack growth rates?
2. The FCGR behavior of 2024-T4 aluminum alloy is given by the following equation,

$$\frac{da}{dN} = 8.5 \times 10^{-12} (\Delta K)^4$$

where *da/dN* is in m/cycle and ΔK in MPa$\sqrt{m}$. A tension arm that activates a relay switch in a circuit breaker is made from this alloy and is under a uniform stress of 120 MPa due to spring forces. This arm has a width of 75 mm and when the relay is activated the stress relaxes completely. The average frequency of activation of the relay is 20/day. During a relative routine inspection of the circuit breaker, a through the thickness edge-crack of 5 mm in length is discovered. For establishing the subsequent inspection interval, you are asked to calculate the crack length as a function of elapsed time in days. The plane strain fracture toughness of this material is given as 33 MPa$\sqrt{m}$. Predict the remaining life of the component. Assume that the *K*-expression for this geometry is given by for $0 < a/W \le 0.5$:

$$K = \sigma \sqrt{\pi a} f\left(\frac{a}{W}\right)$$

where, $F\left(\dfrac{a}{W}\right)$ can be adapted from the K-calibration expression given as part of Example Problem 6.5

3. The following equation is used to ensure dominantly linear elastic conditions during FCGR testing,

$$(W - a) \geq \frac{4}{\pi}\left(\frac{K_{max}}{\sigma_{ys}}\right)^2$$

A compact type specimen of 304 stainless steel is being used to generate FCGR data at a load ratio of 0.5. The width of the specimen is 50 mm and the thickness is 6.25 mm. The loading range being imposed on the specimen is 2×10^3 N. If the yield strength of this material is 180 MPa, at what crack length will the data obtained from this specimen will be invalid according to the above condition? If data are needed up to ΔK values of 30 MPa$\sqrt{m}$, what size specimen will be needed to generate the data assuming a finishing a/W value of 0.6? If the following alternate criterion for specimen size requirement is used,

$$(W - a) \geq \frac{4}{\pi}\left(\frac{K_{max}}{2\sigma_{ys}^c}\right)^2$$

where σ_{ys}^c is the cyclic yield strength of the material and for stainless steel is 1.5 times the yield strength, σ_{ys}. What is the smallest size specimen that will be sufficient for assuming a finishing a/W value of 0.6?
4. Schematically show the influence of (a) single overload and (b) single overload followed by an under-load of equal magnitude on the crack size versus cycles curve during fatigue crack growth testing.
5. You are given the task of generating FCGR data for the material in Problem 3 above, using a ΔK-decreasing technique. Starting from an a/W value of 0.3 and using a compact specimen, design a ΔK-decreasing test for measuring FCGR behavior between $15 \leq \Delta K \leq 30$. Specifically, what size specimen will you choose and the value of "C" that you will specify?
6. A very wide and long plate contains an edge crack of 1 mm length and is subjected to a stress range of 200 MPa. If you are given a choice of three materials, A, B, and C and their corresponding properties, which material gives you the highest resistance to fatigue.

Material	K_{Ic} (MPa (m))$^{1/2}$	C for ΔK in MP (m)$^{1/2}$ and da/dN in m/cycle	n
A	55	5×10^{-14}	4.0
B	70	2×10^{-12}	3.5
C	90	3×10^{-11}	3.0

7. If the exponent in the Paris equation is 4, estimate the time it will take for a crack of length $2a$ in the center of a very wide panel subjected to uniform tension to quadruple relative to the time it takes to double.

8. A plate of 2024-T4 Al (see Problem 1 for FCGR properties) that is 20 cm wide and 5 cm thick has a surface crack in its center is loaded with a cyclic stress of $100\,\text{MPa}\sqrt{m}$. If the surface crack has a depth of 0.5 cm and a surface length of 2 cm, estimate the crack growth rates at the surface and at the deepest point. Comment on how the shape of the crack will change as this crack grows during service.

9. A cylindrical pressure vessel has radius, R, of 1 m and a wall thickness, t, of 5 cm. It has a long axial surface crack of depth equal to 3 mm on the inside surface. The pressure vessel is inflated to a maximum pressure, p, of 20 MPa which reduces to atmospheric pressure every day and then the vessel is refilled. Calculate the life of the pressure vessel if you are given the following additional information:

- $\dfrac{da}{dN} = 4.2 \times 10^{-12}(\Delta K)^{3}$ where FCGR is in m/cycle and ΔK in MPa (m)$^{1/2}$
- $KI_{c} = 100\,\text{MPa (m)}^{1/2}$
- The K-expression for axially cracked cylinders subjected to internal pressure is given in Table 3.1 in Chapter 3.

NOTE

1 The wake of the growing fatigue crack is the region in the immediate vicinity of the crack tip through which the crack has already passed.

7 Environment-Assisted Cracking

7.1 INTRODUCTION

Crack growth that occurs at K levels below K_{Ic} is known as subcritical crack growth. In Chapter 6, we considered crack growth at K levels below K_{Ic} under cyclic (fatigue) loading in innocuous environments. We observed that subcritical cracks can potentially grow to the critical size and cause sudden failures under normal service conditions. Another type of subcritical crack growth is environment-assisted cracking (EAC) which occurs due to the weakening of the material near the crack tip due to combined effects of stress and environment. This is also referred to as stress corrosion cracking in the literature. When EAC occurs in conjunction with cyclic loading, the phenomenon is known as corrosion-fatigue. We will later consider other forms of subcritical crack growth known as creep crack growth and creep-fatigue crack growth which occur at elevated temperatures. In this chapter, we are primarily concerned with EAC that occurs under sustained and cyclic loading at temperatures below the levels at which creep becomes a significant consideration.

In the 1960s exploratory experiments were performed to document the effects of environment on subcritical crack growth in high strength materials [1]. Experiments were performed using 75 mm wide center crack specimens with a 19.3 mm crack in the middle on a very high strength, 300 M, steel ($\sigma_{ys} = 1,700$ MPa) loaded remotely to a uniform stress level of 517 MPa. The specimens were exposed to various environments consisting of common laboratory chemicals to explore their effects on environment assisted crack growth rates. Table 7.1 lists the results of these experiments. The time to failure in various environments was recorded and it ranged from 0.5 hours in recording ink to 2,247 hours in benzene. There were no failures recorded in air in 100 hours. Thus, it is obvious that EAC depends just as much on environment as it does on applied stress level. The requisites for environment-assisted failure are (a) stress, (b) environment, and (c) environment sensitive material. For example, copper and its alloys are susceptible to ammonia containing solutions and compounds, mild steels are susceptible to alkalis and stainless steels are susceptible to chlorides. Thus, there are several material-environment combinations that are potentially prone to EAC and must be understood to effectively mitigate the associated risks.

7.2 MECHANISMS OF EAC

Figure 7.1 shows the fracture surface of a severely cold worked Zr-2.5 Nb tube used in nuclear reactors for carrying pressurized water for transporting heat from the reactor [2]. The tube was 6 m long and had an internal diameter (ID) of 104 mm and a wall thickness of 4 mm and carries residual tensile stresses in the range of 700 MPa

DOI: 10.1201/9781003292296-7

TABLE 7.1

Failure Times for 300 M Steels Specimens Containing 19.3 mm Long Cracks Loaded to Stress Levels of 517 MPa in Various Environments [Johnson and Paris, 1]

Environment	Time to Failure in Hours
Recording ink	0.5
Distilled water	6.5
Butyl alcohol	28.0
Amyl alcohol	35.8
Butyl acetate	18.0
Acetone	120.0
Lubricating oil	150.0
Carbon tetra chloride	No failure in 20 hours
Benzene	2,247
Air	No failure in 100 hours

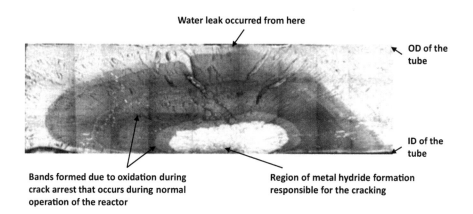

Water leak occurred from here

OD of the tube

ID of the tube

Bands formed due to oxidation during crack arrest that occurs during normal operation of the reactor

Region of metal hydride formation responsible for the cracking

FIGURE 7.1 A through-wall crack in a Zr-2.5 Nb pressure tube of a nuclear reactor. The picture shows crack bands that formed due to oxidation of the fracture surface during crack arrest periods. The crack initiated on the inner surface of the tube in the radial-axial plane and grew by hydride cracking that occurred during reactor shutdown periods and crack arrest during reactor operation periods [2].

from the cold work. During service, the tube leaked water, and as soon as the leak was detected, the reactor was shut down to prevent a more severe event. The crack initiated on the internal surface of the tube by a mechanism called *delayed hydride cracking*. In this mechanism, the metal reacts with the dissolved hydrogen in the water to form hard hydrides that are prone to brittle cracking. Crack growth occurs during reactor shut-down periods when the water temperature is low, driven by the high residual tensile stresses that are locked in the tube from manufacturing. During normal operation of the reactor, the water temperature rises to 250°C so the dissolved hydrogen concentration diminishes to less than 15 ppm which is insufficient to form hydrides, so the crack arrested. The surface of the crack that grew during shutdown oxidized during normal operation of the reactor and left behind bands on the fracture surface that are visible in Figure 7.1. During each subsequent shutdown, the crack grew radially and axially through the tube wall leaving behind a series of bands visible on the fracture surface. When the crack broke through the outer wall of the tube, a leak was detected, and the reactor was shut down.

Hydride cracking has been simulated in Zirconium alloys in the laboratory and has been observed in other hydride forming metals. Another example of hydride cracking was in rail-road tracks. In this instance, the source of hydrogen was trapped hydrogen in the steel itself from the steel making process. The hydrogen was swept into a region containing a manufacturing defect which first grew due to hydride embrittlement and subsequently by fatigue until fast fracture occurred.

The phenomena involved in EAC are not only complex but are also very diverse, so a unified theory for predicting EAC is not possible. Understanding the phenomenon involved in EAC is key to accurately predicting the crack growth behavior. Besides delayed hydride cracking, four other mechanisms of EAC have been observed.

Adsorption mechanism of EAC is said to occur when a specific chemical species is adsorbed on the crack surface at the tip, and it lowers the cohesive strength of the atomic bonds. The embrittling species is frequently liquid metal, such as mercury (Hg), so this mechanism is also known as liquid metal embrittlement (LME) [3]. Examples of LME have been seen in Al, Ni, and Fe alloys. LME occurs in systems where there is little or no tendency between the two species for any reaction to form intermetallic compounds, and there is no tendency to form solid solutions. However, degradation of chemical bonds in the host metal at the crack tip occurs through surface adsorption of the liquid metal. LME is facilitated by the capillary action that draws the liquid metal between the crack faces and transports it to the metal surface near the crack tip. A schematic of the process steps involved was described by Lynch [3] and is shown in Figure 7.2a. Crack growth is seen to occur along the cleavage planes that have the lowest cohesive strength to begin with and are further weakened by the presence of the adsorbed species.

The *oxide film rupture mechanism* of environment assisted crack growth is observed primarily in aqueous environments containing oxidizing agents that form passivating oxide films. The process involves locally rupturing the oxide film in the crack tip region due to high strains at the crack tip and exposing bare metal. The passivating oxide film that forms on the mouth of the crack sets up an active electrochemical cell between the oxygen rich crack mouth region and the oxygen deprived region at the crack tip [4,5]. This cell assists in re-forming the oxide film on the bare metal at the crack tip. Oxygen has high electronegativity so the difference in oxygen concentration sets up the electrochemical cell. The brittle oxide film at the crack tip subsequently ruptures and advances the crack. The continuous process of forming and rupturing of oxide film at the crack tip results in subcritical crack growth. Figure 7.2b schematically shows the steps of film rupture model showing how it leads to subcritical crack growth.

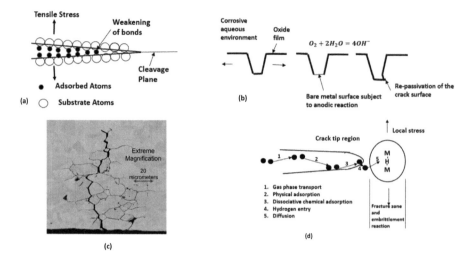

FIGURE 7.2 Schematics of environment assisted cracking processes (a) liquid metal embrittlement [3] (b) anodic dissolution and crack growth by rupture of oxide film [4,5] (c) a picture of grain boundary cracking in 304 stainless steel in an environment containing chloride and (d) the five-step process for hydrogen embrittlement [6].

Environment assisted *grain boundary cracking* occurs because grain boundaries provide an easier path for diffusion of corrosive foreign species and can form compounds that degrade the grain boundary strength. Figure 7.2c shows an example of grain boundary EAC in 304 stainless steel in oxygenated warm water containing chlorides and sulfides that tend to accelerate the cracking rates. Cracks advance by intergranular crack growth through the weakened boundaries. This mechanism is also seen to operate at high temperatures where a combination of favorable conditions exists consisting of (a) high triaxial stress state at the crack tip on grain boundary facets normal to the loading axis, (b) high temperatures to drive diffusion, and (c) a relatively open grain boundary structure providing an easier path for diffusion. This mechanism can also operate at phase boundaries such as boundaries between dendrites in directionally solidified metals where there are fewer grain boundaries available normal to the primary loading direction. Similarly, hydrogen, being the smallest atom in the periodic table, can diffuse into the grain boundaries under a hydrostatic state of stress that exists at crack tips even at ambient temperatures and results in grain boundary cracking by degrading its strength. This mechanism overlaps in characteristics with hydrogen embrittlement described below.

Hydrogen embrittlement is a mechanism of EAC involved in degrading steels, especially high strength steels and Ni base alloys. Atomic hydrogen diffuses into the crack tip region and degrades the strength of atomic bonds [6]. The five-step sequential process shown in Figure 7.2d begins with (a) gas transport to the crack region (b) gas adsorption and migration to the crack tip (c) dissociative reaction that produces atomic hydrogen (d) hydrogen entry into the crack tip and followed by (e) diffusion of hydrogen into the metal and weakening of the metal bonds. Since the process is sequential, the slowest step in the process determines the kinetics of crack

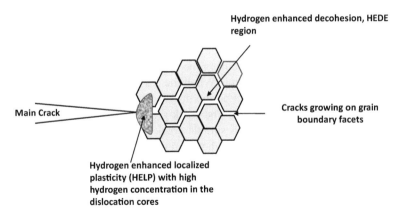

FIGURE 7.3 A schematic showing hydrogen enhanced localized-plasticity (HELP) and hydrogen-enhanced decohesion (HEDE) mechanisms working in conjunction resulting in hydrogen assisted cracking.

growth which is often the diffusion of hydrogen into the metal. This merits further discussion.

Beachem [7] proposed a mechanism that has become known as hydrogen-enhanced localized plasticity (HELP) [8–10]. In this mechanism, hydrogen diffuses into the crack tip region through the core of dislocations assisted by high triaxial stresses present at the crack tip and a high concentration of point defects in the form of vacancies. The diffusion into the core of the dislocations carries hydrogen in sufficient quantity to effectively reduce the lattice resistance to dislocation motion, making it easier for the dislocations to move on their slip planes and contribute to plastic deformation while sweeping hydrogen with it into the grain boundaries. This facilitates hydrogen-enhanced decohesion (HEDE) proposed earlier by Troiano [11], as a mechanism of hydrogen embrittlement. Hydrogen reduces the interface strength between the grain boundary carbides and the surrounding matrix as part of HEDE, promoting crack growth along grain boundary facets as shown schematically in Figure 7.3 along with the zones in which HELP and HEDE are active.

7.3 RELATIONSHIP BETWEEN EAC AND *K* UNDER STATIC LOADS

In this section, we primarily discuss the fracture mechanics approach for measuring and correlating the kinetics of crack growth during EAC, regardless of the underlying mechanism. This approach serves the engineering design community well to assist in ensuring safe operation during service, to test new materials and their performance in corrosive environments, and to develop new materials with better resistance to EAC. But, as mentioned before, full awareness of the underlying mechanisms is necessary for developing models for predicting the EAC behavior.

Like in the case of fatigue crack growth rate and fracture toughness, stress intensity parameter K is an attractive parameter for characterizing the time rate of crack growth, da/dt, during EAC much for the same reasons as for correlating fatigue crack growth rate with ΔK. A schematic of this relationship is shown in Figure 7.4.

The relationship between *da/dt* and *K* shown in Figure 7.4 is unique only after steady-state conditions have been established and the environment and its severity (variables such as partial pressure of the corrosive gas, temperature, PH levels for liquid solutions etc.) are held constant during the test. Further, the term steady-state in this context has the following meaning. When the crack tip is first exposed to stress and environment, there is a transient period during which the environment affected zone develops in the crack tip region. The crack growth rate during this period can either be retarded or there is no measurable crack growth while the environment affected zone develops. In either case, the relationship between *da/dt* and *K* is not unique. It takes crack extension on the order of 0.5 mm from the start of the test for steady-state conditions to develop. This initial data is not included in the steady-state relationship between *da/dt* and *K* because it depends on factors other than the material and environment. The important characteristics of steady-state crack growth data as in Figure 7.4 are as follows:

- K_{IEAC} or K_{EAC} is the threshold level of the stress intensity parameter for EAC. K_{IEAC} is used when the measured threshold is from specimens that meet the size requirements for plane strain while K_{EAC} is used when the specimen only meets the conditions for linear elasticity but is not thick enough to meet requirements for plane strain. If the applied *K* is below the level of this threshold, environment assisted crack growth is not expected to occur. Thus, for structures operating in harsh environment, this is an important design parameter that establishes a combination of maximum allowable stresses and maximum initial crack size to avoid EAC. The test methods, therefore, have focused on determining the K_{IEAC} and K_{EAC}.
- The *da/dt* versus *K* behavior is a material property but is also specific to the environment and can be divided into three regions.
  - Region I is characterized by a strong dependence of *da/dt* on *K*. This region is also referred to as the near-threshold region.
  - In Region II, *da/dt* is only mildly sensitive to *K* but it varies significantly with the intensity of the environment as characterized by temperature

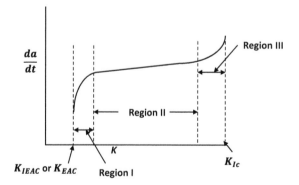

FIGURE 7.4 Schematic relationship between environment assisted crack growth rate and the stress intensity parameter, *K*.

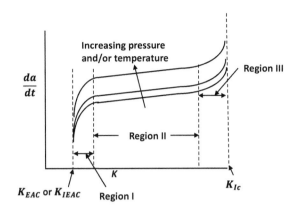

FIGURE 7.5   Schematic of the influence of environment severity on the relationship between *da/dt* and *K*.

and pressure (if the environment is gaseous). This relationship is schematically shown in Figure 7.5.

- In Region III, the crack growth rate is once again very sensitive to the
  applied value of K as it approaches K_{Ic} of the material or the onset of
  fast fracture.

7.4   METHODS OF DETERMINING K_{IEAC}

Three methods are used for determining K_{IEAC} and K_{EAC} of materials. Two of these
methods are coded in ASTM Standard E-1681 [12] using fatigue pre-cracked specimens. Following the standard method ensures consistency in data developed at different laboratories.

K_{IEAC} is defined as the highest level of K for which no crack growth is observed in
a material exposed to a certain environment; the specimen, in this case, also meets
the size requirement specified by ASTM Standard E 399 for plane strain conditions,
equation (5.4) repeated below. K_{IEAC} is then thought of being independent of specimen thickness.

$$B, (W - a), a \geq 2.5 \left(\frac{K_{IEAC}}{\sigma_{ys}} \right)^2 \tag{7.1}$$

K_{EAC}, on the other hand, is also used to describe the highest value of K for which no
crack growth is observed in a material in a specific environment, except the specimen size requirements of ASTM E399 are not met, but the size requirements of
ASTM E647 for fatigue crack growth testing, equation (6.18) below is met.

$$(W - a) = \frac{4}{\pi} \left(\frac{K_{EAC}}{\sigma_{ys}} \right)^2 \tag{7.2}$$

K_{EAC} may vary with specimen thickness. The three methods, (a) constant load method, (b) constant displacement method, and (c) rising load method are described next. The first two, as mentioned before, are coded in ASTM Standard E-1681 [12] and the third is often used as quick method to determine a conservative value of K_{IEAC} or K_{EAC} as the case may be. All three are described below.

Constant load method: In this method, a series of identical specimens are prepared and pre-cracked using fatigue loading. The specimen geometries are like single-edge-crack specimens that are gripped rigidly on one end and are subjected to a cantilever load on the other, Figure 7.6. Thus, the loading is dominantly bending. The region of the specimen containing the crack is immersed in a chamber that holds the environment. Several specimens are loaded to different load levels (or K levels) and the time to failure is recorded. The time to failure is correlated with the applied value of K and the relationship between K and time to failure is used to determine K_{IEAC} or K_{EAC} values as shown in Figure 7.6. The value of K below which no failure is expected is K_{IEAC} or K_{EAC}, depending on the size requirements met by the specimen. The preferred value is K_{IEAC} because it does not depend on specimen thickness, so it classifies as a material property, although only for a specified material-environment combination.

The stress intensity parameter in this case is calculated by equation (7.3) below.

$$K = \frac{P}{BW^{1/2}}\left(\frac{L}{W}\right)f\left(a/W\right) \tag{7.3a}$$

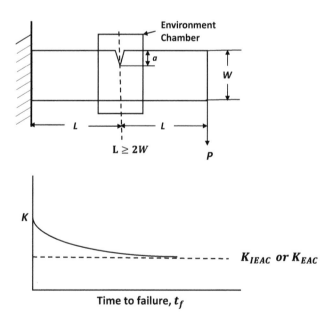

FIGURE 7.6 (a) Schematic of a constant load test and (b) the relationship between applied K and the time to failure.

$$f(a/W) = \frac{6(a/W)^{1/2}}{\alpha^{3/2}}\left[1.9878 - 1.3253\left(\frac{a}{W}\right) + \alpha(a/W)\left\{-3.8308 + 10.1081\left(\frac{a}{W}\right)\right.\right.$$

$$\left.\left. - 17.941\left(\frac{a}{W}\right)^2 + 16.8282\left(\frac{a}{W}\right)^3 - 6.2241(a/W)^4\right\}\right]$$

(7.3b)

In the above equation, the various terms are defined as below:

$P=$ the applied cantilever force on the specimen.
$L=$ Distance between the crack plane and the applied load

$$\alpha = (1 - a/W)$$

Single-edge-crack specimens subject to cantilever load have a high K-gradient so, once the crack initiates, the crack growth rate rises rapidly and the specimen fails; little to no crack growth rate data can be collected from these tests. Also, the K_{IEAC} or K_{EAC} measured by this method is a measure of the threshold K level at which crack growth begins to initiate; this value can be different from the K level at which a growing crack comes to arrest in a corrosive environment. The crack arrest threshold is measured by a different test method that is conducted under constant displacement conditions as described below.

   Constant Displacement Method for Determining K_{IEAC} or K_{EAC}: This method utilizes a wedge opening load specimen that is a modified C(T) specimen as shown in Figure 7.7. The height of the specimen specified by H is less than that of the C(T) specimen. Also, there is provision for a bolt that can be cranked to apply a fixed displacement along the load-line of the specimen. The crack mouth displacement is measured while the bolt is cranked to the specified displacement level. After the

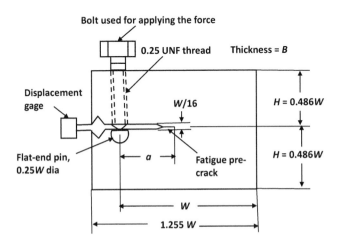

FIGURE 7.7   Schematic of a bolt-loaded compact type specimen, also known as wedge opening load (WOL) specimen, used for constant displacement EAC testing.

load is applied, the displacement gage can be removed and the specimen can be immersed in the environment. Crack size measurements are periodically made by briefly removing the specimen and observing the cracks on both sides of the specimen under good lighting. As the test progresses, the crack grows but the displacement along the load-line remains fixed. To accommodate the increasing compliance of the specimen as the crack grows, the load must decrease. With decreasing load, the K also decreases and the crack growth rate slows down until it comes to arrest when the K_{IEAC} or K_{EAC} values are reached. Thus, the test can be started at a high K level and then continued until the crack has arrested. The measured crack growth rates yield the da/dt versus K data and the K level corresponding to arrest yields K_{IEAC} or the K_{EAC} value. Usually, the K_{IEAC} or K_{EAC} value measured by this technique is lower than the value obtained from the constant load tests. This is the threshold value at which a crack growing in the environment will come to arrest.

The K-value in the constant displacement test is given by the measured displacement using the clip-on gage on the front-face of the specimen. This relationship is given in equation (7.4) below:

$$K = \frac{V_m E}{W^{1/2}} f_1(a/W) \tag{7.4a}$$

For $0.45 \le a/W \le 0.8$

$$f_1(a/W) = \left[1 - \frac{a}{W}\right]^{1/2}\left[0.654 - 1.88\left(\frac{a}{W}\right) + 2.66\left(\frac{a}{W}\right)^2 - 1.233(a/W)^3\right] \tag{7.4b}$$

Outside of the above a/W range, this expression has an error of more than 0.5%.

V_m = crack mouth opening displacement, and E = elastic modulus of the material. If the applied load P is measured or known, the K-expression that is valid for any $\frac{a}{W} \ge 0.2$ is given in equation (7.5) [13]:

$$\frac{K}{P}BW^{1/2} = \frac{2 + a/W}{(1 - a/W)^{3/2}}\left[0.8072 + 8.858(a/W) - 30.23\left(\frac{a}{W}\right)^2\right.$$
$$\left. + 41.088\left(\frac{a}{W}\right)^3 - 24.15\left(\frac{a}{W}\right)^4 + 4.951(a/W)^5\right] \tag{7.5}$$

The bolt is of sufficient diameter that it can be assumed to be rigid relative to the specimen arms and does not contribute significantly to displacement. The relationship between front-face displacement, V_m, and the applied load and crack size is given by equation (7.6) [13] and is valid for any $\frac{a}{W} \ge 0.2$.

$$\frac{BEV_m}{P} = \left(\frac{1 + a/W}{1 - a/W}\right)^2$$
$$\left[7.2616 + 16.488(a/W) - 45.477\left(\frac{a}{W}\right)^2 + 26.956(a/W)^3\right] \tag{7.6}$$

Example Problem 7.1

From equations (7.5) and (7.6) for WOL specimens, derive a relationship for estimating K from measured values of the front-face displacement, V_m, and compare those values with those derived from equation (7.4) and calculate the percentage difference in the values from the two equations as a function a/W.

Solution

Dividing equation (7.5) with equation (7.6) and rearranging the terms, gives us:

$$\frac{\left(\dfrac{K}{P}BW^{\frac{1}{2}}\right)}{\left(\dfrac{BEV_m}{P}\right)} = \frac{W^{\frac{1}{2}}K}{EV_m} = f_2\left(a/W\right)$$

$$= \frac{\dfrac{2+\dfrac{a}{W}}{\left(1-\dfrac{a}{W}\right)^{\frac{3}{2}}}\left[0.8072+8.858\left(\dfrac{a}{W}\right)-30.23\left(\dfrac{a}{W}\right)^2+41.088\left(\dfrac{a}{W}\right)^3-24.15\left(\dfrac{a}{W}\right)^4+4.951\left(\dfrac{a}{W}\right)^5\right]}{\left(\dfrac{1+\dfrac{a}{W}}{1-\dfrac{a}{W}}\right)^2\left[7.2616+16.488\left(a/W\right)-45.477\left(\dfrac{a}{W}\right)^2+26.956\left(\dfrac{a}{W}\right)^3\right]}$$

Or, $K = \dfrac{EV_m}{W^{1/2}} f_2\left(a/W\right)$

Comparing the above equation with equation (7.4a) suggests that $f_2(a/W)/f_1(a/W)$ must be equal to 1 for the K-value from equations (7.4) and (7.5) to be the same. These values are computed and plotted in Figure 7.8. It is observed that the K-values are within 5% of each other for $a/W \geq 0.45$ but are quite different at smaller a/W values. $f_2(a/W)$ is given by a simpler polynomial equation that is valid for $0.2 \leq a/W \leq 1$:

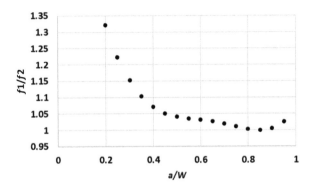

FIGURE 7.8 The differences between K values estimated from equations (7.2) and (7.4).

$$f_2(a/W) = (1-a/W)^{1/2} \left[7.424 - 83.601\left(\frac{a}{W}\right) + 468.17\left(\frac{a}{W}\right)^2 - 1356.1\left(\frac{a}{W}\right)^3 \right.$$

$$\left. + 2134.7(a/W)^4 - 1724.3\left(\frac{a}{W}\right)^5 + 560.79(a/W)^6 \right]$$

Continuously rising load tests: The constant load and constant displacement methods require 1,000–10,000 hours of testing time and multiple specimens must be tested to obtain K_{IEAC} or K_{EAC}. The rising load method, on the other hand, quickly yields a conservative value of K_{IEAC} or K_{EAC}. Therefore, this method is frequently used to screen materials and rank their ability to resist crack growth in an environment. In this technique, specimens are monotonically loaded at a slow rate and the K value at which the crack begins to grow is noted and reported as an estimated K_{IEAC}. Continuously increasing force is applied on the specimen at a slow rate and the force-displacement $(P-V)$ behavior is recorded. Deviation from the linear trend in the $P-V$ behavior is expected to occur as the crack begins to grow. If the material is susceptible to environment, this crack growth begins at K levels that are substantially below the K_{Ic} of the material. The K corresponding to the first point of deviation from the linear $P-V$ behavior is used to estimate K_{IEAC}.

There are no established rates at which the force should be applied so, it is best to use multiple loading rates and the lowest K level observed at which cracking initiates be used as the estimated K_{IEAC}. The K_{IEAC} measured from the rising load tests is slightly different from those measured by the constant load and constant displacement tests. To distinguish this value from "true" K_{IEAC}, it is often referred to as K_{IH}. The subscript H is used because this test is mostly used to test materials in hydrogen containing environments such as high-pressure hydrogen or hydrogen sulfide gases.

7.5 RELATIONSHIP BETWEEN K_{IEAC} AND YIELD STRENGTH AND FRACTURE TOUGHNESS

The relationship between K_{IEAC} and properties such as yield strength and fracture toughness is important in selecting materials because those are the primary properties engineers use in preliminary selection of materials. Based on empirical evidence, it is established that K_{IEAC} decreases with increasing yield strength for materials that are susceptible to EAC. For example, see Figure 7.9 showing the relationship between K_{IEAC} and fracture toughness with yield strength in AISI 4340 steel. The *EAC* tests were conducted in an aqueous environment consisting of sea water [14]. The different yield strengths in these experiments, ranging from 827 to 1,550 MPa, were produced by subjecting the steel to varying quenching and tempering heat treatments. In these materials, K_{Ic} is higher than K_{IEAC} indicating the susceptibility of the material to EAC in sea water. If K_{Ic} and K_{IEAC} are equal, one can conclude that there is no effect of environment on the material. It is also seen in this figure that the gap between K_{Ic} and K_{IEAC} increases as the yield strength increases, particularly past 965 MPa, indicating that increasing the yield strength beyond 965 MPa accelerates the deleterious effects of environment. In aluminum alloy 7075 with T6 temper that is optimized to produce very high strengths (>500 MPa), the K_{IEAC} is shown to reduce to 6.6 MPa$\sqrt{m}$ while

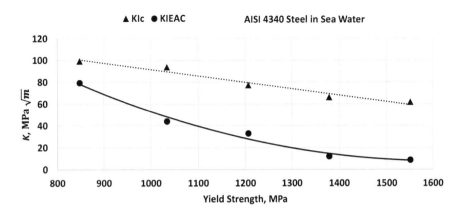

FIGURE 7.9 K_{IEAC} and K_{Ic} as a function of yield strength in AISI 4340 steel in sea water (Figure adapted from Marsh and Gerberich [14]).

the fracture toughness is about $30\,\mathrm{MPa}\sqrt{m}$. If one uses a different temper (T73) that overages the material and decreases the yield strength to 400 MPa, K_{IEAC} and K_{Ic} become equal and rise to a value of about $38\,\mathrm{MPa}\sqrt{m}$. Not recognizing the susceptibility to *EAC* in ultra-high strength aluminum alloys optimized only for strength was one of the primary causes of mid-air explosion of Comet aircrafts in the 1950s. A simple change in heat treatment lowering the yield strength and avoidance of sharp corners near the windows to reduce stress concentrations alleviated the problem. These crashes also led to development of several other more *EAC* resistant aluminum alloys.

Figure 7.10 shows the *da/dt* versus *K* from in AISI 4340 steel with yield strengths of 1,400 MPa with a martensitic microstructure and 1,310 MPa with a bainitic microstructure [15]. Despite the small difference in yield strength between the two materials and identical chemistries, there are significant differences in the overall *EAC* behavior of the two steels. The Bainitic microstructure provides a higher resistance to *EAC* compared to the martensitic microstructure of somewhat similar strength.

Figure 7.11 plots the tendency for EAC with yield strength across several steels and other alloys in gaseous hydrogen environment and no unique correlation is observed between those two properties across different materials [16]. In this figure, the *EAC* resistance is measured by the ratio, $\beta = (RA)_{H2}/(RA)_{He}$, the ratio of the reduction in area in hydrogen environment to that in helium environment in cylindrical specimens containing sharp identical notches machined along the circumference of the specimens. Such specimens promote a triaxial stress like crack tips in planar specimens. The tests were conducted to quickly provide a quantitative measure of the influence of hydrogen in promoting brittle behavior in these materials. Failures in the specimens with lower reductions in area in hydrogen compared to helium, is an index of sensitivity to environment.

In Figure 7.11, the face-centered cubic and hexagonal close packed materials with the highest atomic packing density, such as aluminum alloys, copper, titanium, and stainless steel seem to have none to little susceptibility to hydrogen cracking.

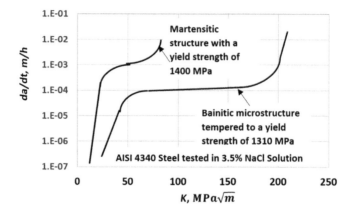

FIGURE 7.10 Crack growth rates versus stress intensity parameter for AISI 4340 steel in martensitic condition with yield strength of 1,400 MPa and in Bainitic condition with a yield strength of 1,310 MPa [15].

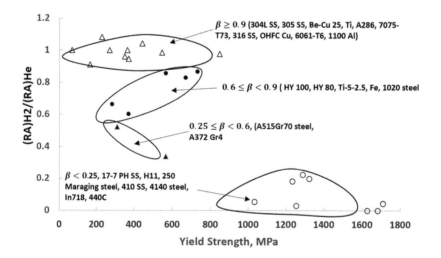

FIGURE 7.11 Plot of EAC resistance in several alloys as a function of their yield strength. The ratio β represents the ratio of reduction in areas of circumferentially notched cylindrical specimens in hydrogen to those in helium environments [16].

This cluster of materials are characterized by $\beta \geq 0.9$. The other extreme is a cluster of materials that includes primarily martensitic steels with values of $\beta < 0.25$, indicating extreme sensitivity to environmental cracking. The cluster consisting of $0.25 \leq \beta < 0.6$ includes bainitic microstructure showing a clear dependence of β on the yield strength across two types of steels. There is also a cluster of ferritic steels with $0.6 \leq \beta < 0.9$ of different types. It appears that among steels, once the yield strength exceeds about 900 MPa, *EAC* becomes an important concern. These are important general guidelines to keep in mind in choosing materials for structural components for applications in deleterious environments. Such considerations supplement fracture mechanics test results.

Besides the yield strength of materials and microstructure, there are other secondary chemical species present in steels that are deleterious to *EAC* resistance. For example, impurity content in the form of sulfur (S) and phosphorus (P) in steels can significantly degrade *EAC*. If these elements migrate to the grain boundaries and form compounds with other elements that also segregate to the grain boundaries during cooling from the austenite region and during tempering heat treatments, they can interact with atomic hydrogen and significantly reduce the cohesive strength of grain boundaries. Therefore, depending on the application, sulfur and phosphorus content in steels is strictly controlled in ferritic and bainitic steels.

Example Problem 7.2

A metallurgical engineer in a ship building company has developed a new steel that is more resistant to EAC by tailoring the microstructure and better control of the material chemistry via reduced impurities in the form of sulfur (S) and phosphorus (P). The yield strength and fracture toughness of the new material is the same as the current material with $\sigma_{ys} = 700$ MPa and a K_{Ic} of 90 MPa$\sqrt{m}$. The material is scheduled to be used in the form of 12.5 mm thick plate in a new ship that will go into production in 4 years for which the material must be ordered 2 years in advance to allow sufficient time for production. You have been hired as a consultant to quantify reduction in the risk for fracture offered by the new steel that comes at a cost. Your first task is to provide a plan for testing that can be completed in 15 months to quantify risk-reduction. What will your plan look like?

Solution

This is a problem that does not have a unique answer and will require judgements, aided by fracture mechanics methodology, to be made while proposing a plan. Since time is of essence, we must only propose essential tests that can be completed in the allocated time.

The yield strength and the K_{Ic} of the new material are the same as the current material so, the advantage of the new material is primarily in the form of a higher K_{IEAC} or K_{EAC}. Let us first perform a quick calculation to determine whether we will pursue measurement of K_{IEAC} or K_{EAC}. Recall that the difference between the two is the size requirements stated in equations (7.1) and (7.2). The highest possible value of K_{IEAC} or K_{EAC} can be 90 MPa$\sqrt{m}$ that is equal to the K_{Ic}. A check to determine the minimum thickness required based on equation (7.1) yields that $B \geq 41.3$ mm. The plate thickness is only 12.5 mm, so the measurement of valid K_{IEAC} is not possible. We must then settle for measuring K_{EAC}. Next, we check the size requirement specified by equation (7.2) and find that the minimum $(W - a)$ must be ≥ 21 mm for a valid K_{EAC} measurement. If we assume that the initial a/W will be 0.5 to make the crack size and uncracked ligament size equal, the minimum required value of $W = 42$ mm. Thus, a specimen with $W = 50$ mm, commonly used in fracture mechanics testing, will meet our needs.

The next set of choices involve using either a constant load tests or a constant displacement test. In this application that involves designing for a very long life (40+ years), we must avoid the possibility of any crack growth during service

TABLE 7.2

Estimated K and Load Levels to be Used for Tests for Measuring K_{EAC} of the Currently Used and the New Steel

Applied K Level		Applied Cantilever Load (kN)	Time to Fracture, Hours to Be Recorded
Fraction of K_{Ic}	K MPa$\sqrt{m}$		
0.90	81	5.026	
0.80	72	4.468	
0.75	67.5	4.189	
0.7	63	3.910	
0.65	58.5	3.630	

making crack growth rate information secondary; thus, the critical parameter is the K_{EAC}. If constant load test frames are available, those tests will be most efficient. These tests will involve test samples with a $W = 50$ mm and a specimen length of 200 mm (see Figure 7.6). Using a WOL specimen (Figure 7.7) will require a sample that is only 50×63 mm in dimension, so the material requirements are considerably less. Assuming that material availability and test frames are not a limitation, we choose to conduct the constant load tests using single edge crack specimens subjected to cantilever loading.

Since the K_{EAC} for the current material is not known, we should also propose to measure that value for comparison with the new material to quantify risk-reduction. We will thus plan six specimens to be machined from each of the two materials. The laboratory where the testing is to be performed has six test stands available to dedicate to this project. Since the specimens will be pre-cracked using fatigue loading, we must specify a machined notch that extends to $0.45W$ (or 22.5 mm), and then pre-crack using fatigue loading as specified by ASTM Standard E1681 to extend the crack by 2.5 mm. Next, we need to specify the K levels (or the load levels) for the six tests that we propose to perform on each of the two materials. The proposed starting K levels are specified in terms of the fraction of the K_{Ic} values as in the Table 7.2 below. The load levels are chosen such that they are always below K_{Ic} but starting at values that are 90% of K_{Ic}. The necessary load to achieve the desired K levels are calculated from equation (7.3) and listed in Table 7.2.

Considering the time needed to procure test material and to machine specimens and fatigue pre-crack them prior to testing, the longest test time available is 1 year that is equal to 8,760 hours. The time to failure at various K levels can be recorded from which an estimated value of K_{EAC} can be obtained.

In making test plans, it is important to make the best use of equipment resources available. Since only six test stands are available, we must plan optimal use of the equipment. We can dedicate three test stands for each material. One of the three should run the test at the lowest K level among those selected K levels for testing in Table 7.2, i.e., 58.5 MPa$\sqrt{m}$. These tests will yield the longest time to fracture so should be started first. In a second test stand, we can begin with a test with the highest K-value of 81 MPa$\sqrt{m}$. If the specimen fails quickly, we can start the test at the next highest K level, 72 MPa$\sqrt{m}$. In the third test stand, we begin with a test at the third highest K level of 67.5 MPa$\sqrt{m}$. The test at 63 MPa$\sqrt{m}$ can go into the test stand that becomes available the soonest. The sixth specimen can be retained as spare and if time permits, can be used to fill-in data gaps, if any or as a duplicate test.

We expect that the K_{EAC} for the new material will be higher than the K_{EAC} for the current material. The maximum permissible flaw sizes can be estimated for a semi-infinite panel under a uniform stress as follows:

$$\frac{(a_i)_{new}}{(a_i)_{current}} = \left(\frac{(K_{EAC})_{new}}{(K_{EAC})_{current}} \right)^2$$

where, a_i refers to acceptable flaw for the current and new materials. As an example for illustration, if the K_{EAC} for the new material is 20% higher than the K_{EAC} for the current material, the flaw tolerance is expected to be 1.44 times larger. The risk of fracture will have been significantly reduced by using a material with higher K_{EAC}.

7.6   ENVIRONMENT ASSISTED FATIGUE CRACK GROWTH

During cyclic loading in the presence of corrosive environment, three variables which were previously labelled in Chapter 6 as either secondary or not significant, become important. These are (a) load ratio, particularly in Region II of the da/dN versus ΔK relationship, (b) loading frequency, and (c) cyclic loading waveform. The influence of these variables on fatigue crack growth rates in corrosive environments is considered using observed trends in experimental data. This discussion will then be followed by simple phenomenological models that can be used to account for cycle and time-dependent effects.

Load ratio, R, effects are significantly enhanced in corrosive environments. In the Paris regime, the data for high and low load ratios can be clustered in a narrow scatter band in benign environments. The crack growth rates can increase significantly in corrosive environments as schematically shown in Figure 7.12. Note that going from low load ratios of 0.1 to high load ratios of 0.5 or greater can yield significantly different fatigue crack growth rates in corrosive environments as schematically shown in Figure 7.12.

High strength low alloy (HSLA) steels are used in pipelines and pressure vessels for storing or transporting hydrogen at high pressures. Figure 7.13 shows the effects of hydrogen at high pressure on fatigue crack growth rates in HSLA steels (yield strength ranging from 480 to 800 MPa). The data includes da/dN versus ΔK behavior at a load ratio of 0.2 and 0.5 in high pressure hydrogen (102 MPa) for A372 Grade J Class 70 steels [17–19]. For comparison, we have also included data at $R=0.2$ in air. The trend for $R=0.5$ in air is expected to be the same as at $R=0.2$. These data show that load ratio is a significant parameter in determining the fatigue crack growth behavior of this steel (and others) in corrosive environments.

Cyclic loading frequency was shown to not be important in benign environments. In corrosive environments, the crack growth rates at the same ΔK can increase systematically with decreasing frequency as shown in Figure 7.14a. This is because more

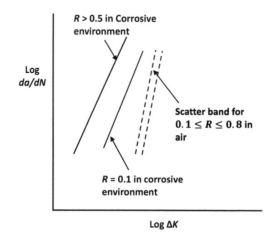

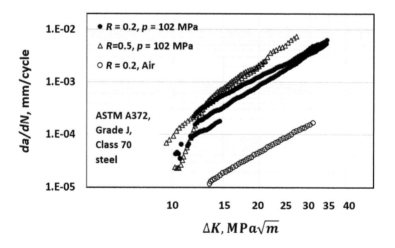

FIGURE 7.12 A schematic showing the fatigue crack growth behavior in steels at different load ratios in benign and corrosive environments.

FIGURE 7.13 Fatigue crack growth behavior of ASTM A372 Grade J, Class 70 steels at two load ratios in high pressure hydrogen (102 MPa) and compared with the behavior in air.

time is available during the loading cycle for environmental degradation to occur in the crack tip region resulting in higher crack growth rates. However, the amount of increase/decrease in crack growth rates with loading frequency can vary with material/environment combinations. In Figure 7.14b for ASTM A372 Grade J Class 70 steel in 10 MPa gaseous hydrogen environment, the crack growth rates per cycle are plotted at various ΔK levels over a wide range of frequencies. In this material/environment combination, the effect of frequency is negligible but that is not always the case. Thus, in generating fatigue crack growth data in corrosive environments for use in design, this variable must be considered. For example, from the data in

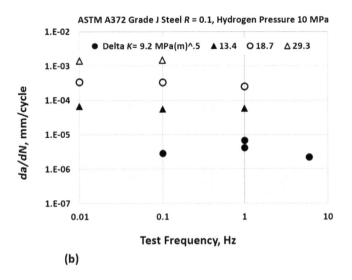

ASTM A372 Grade J Steel R = 0.1, Hydrogen Pressure 10 MPa

(b)

FIGURE 7.14 (a) Schematic representation of how the fatigue crack growth rates vary with loading frequency in corrosive environments and (b) variation in fatigue crack growth rates as a function of loading frequency in A372 Grade J Class 70 steels in hydrogen gas at a pressure of 10 MPa [19].

Figure 7.14b, it can be concluded that data generated at a loading frequency of 1 Hz, is representative of behavior at frequencies that at are two orders of magnitude lower. By testing at 1 Hz instead 0.01 Hz, the test can be speeded up by a factor of 100. As an example, let's consider an application where the frequency of loading is close to 0.01 Hz. At 0.01 Hz, the tests might take an estimated time of a full year to complete. Whereas, if we can use 1 Hz as the testing frequency, the tests can be completed in 3–4 days without affecting the validity of the data for simulating service conditions.

The loading waveform was another variable not considered to be important in benign or innocuous environments but not so in corrosive environments. More time during the loading portion of the cycle has been observed to be more damaging than the time during unloading. This is another variable that should be considered while generating fatigue crack growth data for use in design of components subjected to corrosive environments.

7.7 MODELS FOR ENVIRONMENT ASSISTED FATIGUE CRACK GROWTH BEHAVIOR

The synergistic effects of fatigue and EAC for a constant ΔK are schematically illustrated in Figure 7.15. In Region I, at low loading frequencies the time-dependent effects due to environment are presumed to be dominant. Since ΔK is constant, da/dt must also be constant. So, the cyclic crack growth rate, da/dN is given by equation (7.7):

$$\frac{da}{dN} = \frac{1}{v}\frac{da}{dt} \qquad (7.7)$$

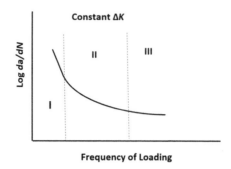

FIGURE 7.15   Synergistic effects of fatigue and environment assisted cracking.

where v = loading frequency.

In Region III, the frequency is too high to allow any time-dependent effects to become significant. The crack tip outpaces the region affected by the environment. Thus, at constant ΔK, da/dN must remain constant. We can write that as, equation (7.8):

$$\frac{da}{dN} = \left(\frac{da}{dN}\right)_0 \tag{7.8}$$

In Region II, we have potential for interaction between EAC and fatigue cracking. The Fatigue cracking is almost always transgranular while EAC can be either transgranular or intergranular or even a mixture of the two modes. Synergistic effects of environment and fatigue could occur if EAC is transgranular or a mixture of intergranular and transgranular. In the case where the fatigue cracking is transgranular and the EAC is only intergranular, there is little interaction that can occur between the two mechanisms. The crack growth rates are given by equation (7.9) below:

$$\frac{da}{dN} = max\left|\left(\frac{da}{dN}\right)_o, \quad \frac{1}{v}\frac{da}{dt}\right| \tag{7.9}$$

The above hypothesis is known as the dominant-damage hypothesis because crack growth rates are dictated by fatigue or by environment, depending on which of the two leads to a higher crack growth rate. A couple of examples of models for predicting crack growth rates in Region II are discussed next.

7.7.1   LINEAR SUPERPOSITION MODEL

Linear superposition model first proposed by Wei and Landes [20] algebraically adds the effects of fatigue and environment and assumes that both mechanisms operate independently, thus one does not enhance/annihilate the damage caused by the other. At a constant ΔK, the time-dependent crack growth rate, da/dt, and the cycle-dependent crack growth rate, da/dN, is thus given by equation (7.10):

$$\frac{da}{dN} = \left(\frac{da}{dN}\right)_0 + \frac{1}{v}\frac{da}{dt} \qquad (7.10)$$

Note that at high frequencies, the second term on the right-hand side becomes negligible so the first term dominates. At low frequencies, the opposite of that happens, so the second term dominates. Thus, the model can effectively be used in all three regions to estimate the environment assisted fatigue crack growth rates. However, this model is expected to accurately represent the behavior when both fatigue and environment result in transgranular crack growth such as in aqueous environments when the two mechanisms operate concurrently. When fatigue crack growth is transgranular and the environment assisted crack growth is intergranular, the model is not expected to yield accurate predictions, and dominant damage mechanism is more appropriate.

An example of implementing the linear superposition model on a titanium alloy (Ti-6 Al-4 V) is shown in Figure 7.16, [21] tested at frequencies of 10, 5, and 1 Hz in a 3.5% NaCl solution compared with the predictions from the model showing reasonable agreement between the predicted and observed crack growth rates. Some discrepancies between predicted and observed crack growth rates are found in the intermediate ΔK range. If necessary, a third synergistic term can be added in equation (7.10) to account for enhancement/annihilation of damage caused by one mechanism over the other. This will improve the correlation between the predicted and actual environment assisted crack growth rates.

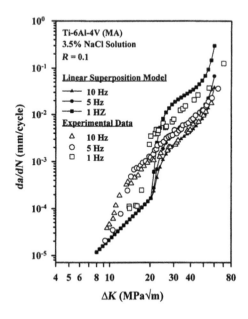

FIGURE 7.16 Prediction of crack growth rates using linear superposition model of Wei and Landes [20] for a Ti-6 Al-4 V alloy in 3.5% NaCl water at a load ratio of 0.1 and frequencies of 0.1, 5, and 10 Hz.

7.7.2 A Model for Predicting the Effects of Hydrogen Pressure on the Fatigue Crack Growth Behavior

Low alloy ferritic steels are used in fabricating cylinders for hydrogen storage at high pressures. A phenomenological model [22,23] is described for predicting the effects of hydrogen pressure ranging from 0.02 to 102 MPa on the kinetics of hydrogen assisted fatigue crack growth (HA-FCG) in SA372 Grade J Class 70 steels. The crack growth kinetics in these steels exhibit a two-region behavior described as transient and the steady-state regions. These regions are characterized by distinct power-law exponents in the relationship between fatigue crack growth rate, da/dN, and the cyclic stress intensity parameter, ΔK.

The phenomenological framework for the model consists of invoking crack tip phenomena such as HELP and HEDE previously discussed in this chapter. A common phenomenon among these mechanisms is that the damage caused is directly related to the concentration of hydrogen, regardless of how the hydrogen gets there [22,23].

The kinetics of hydrogen diffusion in the crack tip region that forms the process zone is expected to be the rate controlling step for environment assisted crack growth, but it must be preceded by surface adsorption of atomic hydrogen to ensure adequate supply of hydrogen. Further, the diffusion of hydrogen in the vicinity of the crack tip is assisted by high hydrostatic stresses that exist in its vicinity. Rate of adsorption is expected to depend on hydrogen pressure and its kinetics are expected to increase initially with pressure and then saturate.

The process zone, that combines the regions of HELP and HEDE in front of the crack tip is defined as a region where there is sufficient hydrogen concentration to weaken the atomic bonds to enhance crack growth rates. The size of this zone is termed as X_{tr}, see Figure 7.18a. When the crack growth rate per cycle, da/dN, is less than X_{tr}, the crack growth during the fatigue cycle occurs entirely within the embrittled zone and therefore, its rate of growth is a strong function of the cyclic stress intensity parameter, ΔK. This leads to a high exponent in the Paris-type relationship between da/dN and ΔK (>6). In the steady-state region, where $da/dN > X_{tr}$, the crack growth per cycle outpaces the increment in size of the process zone, as schematically shown in Figure 7.18b. In this region, the dependence of da/dN on ΔK has a slope equal to that in air, as seen schematically in Figure 7.17b. The hydrogen-degraded process zone forms during a fatigue cycle and the fatigue crack growth during the subsequent cycle occurs in the degraded material within the process zone, or beyond, depending on the applied ΔK. Thus, the overall crack growth rate relationship in hydrogen, $(da/dN)_H$ can be expressed by equation (7.11), where $(da/dN)_{ss}$ is the crack growth rate in the steady state region and $\left(\dfrac{da}{dN}\right)_{tr}$ is the crack growth rate in the transient region. This form of equation allows one of the two mechanisms to dominate the fatigue crack growth rate, depending on ΔK.

$$\left(\frac{da}{dN}\right)_H = \left[\left(\frac{da}{dN}\right)_{tr}^{-1} + \left(\frac{da}{dN}\right)_{ss}^{-1}\right]^{-1} \qquad (7.11)$$

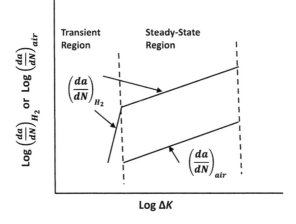

FIGURE 7.17 A schematic representation of the two-region corrosion fatigue behavior of steels in high pressure hydrogen environment [22,23].

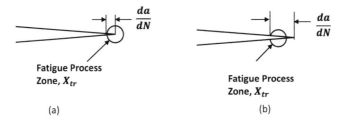

FIGURE 7.18 Schematic representation of the environment assisted fatigue crack growth behavior in (a) the transient region and (b) in the steady-state region.

Both terms on the right-hand side of equation (7.11) can each be written in the following generic form:

$$\left(\frac{da}{dN}\right)_{tr}, \left(\frac{da}{dN}\right)_{ss} = F\left[\Delta K, P_H, R, v\right] \tag{7.12}$$

where P_H is the partial pressure of hydrogen, R is the ratio of the minimum to maximum stress intensity parameter during the fatigue cycle, and v is the loading frequency. The dependence on loading frequency in A372 Grade J Class 70 steel is small as seen in Figure 7.14 in comparison to dependence on other parameters, so it is assumed to be constant. All data used to evaluate the model were generated at a loading frequency of 0.1–1 Hz. The dependence on the load ratio, R, has been modeled [24] with a multiplicative term that can be separated from the dependencies on P_H and ΔK. Thus, for constant R, equation (7.12) can be re-written as in equations (7.13) and (7.14), respectively for the transition and steady state regions.

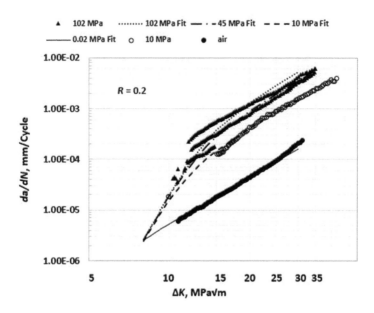

FIGURE 7.19 Environment assisted fatigue crack growth behavior of ASTM A372, Grade J, Class 70 steels in hydrogen at various pressures at a load ratio, R, of 0.2. The predicted growth rates from the model are also shown in the figure.

$$\left(\frac{da}{dN}\right)_{tr} = A_1 \Delta K^{B_1} f(P_H) \tag{7.13}$$

$$\left(\frac{da}{dN}\right)_{ss} = A_2 \Delta K^{B_2} \left(P_H^{m_2}\right)^{d_2} = A_2 \Delta K^{B_2} f_1(P_H) \tag{7.14}$$

We note that the crack growth rate trend in Figure 7.19 in the transient region does not show dependence on hydrogen pressure, so $f(P_H) = 1$ in equation (7.13). If needed, $f(P_H)$ can be a different function. The functional form of equation (7.15) is found to be most suited for representing the pressure dependence of da/dN in the steady-state region.

$$f_1(P_H) = \left(1 + C_2\left(\ln\left(\frac{P_H}{P_{H0}}\right)\right)^{n_2}\right) \tag{7.15}$$

The above form is consistent with the notion that pressure is more likely to affect the hydrogen delivery to the crack tip and its effect on crack growth rate is expected to saturate with increasing pressure. In equation (7.15), P_{H0} = threshold hydrogen pressure below which the effects of hydrogen on the HA-FCG are negligible. In other

TABLE 7.3

HA-FCGR Constants for SA372 Grade J Class 70 Steels in Gaseous Hydrogen

R	A_1 $\left(\dfrac{\text{mm}}{\text{cycle}}(\text{MPa})^{-B_1}\right)$	B_1	A_2 $\left(\dfrac{\text{mm}}{\text{cycle}}(\text{MPa})^{-B_2}\right)$	B_2	P_{H0} (MPa)	C_2	n_2
0.2	9×10^{-15}	9.39	4×10^{-9}	3.116	0.02	0.0045	4.128
0.5	2×10^{-18}	12.8	6×10^{-10}	3.75	0.02	5.0	1.0

words, P_{H0} is the pressure at which the $\left(\dfrac{da}{dN}\right)_{ss} = \left(\dfrac{da}{dN}\right)_{air}$. In equation (7.15) that

occurs when $\dfrac{P_H}{P_{H0}} = 1$. Equation (7.15) then reduces to:

$$\left(\frac{da}{dN}\right)_{ss} = \left(\frac{da}{dN}\right)_{air} = A_2\Delta K^{B_2}$$

P_{H0} for ferritic steels is chosen to be 0.02 MPa. The values of all constants in equations (7.13) and (7.15) for the transient and steady state are given in Table 7.3. The fitted lines are also plotted in Figure 7.19 and appear to fit the data well at different pressures.

The two examples of models described in this section are for applications involving specific materials and specific environments. They describe an approach that one can use to develop corrosion-fatigue models for other material-environment combinations. As mentioned in the beginning of the chapter, no unified models or theories are available for application to EAC, so the discussion on models has focused on methodologies for developing material-environment specific models.

Example Problem 7.3

A pressure vessel made from A372 Grade J Class 70 steel is designed to store hydrogen at a maximum allowable pressure of 350 bar (35 MPa). The outer diameter of the vessel is 400 mm, and it has a wall thickness of 25 mm along the main body of the vessel. Calculate the fatigue life of the vessel if it is cycled between 300 and 60 bar 10 times/day. This vessel is inspected ultrasonically that can very reliably detect flaws that are 1 mm deep and lie on the radial-axial plane on the inside surface of the vessel. These flaws have been shown to be life-limiting for the vessel. Estimate the safe design life/inspection interval during service.

The maintenance engineer states that the vessel is expected to be cycled in service only between 200 and 40 bar so, the expected life should be longer. You are asked to estimate the increase in life/inspection interval of the vessel if the operating pressures are decreased to 200 bar maximum and 40 bar minimum.

The fatigue crack growth behavior of A372 Grade J Class 70 steel in high pressure hydrogen is given in Table 7.3. In addition to these properties, the K_{IEAC} of this material in high pressure hydrogen is given as 40 MPa$\sqrt{m}$ and the $K_{Ic} = 90$ MPa$\sqrt{m}$.

Solution

The first step is to determine the K-expression for this geometry. From Table 3.1 from which the following equations are identified.

$$K = \frac{2pR_o^2}{R_o^2 - R_i^2}\sqrt{\pi a}\,f(a/t)$$

$$f(a/t) = 1.1 + A\left[4.952(a/t)^2 + 1.093(a/t)^4\right]$$

$$A = \left(\frac{0.125}{t/R_i} - 0.25\right)^{0.25} \quad \text{for} \quad 0.1 \le t/R_i \le 0.2$$

$$\Delta K = \frac{2\left(p_{max} - p_{min}\right)R_o^2}{R_o^2 - R_i^2}\sqrt{\pi a}\,f(a/t)$$

$R_o = 200\,mm$, $R_i = 175\,mm$, $t = 25\,mm$, $p_{max} = 30\,MPa$, $p_{min} = 6\,MPa$ for the design operating pressure and $p_{max} = 20\,MPa$, and $p_{min} = 4\,MPa$. The load ratio in both loading conditions, $R = 0.2$. Thus, the following constants in equations (7.13)–(7.15) describe the crack growth behavior.

R	A_1 $\left(\dfrac{mm}{cycle}(MPa)^{-B_1}\right)$	B_1	A_2 $\left(\dfrac{mm}{cycle}(MPa)^{-B_2}\right)$	B_2	P_{H0} (MPa)	C_2	n_2
0.2	9×10^{-15}	9.39	4×10^{-9}	3.116	0.02	0.0045	4.128

The second step is to estimate the critical crack size for maximum design pressure based on the $K_{Ic} = 90\,MPa\sqrt{m}$ and the critical crack size for maximum operating pressure based on the $K_{IEAC} = 40\,MPa\sqrt{m}$. We will then choose the lower of the two critical crack sizes to define the end of life. We write an excel program to estimate the K as a function of crack size for pressure levels of 35, 30, and 20 MPa. These results are shown in Figure 7.20. The Table below shows the critical flaw size based on KIc and the maximum pressure of 30 MPa and the critical flaw sizes based on KIEAC for pressures of 30 and 20 MPa. The critical flaw sizes based on KIEAC are smaller and are used for determining the design life.

Criterion for Critical Flaw Size	Critical Flaw Size (mm)
Based on K reaching K_{Ic} at the maximum design stress, a_{cr1}	9.5
Based on K_{max} reaching K_{IEAC} at maximum operating stress of 300 MPa	5.0
Based on K_{max} reaching K_{IEAC} at maximum operating stress of 200 MPa	7.5

Thus, the critical crack size for the design maximum operating pressure of 300 MPa is 5.0 mm and for reduced maximum pressure of 200 MPa is 7.5 mm. We now need to estimate the number of pressure cycles it will take to grow the crack to 5.0 mm at

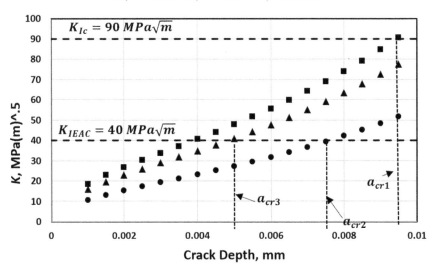

FIGURE 7.20 K as a function of crack depth for an internally pressurized cylinder in Example Problem 7.3 at different vales of pressure. The figure is used to establish the critical crack depth representing the end of useful design life.

a maximum pressure of 30 MPa and at 20 MPa to grow the crack to 7.5 mm. For this we need to formulate the equation for estimating the crack growth rates in hydrogen. For 30 and 20 MPa hydrogen pressure, equation (7.15) reduces to:

$$\text{For 30 MPa hydrogen pressure}: \quad \left(\frac{da}{dN}\right)_{ss} = 4.13 \times 10^{-9} \, \Delta K^{3.116}$$

$$\text{For 20 MPa hydrogen pressure}: \quad \left(\frac{da}{dN}\right)_{ss} = 4.083 \times 10^{-9} \, \Delta K^{3.116}$$

$$\text{In the transient region for both pressures} \left(\frac{da}{dN}\right)_{tr} = 9 \times 10^{-15} \, \Delta K^{9.39}$$

From equation (7.11), the crack growth rates at the two operating maximum pressures are given by:

$$\left(\frac{da}{dN}\right)_{p=30} = \left[\frac{1}{9 \times 10^{-15} \, \Delta K^{9.39}} + \frac{1}{4.13 \times 10^{-9} \, \Delta K^{3.116}}\right]^{-1}$$

$$\left(\frac{da}{dN}\right)_{p=20} = \left[\frac{1}{9 \times 10^{-15} \, \Delta K^{9.39}} + \frac{1}{4.083 \times 10^{-9} \, \Delta K^{3.116}}\right]^{-1}$$

The number of fatigue cycles to grow the crack to a crack size, a is given by:

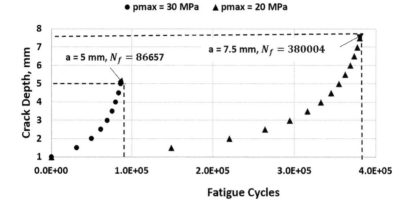

FIGURE 7.21 Crack depth as a function of applied fatigue cycles of a crack in the radial axial plane for maximum operating pressures of 30 and 20 MPa in Example Problem 7.3. The minimum pressure in both cases were 0.2 of the maximum pressures.

$$\left(\int_0^N dN \right)_{p=30} = N_{p=30} = \int_1^5 \left[\frac{1}{9\times10^{-15}\,\Delta K^{9.39}} + \frac{1}{4.13\times10^{-9}\,\Delta K^{3.116}} \right] da$$

$$\left(\int_0^N dN \right)_{p=20} = N_{p=20} = \int_1^{7.5} \left[\frac{1}{9\times10^{-15}\,\Delta K^{9.39}} + \frac{1}{4.083\times10^{-9}\,\Delta K^{3.116}} \right] da$$

We can program the above equations in an excel sheet to first calculate ΔK as a function of crack size and then substitute it in the above integrals to calculate the cycles as a function of crack size. Note that in the above equations, ΔK is in MPa$\sqrt{m}$ and da/dN is mm/cycle. Thus, while calculating ΔK, the crack size must be in meters but while calculating life, the crack size in integration limits is taken in mm. The predicted crack size versus cycles relationship is shown in Figure 7.21. For maximum operating pressure of 30 MPa, the design life is 86,657 cycles. At 10 cycles/day, it is 23.7 years. For a maximum operating pressure of 20 MPa, the design life increases to 380,004 cycles or about 104 years. These calculations can be used to choose appropriate inspection intervals during service or make a decision that the vessel need not be inspected during service because the predicted design life exceeds the service period of the vessel.

7.8   SUMMARY

In this chapter, we have described the fracture mechanics approach for predicting environment assisted subcritical crack growth in structural metallic alloys commonly used in construction of bridges and highways, aerospace structures, pressure vessels, nuclear reactor components, and chemical reactors. The mechanisms for EAC under sustained and cyclic loading are discussed. It was noted that there is considerable variety in the underlying mechanisms that cause EAC therefore, a unified theory that predicts the rates at which the cracks grow in corrosive environments does not currently exist. The mechanisms described include (a) delayed hydride cracking, (b) adsorption of chemical species at the crack tip and weakening the atomic bonds, (c) oxide film rupture mechanism, (d) grain boundary cracking due to embrittlement, and (e) hydrogen embrittlement. The crack growth rates are determined by the slowest process step in the sequence of processes leading to decrease in fracture strength due to environment-caused degradation.

Test methods are described in this chapter to characterize the crack growth rates in terms of crack tip parameters K and ΔK for sustained and cyclic loading, respectively. These measured crack growth rates can be used to estimate the design life of components in service. The experimental techniques for measuring the threshold levels of K for environment assisted crack growth were described. These consisted of the constant load tests, the constant-load-line displacement tests, and the rising load tests. The constant load tests primarily provide the threshold value of K below which no EAC occurs. The constant displacement tests measure the crack growth rates as function of applied K and the threshold K below which a growing crack will stop growing. The constant load and constant displacement techniques have been standardized in American Society for Testing and Materials (ASTM) standard E-1681.

The fracture mechanics approach for predicting subcritical crack growth due to cyclic loading in corrosive environments are discussed. Two models were described that are used to predict corrosion-fatigue behavior in specific material/environment combinations. The linear superposition model is most useful for predicting corrosion fatigue in aqueous environments and the other model is suitable for describing crack growth behavior in gaseous hydrogen environments. The use of these models was illustrated with examples.

REFERENCES

1. H.H. Johnson and P.C. Paris, "Sub-critical Flaw Growth", *Engineering Fracture Mechanics*, Vol. 1, 1968, pp 3–45.
2. A. Alverez et al., Delayed Hydride Cracking in Zirconium Alloys in Pressure Tube Nuclear Reactors, IAEA-TECDOC-1410, ISBN 92-0-110504-5, International Atomic Energy Agency (IAEA), Vienna, 2004, https://www-pub.iaea.org/MTCD/Publications/PDF/te_1410_web.pdf
3. S.P. Lynch, "Mechanisms and Kinetics of Environmentally Assisted Cracking: Current Status, Issues, and Suggestions for Future Work", *Metallurgical and Materials Transactions A*, Vol. 44A, 2013, pp. 1209–1229.
4. P.L. Andresen and P. Ford, "Life Prediction by Mechanistic Modeling and System Monitoring of Environmental Cracking of Fe and Ni Alloys in Aqueous System", *Materials Science and Engineering A*, Vol. 103, 1988, pp. 167–184.
5. M. Alkateb, S. Tadic, A. Sedmark, I. Ivanovic, and S. Markovic, "Crack Growth Analysis of Stress Corrosion Cracking", *Technical Gazette*, Vol. 28, 2021, pp. 240–247.
6. R.P. Wei and G. Shim, in *Corrosion-Fatigue: Mechanics, Metallurgy and Chemistry, American Society for Testing and Materials,* Philadelphia, PA.
7. C. Beachem, "A New Model for Hydrogen Assisted Cracking (Hydrogen Embrittlement)", *Metallurgical Transactions B*, Vol. 3, 1972, pp. 437–451.
8. M.M. Martin, M. Dadafrina, A. Nagao, S. Wang, and P. Sofronis, "Enumeration of Hydrogen-Assisted Localized Plasticity Mechanism for Hydrogen Embrittlement in Structural Materials", *Acta Materialia*, Vol. 165, 2019, pp. 734–750.

9. A. Nagao, M. Dadfrina, B.P. Someday, P. Sofronis, and R.O. Ritchie, "Hydrogen Enhanced Plasticity Mediated Decohesion for Hydrogen-Induced Intergranular and Quasi-Cleavage Fracture of Lath Martensite", *Journal of Mechanics and Physics of Solids*, Vol. 112, 2018, pp. 403–430.

10. R.P. Gangloff, "H-Enhanced Deformation and Fracture in Crack Tip Process Zone", Proceedings of the International Hydrogen Conference, B. P. Somerday and P. Sofronis Editors, ASME—861387, 2017, pp. 1–35.

11. A.R. Troiano, "The Role of Hydrogen and Other Interstitials in the Mechanical Behavior of Metals", *Transactions of ASM*, Vol. 53, 1960, pp. 54–80.

12. ASTM E-1681-03, *Standard Test Method for Determining Threshold Stress Intensity Factor for Environment Assisted Cracking of Metallic Materials*, American Society for Testing and Materials, W. Conshohocken, PA, 2013.

13. A. Saxena and S.J. Hudak, Jr., "Review and Extension of Common Crack Growth Specimens", *International Journal of Fracture*, Vol. 14, 1978, pp. 453–468.

14. P.G. Marsh and W.W. Gerberich, "Stress-Corrosion Cracking of High Strength Steels (Yield Strengths Greater than 1240 MPa)", in *Stress Corrosion Cracking-Material Performance and Evaluation*, 1992, ASM International, Ohio, pp. 63–90.

15. M.T. Wang and R.W. Staehle, *Hydrogen in Metals*, Pergamon Press, Oxford, England, 1972, p. 342.

16. R.P. Jewett, R.J. Walter, W.T. Chandler, and P.P. Fromberg, "Hydrogen Embrittlement of Materials", NASA Technical Report CR-2163, 1973.

17. K.A. Nibur, B.P. Somerday, C. San Marchi, J.W. Foulk, M. Dadfarina, and P. Sofronis, "The Relationship Between Crack Tip Strain and Subcritical Cracking Thresholds in Steels in High Pressure Hydrogen Gas", *Metallurgical and Materials Transactions A*, Vol. 44, 2013, pp. 248–269.

18. B.P. Somerday, P. Bortot, and J. Felbaum, "Optimizing Measurement of Fatigue Crack Growth Relationships for Cr-Mo Pressure Vessel Steels in Hydrogen Gas", PVP2015–45424, Proc. ASME PVP Conference, July 19–23, Boston, MA, 2015.

19. A. Saxena, K.A. Nibur, and A. Prakash, "Applications of Fracture Mechanics in Assessing Integrity of Hydrogen Storage Systems", *Engineering Fracture Mechanics*, Vol. 187, 2018, pp. 368–380.

20. R.P. Wei and J.D. Landes, "Correlation Between Sustained-Load and Fatigue Crack Growth in High Strength Steels", *Materials Research and Standards*, Vol. 9, 1969, pp. 25–27.

21. E. Richey, III, "Empirical Modeling of Environment -Enhanced Fatigue Crack Propagation in Structural Alloys for Component Life Prediction", NASA Contractor Report, 198231, 1995.

22. R.L. Amaro, N. Rustagi, K.O. Findley, E.S. Drexel, and A.J. Slifka, "Modeling the Fatigue Crack Growth of X-100 Pipeline Steel in Gaseous Hydrogen", *International Journal of Fatigue*, Vol. 59, 2014, pp. 262–271.

23. A. Saxena and K.O. Findley, "Modeling the Effects of Hydrogen Pressure on the Fatigue Crack Growth Behavior in SA372 Pressure Vessel Steels" ASME Pressure Vessel and Piping Conference, July 17–22, 2022, Las Vegas, NV, PVP 2022–83958.

24. C. San Marchi, J. Ronevich, P. Bortot, Y. Wada, J. Felbaum, and M. Rana, "Technical Basis for the Master Curve for Fatigue Crack Growth of Ferritic Steels in ASME Section VIII-3 Code" PVP 2019–93907, Proceedings of the ASME 2019 Pressure Vessel and Piping Conference, July 14–19, San Antonio, Texas, 2019.

HOMEWORK PROBLEMS

1. Is K_{IEAC} a material property? Explain for your answer.
2. It is observed that the rate of EAC in Region II of the da/dt versus K relationship was dependent on the pressure and temperature, however, the value of K_{IEAC} was not. Can you speculate on the reasons for this behavior?
3. What are the steps involved in hydrogen embrittlement and what is the rate controlling step?
4. Describe the advantages and disadvantages of constant load versus constant displacement tests for measuring EAC.
5. A cylindrical pressure vessel for containing hydrogen has a radius, R, of 0.5 m and a wall thickness, t, of 5 cm. It has a long axial surface crack of depth equal to 3 mm on the inside surface. The pressure vessel is inflated to a maximum pressure, p, of 30 MPa which reduces to atmospheric pressure every day and then the vessel is refilled. Calculate the life of the pressure vessel if you are given the following additional information:

   - $\dfrac{da}{dN} = 8.0 \times 10^{-12} (\Delta K)^3$ in the H_2 environment where FCGR is in m/cycle and ΔK in MPa$\sqrt{m}$
   - $K_{Ic} = 60$ MPa$\sqrt{m}$
   - $K_{IEAC} = 40$ MPa$\sqrt{m}$
   - The K-expression is given by $K = \dfrac{pR}{t}\sqrt{\pi a}\left[F(a/t)\right]$ where, $F(a/t)$ is given in Table 3.1 for this geometry.

6. Constant displacement tests using bolt loaded WOL specimens are being planned from the material in Problem 5 to measure environment assisted crack growth rates as a function of K. If the width of the specimen is 50 mm and it is 25 mm thick, and the initial crack size after fatigue pre-cracking is 20 mm, estimate the displacements to be applied to the specimen along its load-line by cranking the bolt such that the specimen is loaded to K levels of 55 and 50 MPa$\sqrt{m}$, respectively in separate tests. Also estimate the crack size at which the cracks are expected to arrest. Assume an elastic modulus of 210 GPa.
7. Using the data in Example Problem 7.3, estimate the design life of the pressure vessel for storing hydrogen at a maximum operating pressures of 28 and 24 MPa with minimum pressures of 5.6 and 4.8 MPa and plot the relationship between design life and the maximum operating pressure.
8. If the pressure vessel Example Problem 7.3 is cycled between a maximum operating pressure of 35 and 15 MPa, estimate the design life of the vessel.

8 Fracture under Mixed-Mode Loading

8.1 INTRODUCTION

In the previous chapters, we have developed a theory of fracture that is suitable for elastic, isotropic and homogeneous materials in which the cracks were assumed to grow in pure Mode I. In this mode of loading, the crack faces displace in a direction normal to the crack surface. In Chapter 3, we had also briefly defined two other modes of loading, Modes II and III as previously shown in Figure 3.1. When crack surfaces slide relative to each other on the same plane, Mode II loading is said to occur, and Mode III loading occurs when the crack faces slide relative to each other in the transverse direction. Mixed-mode loading occurs when more than one mode is present such as when the crack surfaces displace in a direction normal to each other and simultaneously also slide relative to each other, either in-plane or out-of-plane. Metals, that are ductile, typically have a much higher resistance to fracture under shear loading than under tensile loading which explains why Mode I fracture has been studied more extensively in metals. Mixed mode fracture is more common in brittle materials so, the examples used in this chapter to describe mixed-mode fracture are primarily from such materials.

Consider a crack that is oriented at an angle to the maximum principal stress direction as shown in Figure 8.1a. If the material deformation and fracture resistance is homogeneous and isotropic, the crack is still expected to propagate in a direction normal to the direction of maximum principal stress as shown in Figure 8.1b and be classified as Mode I crack growth. The strain energy release rate is maximum when the crack grows in a direction normal to the applied first principal stress [1]. On the other hand, if material resistance to fracture is low in another direction(s), crack could propagate in those directions and be classified as mixed mode fracture.

When we consider fracture and crack growth in composites, bones, and other brittle materials such as ceramics and concrete in which resistance to fracture may vary with direction because of distribution of porosity and other defects or due to weak interfaces in the case of composites, the limitations of Mode I fracture theories are exposed and formulation of a theory of mixed mode fracture becomes necessary. For example, see Figure 8.2 where the damage preceding fracture in a carbon fiber/epoxy composite is dominantly due to delamination along carbon fiber/epoxy interface. The Mode I fracture mechanics methods are inadequate in describing such interlaminar fracture.

The primary application of mixed-mode fracture mechanics is to provide laboratory test methods to evaluate these materials, particularly their fracture resistance along the weak planes such as interfaces. These materials are used in spite of their generally poor fracture resistance because of their superior resistance to high

DOI: 10.1201/9781003292296-8

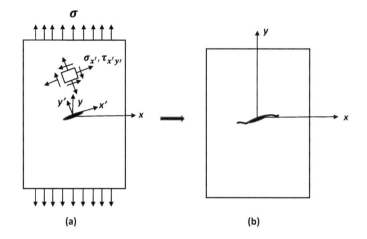

FIGURE 8.1 Example of a crack subject to mixed-mode loading but the crack grows in Mode I.

FIGURE 8.2 Delamination fracture in a composite beam subjected to compressive loading (Photograph reproduced from https://en.wikipedia.org/wiki/Delamination#Inspection_methods).

temperatures and their ability to resist extreme environments. The properties measured from these test methods are used to design stronger and more damage resistant materials and are therefore useful. It must be noted that fracture in composite materials is a much more complex subject and requires many more analytical tools than provided by fracture mechanics to address these complexities. That discussion is outside the scope of this book but, there are other books that describe fracture in composite materials more comprehensively. Prior to building the analytical framework for addressing mixed mode fracture, it is important to understand the characteristics of the materials that require such treatment.

8.2 STRESS ANALYSIS OF CRACKS UNDER MIXED MODE LOADING

As derived in Chapter 3, the stresses under pure Mode I load are given by,

$$\lim_{r \to 0} \sigma_{ij}^{I} = \frac{K_{I}}{\sqrt{2\pi r}} f_{ij}^{I}(\theta) \tag{8.1}$$

where, σ_{ij}^{I} are the components of stress associated with Mode I load, and $f_{ij}^{I}(\theta)$ are angular functions in equations (3.15). Now, let us consider a cracked body that is subjected to remote loading of the type shown in Figure 8.3 consisting of a remote normal stress σ and a shear stress, τ. This produces combined Modes I and II loading conditions at the crack tip. In this discussion we will concern ourselves with combined Modes I and II loading because it is encountered more commonly in fracture of brittle materials and composites. The out-of-plane shear mode is uncommon, so it

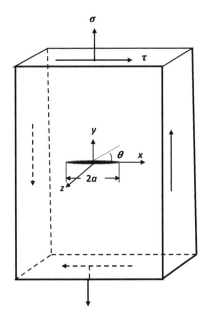

FIGURE 8.3   A crack in a semi-infinite panel subjected mixed Mode I and Mode II loading.

has not been studied much. For Mode II by itself, the crack tip stress fields are given by equation (8.2) [1]:

$$\lim_{r \to 0} \sigma_{ij}^{II} = \frac{K_{II}}{\sqrt{2\pi r}} f_{ij}^{II}(\theta) \qquad (8.2)$$

where,

$$f_{xx}^{II} = \sin\frac{\theta}{2}\left[2 + \cos\frac{\theta}{2}\cos\frac{3\theta}{2}\right] \qquad (8.3a)$$

$$f_{yy}^{II} = \sin\frac{\theta}{2}\cos\frac{\theta}{2}\cos\frac{3\theta}{2} \qquad (8.3b)$$

$$f_{xy}^{II} = \cos\frac{\theta}{2}\left[1 - \sin\frac{\theta}{2}\sin\frac{3\theta}{2}\right] \qquad (8.3c)$$

$$f_{zz}^{II} = 0, \text{for plane stress} \qquad (8.3c)$$

$$f_{zz}^{II} = v\left(f_{xx}^{II} + f_{yy}^{II}\right), \text{for plane strain} \qquad (8.3d)$$

$$\text{Also}, K_I = \sigma\sqrt{\pi a} \qquad (8.4a)$$

$$K_{II} = \tau\sqrt{\pi a} \qquad (8.4b)$$

For $\theta = 0$, the only nonzero component of f_{ij}^{II} is f_{xy}^{II}, and it is $= 1$. Thus, for mixed Mode I and II, along $\theta = 0$, the normal stress components are due only to K_I but, the in-plane shear, which was 0 in pure Mode I, is given by equation (8.5).

$$\tau_{xy} = \frac{K^{II}}{\sqrt{2\pi r}} \qquad (8.5)$$

It is important to point out that stress intensity parameters for two of more modes are not additive. However, normal and shear stresses resulting from Modes I and II loading are additive as given by equations (8.7):
    Thus,

$$K_{\text{total}} \neq K_I + K_{II} \qquad (8.6)$$

$$\sigma_{xx} = \frac{K_I}{\sqrt{2\pi r}}\cos\frac{\theta}{2}\left(1 - \sin\frac{\theta}{2}\sin\frac{3\theta}{2}\right) - \frac{K_{II}}{\sqrt{2\pi r}}\sin\frac{\theta}{2}\left(2 + \cos\frac{\theta}{2}\cos\frac{3\theta}{2}\right) \quad (8.7a)$$

$$\sigma_{yy} = \frac{K_I}{\sqrt{2\pi r}} \cos\frac{\theta}{2}\left(1 + \sin\frac{\theta}{2}\sin\frac{3\theta}{2}\right) + \frac{K_{II}}{\sqrt{2\pi r}}\sin\frac{\theta}{2}\cos\frac{\theta}{2}\cos\frac{3\theta}{2} \qquad (8.7b)$$

$$\tau_{xy} = \frac{K_I}{\sqrt{2\pi r}} \cos\frac{\theta}{2}\sin\frac{\theta}{2}\sin\frac{3\theta}{2} + \frac{K_{II}}{\sqrt{2\pi r}}\cos\frac{\theta}{2}\left(1 - \sin\frac{\theta}{2}\sin\frac{3\theta}{2}\right) \qquad (8.7c)$$

The K expressions given in equation (8.4a and b) are for the simplest geometry of a crack of size $2a$ in a semi-infinite plate subjected to remote uniform normal and shear stress. K_I for several geometries were derived in Chapter 3, and similarly K_{II} for different geometries can be derived and are listed in handbooks of stress intensity parameter [2] for many geometries.

8.3 MIXED MODE CONSIDERATIONS IN FRACTURE OF ISOTROPIC MATERIALS

In single mode fracture, whether Modes I, II, or III, the critical K for fracture can be measured in that mode and compared to the applied K also calculated for the same mode to determine whether fracture will occur. Under mixed-mode loading conditions, a combination of fracture resistance in the operating modes is needed. There are three criteria that have been proposed for fracture under mixed-mode conditions that are described next.

8.3.1 FRACTURE CRITERION BASED ON ENERGY AVAILABLE FOR CRACK EXTENSION

As mentioned earlier, the values of K in the various modes cannot simply be added so we cannot algebraically add the K values from different modes to compute the crack driving force under mixed mode conditions. In other words, the overall crack driving force cannot be written as, $K \neq K_I + K_{II} + K_{III}$ as also previously mentioned. On the other hand, energy available for fracture in the form of the Griffith's energy, G, is additive and can be compared with the critical value G_c measured under similar mixed-mode conditions to determine if fracture is expected or not. Thus, the crack extension force under mixed mode conditions is given by equation (8.8), if only two modes of fracture are present:

$$G = \frac{K_I^2}{E} + \frac{K_{II}^2}{E} \qquad (8.8)$$

Thus, G can be estimated by combining equation (8.8) with equation (8.4). Fracture toughness as a function of mode-mixity is shown in Figure 8.4 for a brittle ceramic, alumina (Al_2O_3) and for human bones [3]. The mode mixity is expressed in terms of $\psi = \tan^{-1}\left(\frac{K_{II}}{K_I}\right)$ and the critical energies, G_c, for fracture are normalized by their values, G_o, in pure Mode I [3,4]. Note that $\psi = 0$ corresponds to pure Mode I

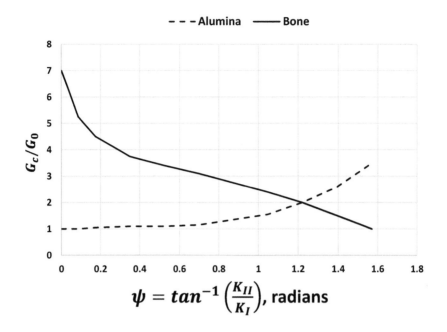

$$\psi = tan^{-1}\left(\frac{K_{II}}{K_{I}}\right), \text{ radians}$$

FIGURE 8.4 Critical fracture energy of alumina as a function of mode mixity normalized by the critical fracture energy in Mode I [3,4].

while $\psi=\pi/2$ corresponds to pure Mode II and all in-between mixed-mode states for $0\leq\psi\leq\pi/2$. In alumina, the fracture toughness is seen to increase with mode mixity. Thus, Mode II and the mixed-mode fracture toughness values are higher than the fracture resistance in Mode I. On the other hand, the fracture toughness of bones is seen to decrease as a function of mode-mixity. The remote stress and the initial direction of crack propagation in this theory is thus determined by the direction in which the energy release rate exceeds the fracture resistance.

Metallic materials typically exhibit fracture resistance that is higher in the shear direction resulting in higher K_{IIc} than K_{Ic}, like the behavior of alumina shown in Figure 8.4. To measure K_{IIc} and the mixed-mode fracture toughness requires that the fracture occur in the direction of the original crack plane. Recall that while measuring fracture toughness in Mode I in Chapter 5, the fracture was expected to propagate along the plane of the original crack to qualify as a valid K_{Ic} measurement. In metallic materials with varying resistance to fracture under mixed-mode conditions, it cannot be assumed that fracture will proceed along the original crack plane if the loading is mixed-mode. Typically, the crack will propagate at an angle θ_c from the original crack plane. If the crack does indeed propagate along the original crack plane, θ_c will be 0. We refer to θ_c as the angle of initial crack propagation and measure its value relative to the original crack plane along with the remote force (or stress) for fracture. The following example explains this criterion of fracture further.

Example Problem 8.1

Consider a 2 cm long crack with its plane inclined at an angle β from the loading direction in a semi-infinite body loaded under uniform tension in a high strength aluminum alloy. The critical fracture energy, G_c, as a function of mode mixity is given by the following equation:

$$G_c = 11.7 + 7.453\psi$$

where G_c in kJ/m² and $\psi = \tan^{-1}\left(\dfrac{K_{II}}{K_I}\right)$ is in radians. Estimate the fracture stress in the structure for $\beta = 5°$, 10°, 15°, 20°, 30°, 40°, 45°, 70°, and 90°. The elastic modulus of the alloy is 70 GPa and the Poisson's ratio, $v = 0.3$.

Solution

The K_I and K_{II} for this configuration are given by equation (8.9a and b), respectively.

$$K_I = \sigma\sqrt{\pi a}\,\sin^2\beta \tag{8.9a}$$

$$K_{II} = \sigma\sqrt{\pi a}\,\sin\beta\cos\beta \tag{8.9b}$$

Thus, energy available for fracture is given by:

$$G(\beta) = \left(\frac{K_I^2}{E} + \frac{K_{II}^2}{E}\right)\left(1 - v^2\right) = \frac{\sigma^2\pi a}{E}\left(1 - v^2\right)\sin^2\beta\left(\sin^2\beta + \cos^2\beta\right)$$

$$= \frac{\sigma^2\pi a}{E}\left(1 - v^2\right)\sin^2\beta \tag{8.10}$$

$G(\beta)/G(\beta = \pi/2)$ is plotted in Figure 8.5 for this configuration. For this loading configuration, it is the highest for $\beta = 90°$ and goes to 0 for $\beta = 0$, when the crack is oriented along the normal stress direction.

Fracture is predicted to occur when, $G(\beta) = G_c(\beta)$. This assumes that the direction in which the fracture occurs is along the plane of the original crack. The fracture stress, $\sigma_c =$ is given by the following equation:

$$\frac{\sigma_c^2\pi(0.01)}{70,000}\sin^2\beta\left(1 - 0.3^2\right) = 11.7\times10^{-3} + 7.453\times10^{-3}\tan^{-1}\left(\frac{K_{II}}{K_I}\right)$$

From equations (8.9), $\dfrac{K_{II}}{K_I} = \dfrac{1}{\tan\beta}$

$$\sigma_c = \left[\frac{70,000}{0.01\pi}\frac{10^{-3}\left(11.7 + 7.453\tan^{-1}\left(\cot\beta\right)\right)}{\left(1 - 0.3^2\right)\sin^2\beta}\right]^{1/2}$$

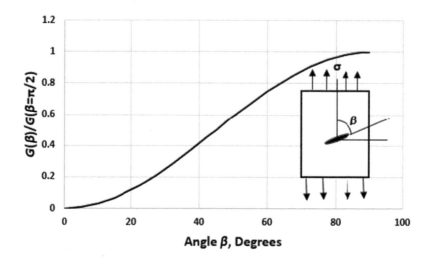

FIGURE 8.5   Variation in the energy available for fracture as a function of crack plane orientation with the loading axis in a semi-infinite body with a remote uniform stress.

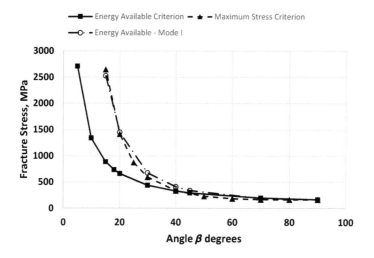

FIGURE 8.6   Predicted stress as a function of the angle between the normal to the crack plane and the loading axis for a semi-infinite panel subjected to uniform stress.

The predicted fracture stress from this theory is shown Figure 8.6. Note that for $\beta=90°$, the problem reduces to that of pure Mode I and indeed the results of fracture stress calculations are also identical to ones expected for pure Mode I and the crack is predicted to grow along the original crack plane that is normal to the direction of the external normal stress and lies along which the strain energy available for crack extension is the maximum. Therefore, this yields the lowest value of σ_c.

Since the fracture toughness is the lowest under Mode I conditions, it is possible for $\beta<90°$, in other words for an inclined crack, that G_I component exceeds

G_{Ic} before $G(\beta)$ exceeds $G_c(\beta)$. This can be tested by equating the Mode I component of G for an angle of inclination β to G_{Ic} by the following equation:

$$\frac{\sigma_c^2 \pi a}{E}(1-v^2)\sin^4\beta = 11.7\times10^{-3}$$

In the above equation, stress and E are in MPa and G_{Ic} in MJ/m^2. Thus,

$$\sigma_c = \sqrt{\frac{11.7\times10^{-3}\times70,000}{\pi\times0.01(1-0.3^2)}}\frac{1}{\sin^2\beta} = \frac{169.3}{\sin^2\beta}$$

The fracture stress, σ_c, as a function of β is plotted in Figure 8.6 and labelled as Energy Available – Mode I criterion. We notice that the predictions from the energy available criterion and the energy available in Mode I criterion are very similar for $90° \le \beta \le 40°$ and then diverge for $\beta < 40°$. For the latter case, the fracture stress predicted by the energy available criterion is less than from the energy available – Mode I criterion. For $\beta \ge 40°$, Mode I fracture dominates and the crack is expected to grow normal to the remote stress direction yielding a value of $\theta_c = \beta - 90$. For $\beta < 40°$, the predicted value of the initial fracture angle θ_c is 0 and fracture will occur at a remote stress level given by the energy available criterion. It is expected that there will be shift from essentially a Mode I fracture to one in which Mode II is also a contributor. In a variation of the above example, if we consider a material such as human bones in which the resistance to Mode II fracture is less than under Mode I conditions, the prediction of when the shift occurs from dominantly Mode I to one in which the Mode II contribution is also a factor will be different from $\beta = 40°$. To prove this is left as an end of chapter exercise.

8.3.2 MAXIMUM CIRCUMFERENTIAL STRESS FRACTURE CRITERION

A second mixed mode fracture criterion was proposed by Erdogan and Sih [5] and is based on maximum normal stress along the crack plane reaching a critical value, σ_c. The magnitude of the critical normal stress is the same as required for fracture in pure Mode I. From equation (3.15) for $\theta = 0$,

$$\sigma_y = \sigma_x = \frac{K}{\sqrt{2\pi x}} = \sigma_c; \quad \text{and} \quad \tau_{xy} = 0; \quad \text{and} \quad \tau_{xy} = 0$$

Let us consider a planar cracked body subjected to a remote uniform normal stress and a remote uniform shear stress as shown in Figure 8.7, so the crack is under combined Mode I and Mode II loading. In this configuration, the normal and shear stresses are best expressed in polar coordinates and are given by equations (8.11):

$$\sigma_r = \frac{K_I}{\sqrt{2\pi r}}\left(\frac{5}{4}\cos\frac{\theta}{2}-\frac{1}{4}\cos\frac{3\theta}{2}\right)+\frac{K_{II}}{\sqrt{2\pi r}}\left(-\frac{5}{4}\sin\frac{\theta}{2}+\frac{3}{4}\sin\frac{3\theta}{2}\right) \quad (8.11\text{a})$$

$$\sigma_\theta = \frac{K_I}{\sqrt{2\pi r}}\left(\frac{3}{4}\cos\frac{\theta}{2}+\frac{1}{4}\cos\frac{3\theta}{2}\right)+\frac{K_{II}}{\sqrt{2\pi r}}\left(-\frac{3}{4}\sin\frac{\theta}{2}-\frac{3}{4}\sin\frac{3\theta}{2}\right) \quad (8.11\text{b})$$

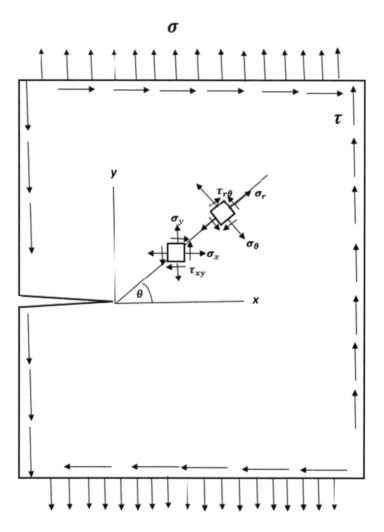

FIGURE 8.7 Crack in a plate subjected to mixed mode loading consisting of a remote uniform normal stress, σ, and a shear stress, τ.

$$\tau_{r\theta} = \frac{K_I}{\sqrt{2\pi r}}\left(\frac{1}{4}\sin\frac{\theta}{2}+\frac{1}{4}\sin\frac{3\theta}{2}\right)+\frac{K_{II}}{\sqrt{2\pi r}}\left(\frac{1}{4}\cos\frac{\theta}{2}+\frac{3}{4}\cos\frac{3\theta}{2}\right) \qquad (8.11c)$$

As per this criterion, fracture occurs at the crack tip along the radial direction for which σ_θ has the maximum value, or in other words, along the maximum principal stress direction, with the corresponding shear stress being zero. The maximum principal stress has a magnitude equal to:

$$\sigma_\theta = \sigma_c = \frac{K_{Ic}}{\sqrt{2\pi r}} \qquad (8.12)$$

Thus, the θ at which $\tau_{r\theta}=0$ in equation (8.11c) can be calculated and labelled as θ_c and can be calculated as follows [4]:

$$K_I\left(\sin\frac{\theta_c}{2}+\sin\frac{3\theta_c}{2}\right)+K_{II}\left(\cos\frac{\theta_c}{2}+3\cos\frac{3\theta_c}{2}\right)$$

$$= K_I\sin\theta_c + K_{II}(3\cos\theta_c-1)=0$$

or

$$\frac{\sin\theta_c}{(3\cos\theta_c-1)}=-\frac{K_{II}}{K_I} \tag{8.13}$$

Solving for θ_c from equation (8.13) and then substituting its value in equation (8.11b) to calculate σ_θ and then substituting it in equation (8.12) leads to:

$$K_I\left(3\cos\frac{\theta_c}{2}+\cos\frac{3\theta_c}{2}\right)-3K_{II}\left(\sin\frac{\theta_c}{2}+\sin\frac{3\theta_c}{2}\right)=4K_{Ic} \tag{8.14}$$

From this criterion, one can then predict the angle θ_c along which the crack is expected to first propagate and the remote critical stress at which fracture is expected to occur. An example of the steps involved are illustrated in Example 8.2.

Example Problem 8.2

Find the fracture stress using the maximum stress criterion for the geometry and loading in Example Problem 8.1. Use the fracture energy given in Example Problem 8.1 to estimate the K_{Ic} for the material.

Solution

Combining equations (8.9) and (8.13) yields:

$$-\frac{\sin\theta_c}{(3\cos\theta_c-1)}=\cot\beta$$

or

$$\beta=\cot^{-1}\left(-\frac{\sin\theta_c}{(3\cos\theta_c-1)}\right)$$

Figure 8.8 shows the relationship between β and θ_c from the above equation and can be approximately rewritten by regression analysis as follows:

$$\theta_c=-48.18-0.826\beta+0.015\beta^2$$

For $\beta=90°$, as expected, $\theta_c=0$ or normal to the Mode I load direction and,

$$K_{Ic}=\sqrt{\frac{EG_{Ic,\,\psi=0}}{1-v^2}}=\sqrt{\frac{70\times10^3\times11.7\times10^{-3}}{1-0.3^2}}=30\text{ MPa}\sqrt{m}.\text{ Substituting the values}$$

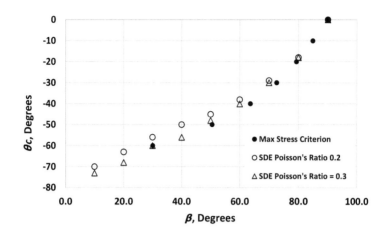

FIGURE 8.8 Figure showing the relationship between the direction of initial crack growth, θ_c, and the angle β. Note that θ_c as measured from the direction of the crack plane has a negative value.

of θ_c for various values of β can set up the relationship for the critical fracture stress, σ_c as a function of β. This relationship is given by:

$$\sigma_c = \frac{4K_{Ic}}{\sqrt{\pi a}\left[\sin^2\beta\left(3\cos\dfrac{\theta_c}{2}+\cos\dfrac{3\theta_c}{2}\right)-3\sin\beta\cos\beta\left(\sin\dfrac{\theta_c}{2}+\sin\dfrac{3\theta_c}{2}\right)\right]}$$

The crack size, $a=0.01$ m from Example Problem 8.1. The predicted angle of initial fracture and the fracture stress, σ_c, are given in Table 8.1. The predicted fracture stress from this theory is plotted in Figure 8.6 as a function of the angle β where a comparison between predictions from the maximum stress and energy available for fracture theories can be made. Between $40\leq\beta\leq90$, the results from the two theories are quite similar, while they diverge for $\beta<40°$ when the Mode II component of the loading increases and the Mode I component decreases. For $\beta<40°$, the predictions from this theory continue to agree with the strain energy available – Mode I estimates. This is entirely expected and left for the reader to ponder why. Figure 8.8 shows the relationship between the angle β and the predicted initial fracture angle θ_c from the maximum stress theory. Note that at $\beta=90°$, we have Mode I conditions so $\theta_c=0°$ and for other values of β, the angle $\beta-\theta_c$ is the angle between the direction of the normal stress on the cracked body and the predicted angle of initial fracture. For $\beta>25°$, this angle is not far from 90° indicating that the predicted direction of initial crack growth is approximately in a direction normal to the direction of the external loading as shown in Figure 8.1b. This is similar to the predictions from the energy available theory.

8.3.3 STRAIN ENERGY DENSITY (SED) AS MIXED MODE FRACTURE CRITERION

A third mixed-mode fracture criterion has been proposed by Sih [7, 8] using SED per unit volume. SED per unit volume, dW/dV, for isothermal conditions is defined in equation (8.15) below.

TABLE 8.1
Estimates of Fracture Stress, the Initial Angle of Fracture, and the Angle β for a Uniformly Stressed Panel with an Inclined Crack from the Maximum Circumferential Stress Theory

β (degrees)	$\dfrac{K_{II}}{K_I} = \cot\beta$	θ_c (degrees)	σ_c (MPa)	$\beta-\theta_c$ (degrees)
90	0.00	0.0	169.3	90
80	0.18	−18.3	167.6	98.3
70	0.36	−32.5	169.1	102.5
60	0.58	−43.8	185.6	103.8
50	0.84	−52.0	230.8	102
40	1.19	−57.2	335.2	97.2
30	1.73	−59.5	593.4	89.5
25	2.14	−59.5	873.4	94.5
20	2.75	−58.7	1,416.2	78.7
15	3.73	−57.2	2,648.6	72.2

$$\text{Strain Energy Density} = SED = \frac{dW}{dV} = \int_{0}^{\varepsilon_{ij}} \sigma_{ij} d\varepsilon_{ij} \tag{8.15}$$

For a 2-d elastic stress fields, SED can be written as:

$$SED = \frac{dW}{dV} = \frac{1}{4\mu}\left[\frac{\kappa+1}{4}\left(\sigma_x + \sigma_y\right)^2 - 2\left(\sigma_x\sigma_y - \tau_{xy}^2\right)\right] \tag{8.16}$$

where $\kappa = 3-4v$ for plane strain and $= \dfrac{3-v}{1+v}$ for plane stress, and $\mu =$ shear modulus. If we define a SED factor, S, such that:

$$\frac{S}{r} = \frac{dW}{dV} \tag{8.17}$$

By combining equations (8.17) and (8.7) leads to the following relationships for cracked bodies subjected to mixed-modes I and II loads:

$$S(\theta) = \frac{1}{\pi}\left[a_{11}K_I^2 + 2a_{12}K_IK_{II} + a_{22}K_{II}^2\right] \tag{8.18}$$

$$a_{11} = \frac{1}{16\mu}(1+\cos\theta)(1+\kappa)$$

$$a_{12} = \frac{1}{16\mu}\sin\theta\left[2\cos\theta - (\kappa-1)\right]$$

$$a_{22} = \frac{1}{16\mu}\left[(\kappa+1)(1-\cos\theta)+(1+\cos\theta)(3\cos\theta-1)\right]$$

Thus, S, the SED factor, can be thought of as the amplitude of the strain energy field that has a $1/r$ type singularity at the crack tip, analogous to $(1/r)^{-1/2}$ type singularity in stress and strain fields. Accordingly, the theory states the crack begins to grow when S reaches a critical value, S_c. However, note that S is direction sensitive. In other words, the magnitude of S varies with angle with respect to the crack plane, unlike K which is not dependent on direction. It is hypothesized that the initial direction of crack propagation is determined by the direction in which the strain energy factor, S, is minimum, leading to the following conditions given by equation (8.19).

$$\frac{\partial S}{\partial \theta} = \frac{1}{\pi}\Big[\big[2\cos\theta-(\kappa-1)\big]\sin\theta K_I^2 + 2\big[2\cos2\theta-(\kappa-1)\cos\theta\big]K_I K_{II}$$

$$+\big[(\kappa-1-6\cos\theta)\sin\theta\big]K_{II}^2\Big] = 0 \tag{8.19a}$$

$$\frac{\partial^2 S}{\partial^2 \theta} = \frac{1}{\pi}\Big[\big[2\cos2\theta-(\kappa-1)\cos\theta\big]K_I^2 + 2\big[(\kappa-1)\sin\theta-4\sin2\theta\big]K_I K_{II}$$

$$+\big[(1-\kappa)\cos\theta-6\cos2\theta\big]K_{II}^2\Big] > 0 \tag{8.19b}$$

Figure 8.8 shows the values of θ_c, the angle of predicted initial fracture propagation direction for various values of β by the SED factor theory for the same problem as described in Example Problem 8.2. The predicted angles of fracture from the SED theory are comparable to those from the maximum circumferential stress criterion for this geometry. Note that the SDE factor depends on the Poisson's ratio, but not strongly.

In summary, the following can be concluded about the various mixed mode fracture theories.

- All three theories, (a) based on maximum energy available for crack extension, (b) maximum circumferential stress criterion, and (c) SED factor predict the stress at fracture and the direction of initial fracture.
- All theories are consistent in their predictions for fracture in pure Mode I. There is considerable similarity in the prediction of fracture stress and direction of initial crack propagation between the three theories when Mode I conditions dominate.
- For cracks inclined at an angle β in a semi-infinite panel loaded by a remote normal stress, the three theories predict that the initial fracture will occur in a direction normal to the applied remote stress if Mode I loading conditions dominate, such as when $\beta > 40°$. The predictions diverge for $\beta < 40°$.
- The maximum circumferential stress theory is the most straight-forward of the three theories and therefore the preferred theory for use in engineering applications.

Example Problem 8.3

Figure 8.9 shows the geometry of a semi-infinite panel of a structural steel with an edge crack that is loaded with a remote uniform tensile stress, σ, and an in-plane uniform shear stress, τ. If the K_{Ic} of the steel is 80 MPa$\sqrt{m}$ and the crack size is 0.01 m, estimate the angle θ_c at which the fracture will initiate and the normal fracture stress, σ_c, for $0 \le \dfrac{\tau}{\sigma} \le 0.5$ at intervals of 0.05. Assume that the elastic modulus, $E = 210$ GPa and the Poisson's ratio is 0.3. The K in the two modes for this geometry are given by:

$$K_I = 1.12\sigma\sqrt{\pi a} \quad \text{and} \quad K_{II} = 1.12\tau\sqrt{\pi a}$$

Solution

From the K-expressions for the two modes, it follows that:

$$\frac{K_{II}}{K_I} = \frac{\tau}{\sigma}$$

Substituting the value of K_{II}/K_I from the above equation into equation (8.13) sets up a relationship between θ_c and τ/σ that is given in the Table 8.2 and shown in Figure 8.10a. Next, we can rewrite equation (8.14) in the following form:

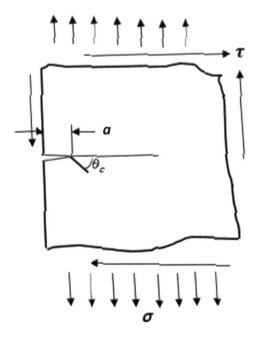

FIGURE 8.9   A semi-infinite panel with an edge crack of length a subjected to remote uniform normal and shear stress causing mixed mode loading conditions at the crack tip.

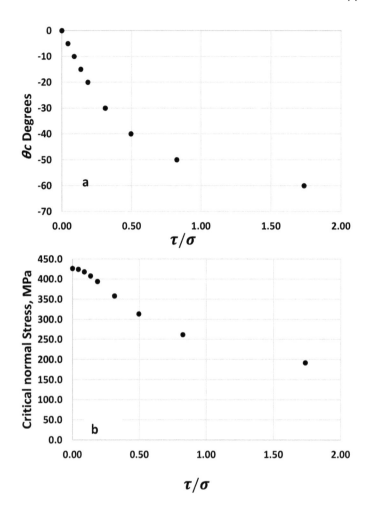

FIGURE 8.10 (a) The angle of initial fracture from the crack plane as a function of mode mixity, and (b) critical fracture stress as a function of mode mixity, in Example Problem 8.3.

$$\sigma_c = \frac{4K_{Ic}}{1.12\sqrt{\pi(0.01)}\left(3\cos\dfrac{\theta_c}{2}+\cos\dfrac{3\theta_c}{2}\right)-3\dfrac{\tau}{\sigma}\left(\sin\dfrac{\theta_c}{2}+\sin\dfrac{3\theta_c}{2}\right)}$$

For the given value of K_{Ic} of 80 MPa$\sqrt{m}$, the value of σ_c can be calculated for various values of $\dfrac{\tau}{\sigma}$ or K_{II}/K_I. The results are listed in Table 8.2 below and are plotted in Figure 8.10b.

These results show that as $\dfrac{\tau}{\sigma}$ or K_{II}/K_I increases, the normal stress required for fracture decreases and the theory can predict the angle at which the initial crack is expected to grow.

TABLE 8.2

Influence of Mode Mixity $\left(\dfrac{K_{II}}{K_I}\right)$ on the Initial Angle of Fracture and the Fracture Stress Using the Maximum Circumferential Stress Theory

$\dfrac{\tau}{\sigma} = \dfrac{K_{II}}{K_I}$	θ_c (degrees)	σ_c (MPa)
0.00	0	426.4
0.04	−5	424.3
0.09	−10	418.0
0.14	−15	407.8
0.19	−20	394.1
0.31	−30	358.1
0.50	−40	313.5
0.83	−50	261.8
1.74	−60	191.8

8.4   FRACTURE TOUGHNESS MEASUREMENTS UNDER MIXED-MODE CONDITIONS

Fracture in materials in which the fracture toughness and elastic properties vary with direction, occurs inherently under mixed mode conditions. The direction of initial fracture in these materials is often dictated by planes along which the fracture toughness is low. It is thus necessary to document the direction of primary loading and the orientation of the crack plane to characterize the fracture toughness of these materials, using the nomenclature recommended by ASTM [8]. The various orientations of the fracture specimens are referenced to a characteristic direction of the product. For example, for a plate, L is the direction of rolling, T is the transverse width direction, and S is the short-transverse direction along the thickness of the plate. For cylindrical material forms, C is the circumferential direction and R the radial direction as seen in Figure 8.13b. The fracture specimen orientation is represented by two letters. For example, in the LT orientation, the first letter, L, refers to the direction of the applied load and the second letter, T, to the direction of crack propagation. Other examples of designation of orientations are shown in Figure 8.11a and b.

8.4.1   FRACTURE IN BONES

Human cortical bone is an example of materials that are inherently brittle and contain defects in the form of porosity and clusters of porosity. Ceramics and concrete are examples of similar man-made materials that are used only in compressive load applications. Unlike metals, brittle materials have intrinsically low fracture toughness because of lack of mechanisms by which plastic deformation can occur. Plastic deformation is attributed to the mobility of dislocations that are present in all crystalline materials. In brittle materials, the dislocations, though present, are unable to glide on planes in which they reside and therefore cannot contribute to toughness of these materials through plastic deformation. The sources of toughness in these

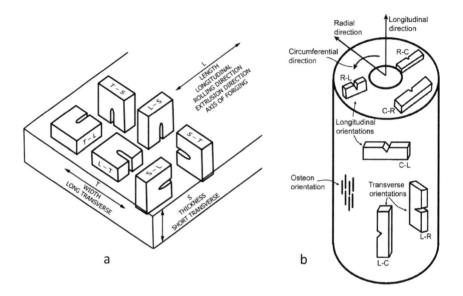

FIGURE 8.11 Fracture specimen orientation terminology (a) for plates and (b) for cylindrical forms of material such as rods and round shafts [8] (Reproduced with permission from ASTM, Conshohocken, PA).

materials are engineered extrinsically by reinforcing them with glass, metal, polymeric, or ceramic fibers. For example, bones are reinforced by fibers called osteons that exist naturally as part of the bone structure, and concrete is reinforced by steel rods.

Microstructural features of the bone participate in providing toughness to the bone. Intrinsic toughening mechanisms include crack deflection at interfaces between microstructural constituents, bridging between micro-cracks via uncracked ligaments, and formation of multiple microcracks that diffuse the intensity of stresses associated with a dominant singular crack tip, as pointed out by Launey, Beuhler, and Ritchie [10]. These phenomena occur ahead of the crack tip. Other extrinsic toughening mechanisms such as fiber bridging occurs behind the crack tip. The "pseudo" fracture toughness of the cortical bone depends on these microstructural factors that have considerable variability from person to person and because bone is a living tissue, it changes and degrades as part of aging [11,12]. Thus, fracture toughness measurement procedures to quantify the effects of aging on bone properties are useful.

For reporting fracture toughness, the long direction of the bone is designated L that is also the direction of fiber orientation and the direction in which the elastic modulus, strength, and toughness are all expected to be the highest. The radial and circumferential directions are designated as R and C, respectively as shown in Figure 8.11b. Ritchie et al. [11] and Nalla et al. [12] have reported fracture toughness in the LC, CL, and CR orientations as $5.3, 3.5, 2.2\,\text{MPa}\sqrt{m}$, respectively. It is noted that the fracture resistance in the two transverse directions, LR and LC are higher because the crack must cut through fibers or osteons, oriented in the longitudinal direction.

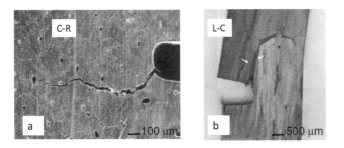

FIGURE 8.12 Fracture propagation in human cortical bone in specimens oriented (a) in the C-R orientation and (b) in the L-C orientation. In both photomicrographs, the crack direction tends to change into the L direction so it can avoid fracturing the osteons (Ritchie et al. [11] and Nalla et al. [12], Reproduced with permission of John Wiley).

Comparatively, the toughness is lower in the CL and the RL directions because the crack can propagate along the fiber directions without having to cut through the osteons. This directional dependence of fracture resistance in bones favors the initial crack propagation to occur along the longitudinal (*L*) direction that is parallel to the osteon fibers as seen in Figure 8.12 [11,12].

In the C-R orientation, the crack is seen to grow nominally under Mode I along its original plane so the fracture toughness measurement of 2.2 MPa$\sqrt{m}$ can be classified as K_{Ic}. The measured fracture toughness in the L-C orientation is a "pseudo value" of K_{Ic} because the crack propagates on a plane different from the original crack plane. It will not meet the requirements for a valid K_{Ic} measurement. Such measures of fracture toughness are only good as qualitative measures and cannot be characterized as valid K_{Ic}, a term which has a precise meaning and requires several conditions to be met, including the requirement that the fracture proceed on the plane of the original crack.

From the results of the test in LC orientation in which the dominant crack growth is in the *L* orientation and the crack growth has the appearance of an in-plane shear fracture, no assessments about K_{IIc} can be made because to measure K_{IIc} the crack must grow under the influence of dominantly Mode II conditions and in the plane of the original crack. This is discussed in more detail in the next section.

8.4.2 Measurement of Fracture Toughness under Mode II (K_{IIc})

In brittle materials such as concrete, rocks, and marble, fracture resistance under Mode II shear loading is significantly higher in comparison to Mode I tensile loading. This makes it difficult to measure K_{IIc} because often under mixed mode loading conditions, the K_I exceeds K_{Ic} while K_{II} remains below K_{IIc}, so the fracture occurs in Mode I. Thus, we need special specimens that create loading conditions in which the K_{II}/K_I is consistently high to promote a Mode II fracture. The shear box specimen is one specimen shown in Figure 8.13 [13] that enables such loading. Mode I is only relevant if it opens the crack, so a compressive load does not contribute to Mode I loading.

In the shear box specimen proposed by Rao et al. [13], Figure 8.13a, the specimen is placed between two beveled dies, which can be set at any inclination angle,

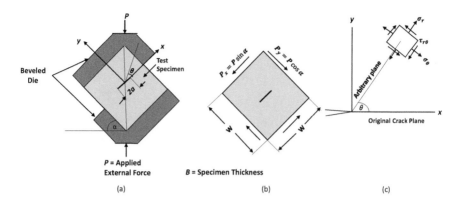

FIGURE 8.13 Shear box test for determining Mode II fracture toughness (a) specimen and loading configuration, (b) resultant loads, and (c) crack tip coordinate system.

α, with the horizontal. A single center crack of length $2a$ is introduced by machining in a square specimen with the planar width, W, and a thickness, B, as shown in Figure 8.13b. The loading normal to the crack plane is compressive to suppress the magnitude of the Mode I stress intensity parameter. Using the coordinate system shown in Figure 8.13c, P_x, the component of the force, P in the direction of the crack plane and P_y, the component of the force in the direction normal to the crack, and the corresponding remote normal and shear stresses they generate in the specimen can be written as:

$$P_x = P\sin\alpha \quad \text{and} \quad \tau_{xy} = \frac{P}{A}\sin\alpha \tag{8.20a}$$

$$P_y = -P\cos\alpha \quad \text{and} \quad \sigma_y = -\frac{P}{A}\cos\alpha \tag{8.20b}$$

In equation (8.20), the area, $A = B(W - 2a)$, the area of the specimen that bears the force, and it is assumed that the frictional force on the specimen is negligible in comparison to the shear force due to the external force. The Mode I and Mode II stress intensity parameters for this geometry as a function of θ are given by [13,14]:

$$K_I(\theta) = \frac{P}{B(W - 2a)}\sqrt{\pi a}$$

$$\times \left[-\cos\alpha\,\cos^3\left(\frac{\theta}{2}\right) - 3\sin\alpha\,\sin\frac{\theta}{2}\cos^2\left(\frac{\theta}{2}\right) + \frac{3\sin\alpha}{2(1 - 2a/W)}\right] \tag{8.21}$$

$$K_{II}(\theta) = \frac{P}{B(W - 2a)}\sqrt{\pi a}$$

$$\times \left[-\cos\alpha\sin\frac{\theta}{2}\cos^2\left(\frac{\theta}{2}\right) + \sin\alpha\cos\frac{\theta}{2}\left(1 - 3\sin^2\frac{\theta}{2}\right)\right] \tag{8.22}$$

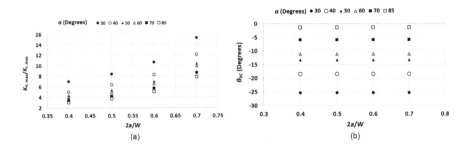

FIGURE 8.14 (a) Ratio of $K_{II,max}/K_{I,max}$ as function of crack size and the angle of inclination, α, and (b) the angle, θ_{IIC} from the original plane of the crack where $K_{II,max}/K_{I,max}$ is the highest for a given α and $2a/W$ value.

The maximum values of $K_I(\theta)$ and $K_{II}(\theta)$ can be determined from the following equations:

$$\frac{\partial K_I(\theta)}{\partial \theta} = 0, \quad \frac{\partial^2 K_I}{\partial \theta^2} < 0 \qquad (8.23a)$$

$$\frac{\partial K_{II}(\theta)}{\partial \theta} = 0, \quad \frac{\partial^2 K_{II}}{\partial \theta^2} < 0 \qquad (8.23b)$$

The ratio $K_{II,max}/K_{I,max}$ can be calculated, and its values are plotted as a function of α and $2a/W$ in Figure 8.14a. Similarly, the value of θ_{IIC}, the angle at which K_{II} is the highest is also calculated and plotted as a function of α and $2a/W$ in Figure 8.14b.

To promote Mode II fracture, $K_{II,max}/K_{I,max}$ must be high, and θ_{IIC} must be nearly $0°$ so the crack grows along its original plane. From Figure 8.14a and b, these conditions are achieved when $60° \le \alpha \le 85°$ and $2a/W > 0.5$. At $\alpha = 60°$ and $2a/W = 0.5$, the angle θ_{IIC} is about $-10°$ and $K_{II}/K_I > 5$. These conditions will favor Mode II fracture.

Two more variations of the shear box specimen are the single notch shear box (SNSB) and double notch shear box specimen (DNSB) that have also been used for determining the Mode II fracture toughness. These are shown in Table 8.3. The expressions for determining K-values for Mode II fracture toughness are [14]:

$$K_{II} = \frac{P\sin\alpha}{B\sqrt{W}} F_1(a/W) \quad \text{for SNSB specimen} \qquad (8.24)$$

$$K_{II} = \frac{P\sin\alpha}{B\sqrt{W}} \sqrt{\pi(a/W)}F_2(2a/W) \quad \text{for the DNSB specimen} \qquad (8.25)$$

where F_1 and F_2 are listed in Table 8.3 for both specimens. The θ_{IIC} for $65° \le \alpha \le 75°$, was in the range of $5°$. Thus, fractures in these specimens are expected to occur under Mode II conditions. Rao et al. [13] have performed several tests on samples of rocks, marble, and granite and have demonstrated that these specimens yield very close to Mode II fracture toughness values.

TABLE 8.3

K_{II} Calibration Functions for the SNSB and DNSB Specimens [13,14]. W = planar width of the specimen

Specimen Geometry	K_{II}-Calibration Function, F

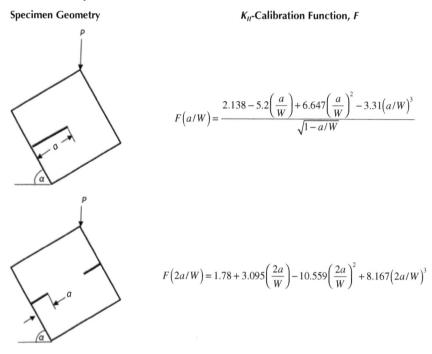

$$F(a/W) = \dfrac{2.138 - 5.2\left(\dfrac{a}{W}\right) + 6.647\left(\dfrac{a}{W}\right)^2 - 3.31(a/W)^3}{\sqrt{1 - a/W}}$$

$$F(2a/W) = 1.78 + 3.095\left(\dfrac{2a}{W}\right) - 10.559\left(\dfrac{2a}{W}\right)^2 + 8.167(2a/W)^3$$

Abbreviations: DNSB, double notch shear box specimen; SNSB, single notch shear box.

TABLE 8.4

Specimen Characteristics and Test Results from Two Specimens of Granite

Specimen ID	a (mm)	α (degrees)	Max Load (kilo-N)	K_{IIc} $\left(\text{MPa}\sqrt{m}\right)$	K_{IIc}/K_{Ic}
A	35	70	86	4.9	2.6
B	42	65	88.6	4.8	2.55

Example Problem 8.4

Single notch shear box specimens from two blocks of granite are tested. For both specimens, $B = W = 70$ mm. The crack growth during fracture occurred at an angle of less than 5° from the original crack plane. The other specimen characteristics and measurements from the tests are in Table 8.4. The K_{Ic} of granite is measured as $1.88\,\text{MPa}\sqrt{m}$. Calculate the value of K_{IIc} and compare it to K_{Ic} of the material and use the result to rationalize that the measurements are accurate values of K_{IIc}.

Solution

Substituting the values of the various variables in equation (8.24), the K_{II} values at fracture can be estimated. These values are listed in Table 8.4 for both specimens. The K_{IIc} values are 2.5–2.6 times that of K_{Ic} that is less than the expected K_{II}/K_I ratio of the specimen which is about 5 as seen in Figure 8.14a. This means that K_{II} reached the critical value before K_I could reach K_{Ic}. Also, a confirmation of that condition is also implied in the result that the fracture proceeded nearly along the original plane of the crack. The values of K_{IIc} obtained from the two tests conducted at different values of α and crack size yield identical values of K_{IIc}. This is another indicator that the measured K_{IIc} is a material property.

8.4.3 Measurement of Interfacial Toughness in Laminate Composites

A common mode of failure in laminated composites is delamination that can be modelled as interlaminar crack propagation of the type shown in Figure 8.2. Laminates are designed to provide high strength and stiffness in the plane of the sheet but delamination between adjacent sheets or planes can occur and limit the flexural strength of the composite beams. Simple beam theory predicts that in a beam subject to bending, shear stress is the highest and the normal stress is zero at the mid-plane of the beam. Thus, delamination under Mode II or mixed mode conditions can potentially occur at or near the mid-plane and limit the load carrying capacity of the beam. Therefore, improving delamination resistance is a goal of composite manufacturers that can be achieved by deploying adhesives that lead to higher fracture resistance.

Mixed-mode fracture mechanics provides an analytical framework for characterizing delamination fracture. There is no reason to believe that the resistance to fracture in adhesive joints in Modes I and II and under a mixture of Modes I and II should be the same. Thus, there is need for developing test methods for accurately measuring not only the critical values of G under pure Mode I, but also under Mode II, and under varying G_{II}/G_I ratios. It is important to state again that because G_I and G_{II} can be added to obtain the overall G, while the corresponding K_I and K_{II} cannot be algebraically added, the use of G is preferred for addressing interlaminar fracture.

In this section, we will consider three test specimens that can be used to characterize the interlaminar toughness in unidirectional composites consisting of high strength and high modulus fibers, such as carbon fibers or glass fibers embedded in a polymeric matrix. The tests are classified as (a) double cantilever beam (DCB) test used for characterizing Mode I fracture toughness, (b) end-loaded split (ELS) test used for characterizing Mode II toughness, and (c) ELS test for mixed-mode fracture toughness.

Figure 8.15 shows a double cantilever beam (DCB) specimen and the applied load to achieve pure Mode I conditions. The specimen consists of a long beam of length, L, thickness, B, and height, $2h$. The beam is fabricated by pressing and curing several laminated unidirectional composite sheets that are bonded together. A starter crack is introduced in the specimen along the direction of the fibers as shown in the figure. The Load, P, is applied on the specimens via blocks that are also adhesively bonded to the specimen as shown in Figure 8.15. The displacement, δ, between the two sides

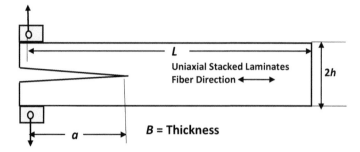

FIGURE 8.15 The double cantilever beam (DCB) specimen used for measuring interfacial fracture toughness of uniaxial composites with the crack propagating in the direction of fibers.

of the crack is measured underneath the load-line to generate the P-δ curves. The crack size is measured visually with the aid of low power telescope and recorded during the test at various points. The test data are analyzed using the following relationships initially derived by Williams [15] and then refined and extended by others [16–18].

Using the simple beam theory that assumes small displacements and beam arms that are long in comparison their height, $2h$, the Mode I value of the crack extension force, G_I, is given by equation (8.26a and b):

$$G_I = \frac{12P^2a^2}{B^2h^3E_{11}}F \qquad (8.26a)$$

Noting that the compliance for this specimen, $C = \dfrac{\delta}{P} = \dfrac{8a^3}{Bh^3E_{11}}$ and substituting it into equation (8.26a) yields an alternate expression for G_I as follows:

$$G_I = \frac{3}{2}\frac{P\delta}{Ba}F \qquad (8.26b)$$

E_{11} = elastic modulus of the substrate material in the longitudinal direction, δ is the total displacement between the points of load application that is measured during the test, and F is a correction factor that accounts for (a) behavior that deviates from predictions of the simple beam theory and (b) estimates the effective crack length that accounts for damage and deformation ahead of the crack tip, similar to the plastically adjusted crack length described in Chapter 4. In general, equation (8.26b) is expected to be more accurate because it is based on directly measured quantities such as P, δ, B, and the crack size, a. The readers are referred to extensive discussion on the accuracy of G estimates from several techniques in references [15–18]. For our purposes here, we will assume $F \approx 1$.

Typical results from tests conducted on unidirectional composites using carbon fibers embedded in three matrix materials, (a) PEEK (ether-ether Ketone), (b) PES (ether sulphone), and (c) epoxy-resin are shown in Figure 8.16 from the work of Hashemi, Kinloch and Williams [16]. The dark circular dots in the P-δ diagram are the points of crack initiation determined visually. Considerable amount of stable

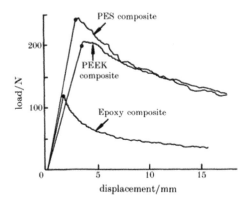

FIGURE 8.16 Load versus load-line displacements from double cantilever beam (DCB) specimens from three types of composites (Figure taken from Hashemi, Kinloch, and Williams [16] reproduced with permission).

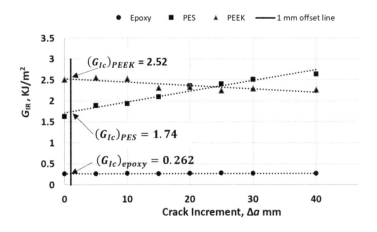

FIGURE 8.17 Mode I crack growth resistance of the three composites (Data taken from reference [16]).

crack growth was observed in all specimens. These data were reduced to obtain G_{IR} versus crack extension relationships for the three adhesives, as shown in Figure 8.17, where G_{IR} is the Mode I Griffith's fracture energy per unit area of crack extension. Since the crack extension is stable, one can assume that the applied value of $G_I = G_{IR}$, the crack growth resistance.

The G_{IR} values for the epoxy resin and PEEK composites seem to be constant with crack extension, Δa, but for the PES composite there is a tendency similar to "R-curve" behavior observed in metals; in the latter, the resistance to fracture increases with stable crack growth. In metals, the increase in crack growth resistance is due to energy spent to overcome strain hardening, and in extending the plastic zone to accommodate the higher K level. Viscous polymers undergo analogous energy dissipative processes that can lead to similar behavior. In the glassy

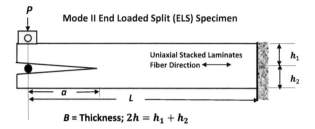

FIGURE 8.18   End-loaded split (ELS) specimen geometry used for measuring Mode II interlaminar fracture resistance.

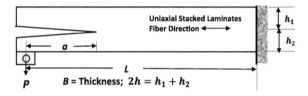

FIGURE 8.19   End-loaded split (ELS) specimen configured for measuring inter-laminar fracture resistance mixed-mode loading.

state, such dissipative processes are unavailable, so the stable crack growth resistance tends to be flat.

For measuring G_{IIc} and fracture resistance under mixed mode conditions, a specimen called the ELS, Figure 8.18, is used [16]. A small roller is placed between the two halves of the specimen as shown in the figure. Both arms are bent in the same sense. If $h_1 = h_2$, the load is shared equally between the two arms that makes $G_I = 0$, and G_{II} is given by the following equation [16,18]:

$$G_{II} = \frac{9P\delta}{2Ba}\left[\frac{a^3}{3a^3 + L^3}\right] \qquad (8.27)$$

The above formula can be modified to account for presence of damage in front of the crack tip [16] to get more accurate expressions for G_{II}.

For mixed-mode loading, the specimen geometry used for conducting the tests is shown in Figure 8.19. For this test configuration, the values of the various components of G are given by equations (8.27) [16].

$$G_I = \frac{6P\delta}{Ba}\left[\frac{a^3}{7a^3 + L^3}\right] \qquad (8.28a)$$

$$G_{II} = \frac{9P\delta}{2Ba}\left[\frac{a^3}{7a^3 + L^3}\right] \qquad (8.28b)$$

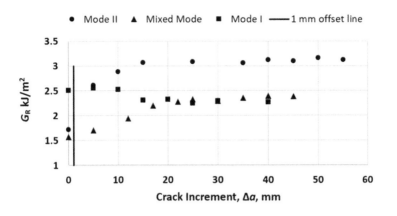

FIGURE 8.20 A comparison between the interlaminar fracture resistance, G_R, as a function of crack extension for a carbon fiber composite utilizing the PEEK matrix subjected to Mode I, Mode II and mixed-mode loading (The data were taken from Hashemi, Kinloch and Williams [16]).

$$G_R = G_I + G_{II} = \frac{21P\delta}{2Ba}\left[\frac{a^3}{7a^3 + L^3}\right] \qquad (8.28c)$$

The mode-mixity in this case, G_I/G_{II}, is given by equation (8.29). It can be varied as needed by choosing different values of h_1 and h_2. For $h_1 = h_2$, $G_I/G_{II} = 1.33$.

$$\frac{G_I}{G_{II}} = \frac{1}{3}\left[\frac{h_1}{h_2}\left(1 + \frac{h_1}{h_2}\right)\right]^2 \qquad (8.29)$$

Figure 8.20 shows the relationship between G_R and Δa for the PEEK carbon fiber composite under Modes I, II and mixed mode with a G_I/G_{II} value of 1.33 corresponding to the condition of $h_1 = h_2$ using data provided in reference [17]. The G_{Ic}, G_{IIc}, and G_c for the condition of $G_I/G_{II} = 1.33$ are 2.52, 2.0, and 1.6 kJ/m² respectively. These are plotted in Figure 8.21 showing that the highest resistance to fracture is under Mode I and it decreases with mode-mixity and then increases again as we move to pure Mode II conditions. This behavior is specifically for PEEK composite and should not be generalized. Ideally, such trends should be derived from multiple tests conducted under several mixed-mode conditions.

Example Problem 8.5

The Mode II fracture resistance of a rubber toughened film adhesive on a carbon fiber composite substrate is being characterized using a Mode II ELS specimen that has a width of 20 mm, a length of 130 mm, and $h = 2$ mm. The initial crack length in the specimen is 65 mm. The measured load, displacement and crack length of the test is given in Table 8.5. Construct the G_{II} fracture resistance curve as a function of crack growth and determine the G_{IIc} value of the adhesive bond.

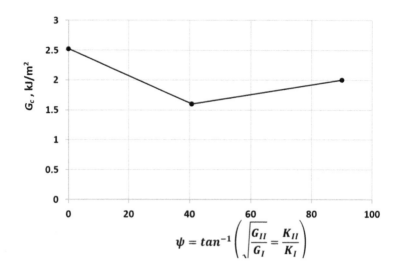

$$\psi = tan^{-1}\left(\sqrt{\frac{G_{II}}{G_I}} = \frac{K_{II}}{K_I}\right)$$

FIGURE 8.21 Inter-laminar fracture resistance of carbon fiber/PEEK composite as a function of mode-mixity.

TABLE 8.5

Data from an ELS Tests Conducted on a Joint Using a Rubber Toughened Film Adhesive on a Carbon Fiber Composite Substrate

P (N)	a (mm)	δ (mm)	G_{IIR} (J/m²)
0	65	0	0
270	65	22	1,869
275	67	23	2,099
300	68	25	2,485
315	70	28	3,052.8
355	75	33	4,245.8
340	80	35	4,590.9
310	90	38	4,897.1
275	100	40	4,762.4

Abbreviations: ELS, end-loaded split

Solution

The load-displacement behavior from the test is plotted in Figure 8.22. From equation (8.27), the G_{IIR} values can be calculated for various values of δ and crack length. These are also listed in Table 8.5 and are plotted as a function of crack extension in Figure 8.23.

The value of G_{IIC} is determined from the plot of the $G_{II,R}$ versus Δa in Figure 8.23 using the data in the range of $0 \leq \Delta a \leq 5$ mm. A straight line is fitted through the data in this interval by regression to obtain the following relationship:

$$G_{IIR} = 1770.3 + 242.4 \Delta a$$

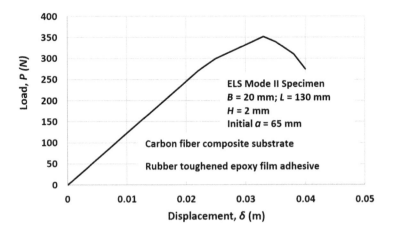

FIGURE 8.22 Load versus displacement diagram from a Mode II ELS test conducted on an adhesively bonded specimen (Data taken from reference [18]).

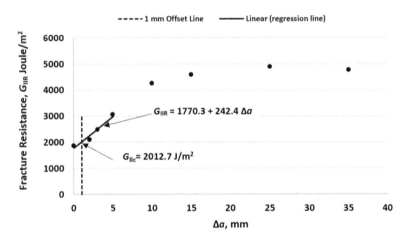

FIGURE 8.23 Fracture resistance, G_{IIR}, as a function of crack extension from a Mode II ELS test conducted on an adhesively bonded specimen.

The G_{IIR} is in J/m² and Δa is in mm. A 1 mm offset line is drawn to determine the value of G_{IIR} corresponding to 1 mm of crack extension. This value is 2,017.7 J/m² and is labelled as G_{IIC}.

8.5 FATIGUE CRACK GROWTH UNDER MIXED-MODE LOADING

Fatigue cracks subjected to mixed-mode loading propagate in a nonself-similar manner because the direction of their growth constantly changes due to changes in ΔK_I, ΔK_{II}, and their ratio $\Delta K_I/\Delta K_{II}$. This is a complex analytical and experimental problem for which adequate experimental data to guide the theoretical developments are limited. Extensions of theories presented in earlier sections of this chapter for predicting

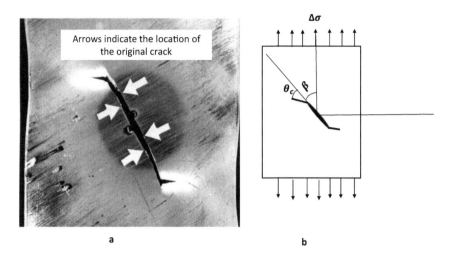

a   b

FIGURE 8.24   (a) Growth of fatigue cracks in panels of an aluminum alloy subjected to uniform cyclic tensile stress containing an inclined crack showing the direction of crack growth (Photograph from the work of Tanaka [19] reproduced with permission). (b) Schematic of the angle of inclination and the angle of crack growth.

fracture under monotonic loading are considered here for predicting fatigue crack growth. These theories are limited in their application to materials that have been used to develop the data while widely accepted and comprehensive theories for predicting mixed-mode fatigue crack growth rates have yet to emerge.

One of the earliest experimental investigation of mixed-mode fatigue crack growth behavior was conducted by Tanaka [19] in which he tested panels of aluminum alloy under uniaxial loading but containing inclined cracks. He measured the crack growth rates and the angle of growth of cracks as shown in Figure 8.24. The tests were conducted at high load ratio (0.65) to avoid rubbing between crack faces. The arrows in the picture point to where the initial crack was located on its two faces. The displacements in the in-plane shear direction between the crack faces is indicative of the presence of Mode II loading. This is even more obvious when one sees the displacement between the two halves of the hole in the center of the specimen.

The maximum circumferential stress theory described in this chapter has been used [19,20] to predict the initial direction of crack propagation. To examine the results from theory, the data collected from Tanaka's experiments are summarized in Table 8.6.

The predicted and actual initial angles of crack propagation, θ_c, as a function of β, the inclination angle, are shown in Figure 8.25. It shows good agreement for $\beta > 45°$ when Mode I is dominant. For smaller angles, the differences between predicted and observed angles increase indicating that the maximum principal stress theory may not represent the behavior. Further examination of the crack tip region in Figure 8.24a shows that there is considerable shear deformation in the crack tip region of this specimen tested with $\beta = 30°$, indicating that the dominant mode maybe transitioning to Mode II where the maximum circumferential stress criterion does

TABLE 8.6

Data Reported by Tanaka [20] from Tests Conducted on Al Alloy Panels

β (degrees)	θ_c (degrees)	ΔK_I (MPa$\sqrt{m}$)	ΔK_{II} (MPa$\sqrt{m}$)	da/dN (mm/cycle)
90	0	2.02	0	5.8×10^{-7}
90	0	1.98	0	7.9×10^{-7}
72	−28	1.76	0.577	7.4×10^{-7}
45	−49	1.18	1.18	7.2×10^{-7}
30	−23	0.85	1.47	1.4×10^{-6}
30	−52	0.775	1.34	2.1×10^{-7}

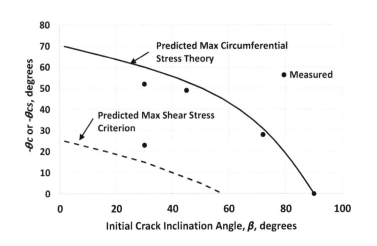

FIGURE 8.25   Plot of the predicted initial angle of crack growth using the maximum circumferential stress criterion (solid line) and the max shear stress criterion (dashed line) as a function if the angle of inclination. The plot also includes the measured angles from tests performed by Tanaka [19].

not apply. Since the shear yield strength is approximately half of the tensile yield strength, shear deformation is expected to accumulate rapidly. The direction of the maximum shear stress is at 45° to the principal stress direction. Thus, based on maximum shear stress direction, the predicted crack growth direction will be at an angle, θ_{cs}, given by equation (8.30) and plotted as the dashed line in Figure 8.25.

$$\theta_{cs} = \theta_c + 45 \tag{8.30}$$

The data and the picture shown in Figure 8.24 support the proposal that for $\beta > 45°$, the fatigue crack growth is dominated by Mode I and then transitions to Mode II domination when $\beta < 45°$ with the transition completing when $\beta < 30°$.

Yan, Du, and Zhang [20] used the maximum principal stress theory to predict the direction of continued crack growth using a specially developed software and concluded the following.

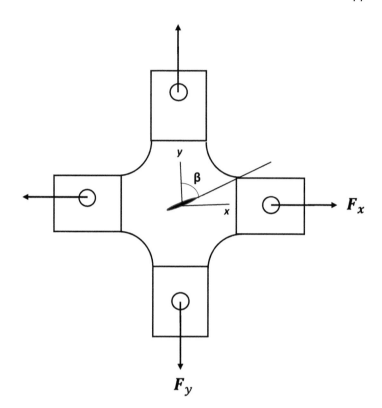

FIGURE 8.26   Specimen used for biaxial fatigue loading.

- Under uniaxial loading, the fatigue crack growth rapidly tends to grow in a direction normal to the applied loading even though it may initially grow at an angle θ_c in Figure 8.24b.
- Under equi-biaxial loading, $F_x = F_y$, (see Figure 8.26 for a specimen used for applying biaxial loading), the cracks are predicted to initiate and continue to grow in their original plane regardless of the value of β.

Predicting the rates of fatigue crack growth under mixed-mode loading are complex, so by necessity the approaches used to accomplish that rely heavily on experimental data. Experiments conducted by Tanaka [19], previously described, to guide the development of mixed-mode fatigue crack growth rate theories were a significant step forward. He proposed that one way to account for the effect of mode-mixity on the rate of fatigue crack growth was to define an effective stress intensity parameter range, ΔK_{eff}, that includes terms involving both ΔK_I and ΔK_{II} to replace its Mode I counterpart, ΔK in the Paris law. Thus, the Paris fatigue crack growth law for mixed-mode loading is written as in equation (8.31).

$$\frac{da}{dN} = C\left(\Delta K_{eff}\right)^n \tag{8.31}$$

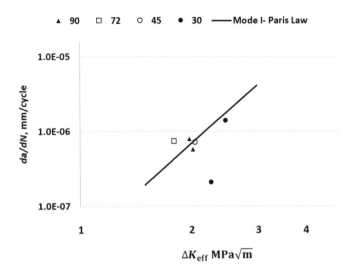

FIGURE 8.27 Mixed-mode fatigue crack growth behavior in an Al alloy correlated with ΔK_{eff} using data from Tanaka [19].

where C and n are constants in the Paris-law for Mode I. To formulate ΔK_{eff}, Tanaka proposed that one can sum the magnitude of crack tip plastic displacements along the direction of the crack attributed to the two modes. He further assumed that the magnitudes of the displacements attributed to the two modes are independent and can be algebraically added. He also assumed that the yield strength of isotropic metals under shear is half of the yield strength under tensile loading and used the previous estimates of the extent of plasticity at the crack tip provided by Bilby, Cottrell, and Swinden [21] and dislocation pileup analysis of Weertman [22] to estimate plastic displacements, to propose the following definition of ΔK_{eff}, equation (8.31).

$$\Delta K_{eff} = \left[\Delta K_I^4 + 8\Delta K_{II}^4 \right]^{1/4}$$ (8.31)

Alternate more complex forms of equation (8.31) that include cross-terms between ΔK_I and ΔK_{II} have also been formulated [20,23] but none have been either experimentally proven to be significantly better than the simple equation (8.31) for predicting mixed-mode fatigue crack growth rates, or have provided a more appealing theoretical justification. Figure 8.27 plots the data summarized in Table 8.6 using equation (8.31) for estimating ΔK_{eff}. With the exception of one data point for $\beta = 30°$, the data are in good agreement with the Mode I Paris law shown in the same figure. The data in the figure is quite limited in the crack growth rates covered and values of β. The one point that lies outside the trend is from the same sample in which the crack growth direction varied significantly from predictions based on maximum circumferential stress theory in Figure 8.25.

Fremy et al. [24] have also pointed out that the validity of the above approach is limited to dominantly linear elastic conditions and when the loading path during cyclic loading is proportional. In other words, it is assumed that the ratio K_{II}/K_I remains constant during the various portions of the fatigue cycle. This is of particular concern when the cyclic loading includes thermal-mechanical stresses. These researchers have provided difficult to obtain fatigue crack growth data that will be useful in the future for developing concepts to address this need.

8.6 SUMMARY

In this chapter, we have considered fracture mechanics approaches for addressing fracture and crack growth problems under the conditions of mixed-mode loading. Mode I is the most prevalent condition under which fracture, and crack growth occurs in structural components but in some cases, the consideration of the influence of Mode II and Mode III loading becomes necessary. The discussion provided in the chapter is limited to mixed Modes I and II to keep the analysis at the level of this textbook and because it also addresses the most common of the mixed-mode fracture and crack growth problems encountered in engineering structures.

The crack tip stress field equations were derived for combined Modes I and II loading under linear elastic conditions. It was noted that the stress intensity parameters K_I and K_{II} cannot be algebraically added to characterize the crack tip stress fields, but the various components of stresses produced by loading in each mode can be summed by the principle of linear superposition. On the other hand, Griffith's energies available for crack extension for different modes can be algebraically summed to estimate the total energy available for crack extension. Three criteria for predicting crack growth and fracture have been discussed to characterize the direction of initial crack growth and fracture under mixed-mode conditions. These are:

- Maximum strain energy release rate in which the fracture is said to occur when the total energy available for crack extension reaches a critical value.
- The maximum circumferential stress criterion predicts that the initial fracture and crack growth occurs in the direction normal to the direction of maximum principal stress at the crack tip. Fracture occurs in that direction when the normal circumferential stress equals the normal stress for fracture in Mode I.
- The minimum SED criterion assumes that fracture and crack growth occur in the direction in which the SED is minimum and when the strain energy factor, S, defined as the amplitude of the singular crack tip SED distribution, approaches a critical value, S_c.

Among the three criteria, the maximum circumferential stress criterion, because of its simplicity and direct relation to crack tip stress, has received the most attention. This criterion is used in the rest of the chapter to address mixed-mode fracture in homogeneous and isotropic materials, and in anisotropic materials such as human bones and laminated composites that contain interfaces that are subject to combined Mode I and II loading. Test methods are described for characterizing interfacial fracture under pure Mode I and pure Mode II, and mixed Mode I and II conditions.

The chapter concludes with a discussion of fatigue crack growth under mixed-mode conditions. The initial direction of crack growth is shown to be predicted by the maximum circumferential stress theory if the loading consists of $\Delta K_{II}/\Delta K_I$ that is less than one. When shear loading dominates, the theory is shown to not be reliable. A definition of effective cyclic stress intensity parameter range, ΔK_{eff}, that combines contributions from ΔK_I and ΔK_{II} in a manner that sums the plastic displacements at the crack tip due to the two modes is used to consolidate mixed-mode fatigue crack growth data. Again, this model works well when $\Delta K_{II}/\Delta K_I$ are less than one. For other values of $\Delta K_{II}/\Delta K_I$ more advanced treatments outside the scope of this book are needed [24,25].

REFERENCES

1. T.L. Anderson, *Fracture Mechanics: Fundamentals and Applications*, 2nd Edition, CRC Press, LLC, Boca Raton, FL, 1995.
2. H. Tada, P.C. Paris, and G.R. Irwin, *The Stress Analysis of Cracks Handbook*, 3rd Edition, American Society for Mechanical Engineers, New York, NY, 2000.
3. D. Olvera, E.A. Zimmermann, and R.O. Ritchie, "Mixed-Mode Toughness of Human Cortical Bone Containing a Longitudinal Crack in Far-Field Compression", *Bone*, Vol. 50, 2012, pp. 331–336.
4. S. Weiner and H.D. Wagner, "The Material Bone: Structural-Mechanical and Functional Relations", *Annual Review of Material Science*, Vol. 28, 1998, pp. 271–298.
5. F. Erdogan and G.C. Sih, "On Crack Extension in Plates Under Plane Loading and Transverse Shear", *Journal of Basic Engineering, Transactions of ASME*, Vol. 85D, 1963, pp. 519–527.
6. E.E. Gdoutos, *Fracture Mechanics: An Introduction*, Kluwer Academic Publishers, Dordrecht/Boston/London, 1993.
7. G.C. Sih, "Energy Density Concept in Fracture Mechanics", *Engineering Fracture Mechanics*, Vol. 5, 1973, pp. 1037–1040.
8. G.C. Sih, "Strain Energy Density Factor Applied to Mixed-Mode Crack Problems", *International Journal of Fracture*, Vol. 10, 1974, pp. 305–321.
9. Standard Terminology Related to Fatigue and Fracture Testing, ASTM Standard E 1823-21, ASTM International, 100 Bal Harbor Drive, West Conshohocken, PA, 2021.
10. M.E. Launey, M.J. Beuhler, and R.O. Ritchie, "On the Mechanistic Origins of Toughness in Bone", *Annual Review of Materials Research*, Vol. 40, 2010, pp. 25–53.
11. R.O. Ritchie, J.H. Kinney, J.J. Kruzic, and R.K. Nalla, "A Fracture Mechanics and Mechanistic Approach to the Failure of Cortical Bone", *Fatigue and Fracture of Engineering Materials and Structures*, Vol. 28, 2005, pp. 345–371.
12. R.K. Nalla, J.S. Stolken, J.H. Kinney, and R.O. Ritchie, "Fracture in Human Cortical Bone: Local Fracture Criteria and Toughening Mechanisms", *Journal of Biomechanics*, Vol. 38, 2005, pp. 1517–1525.
13. Q. Rao, Z. Sun, O. Stephansson, C. Li, and B. Stillborg, "Shear Fracture (Mode II) of Brittle Rock", *International Journal of Rock Mechanics and Mining Sciences*, Vol. 40, 2003, pp. 355–375.
14. D. Sun, Q. Rao, S. Wang, Q. Shen, and W. Yi, "Shear Fracture (Mode II) Toughness Measurement of Anisotropic Rock", *Theoretical and Applied Fracture Mechanics*, Vol. 115, 2021, 103043.
15. J.G. Williams, "On the Calculation of Energy Release Rates for Cracked Laminates", *International Journal of Fracture*, Vol. 36, 1988, pp. 101–119.

16. S. Hashemi, A.J. Kinloch, and J.G. Williams, "The Analysis of Interlaminar Fracture in Uniaxial Fiber-Polymer Composites", *Proceedings of the Royal Society, London*, Vol. A427, 1990, pp. 173–199.
17. B.R.K. Blackman, A.J. Kinloch, and M. Paraschi, "The Determination of Mode II Adhesive Fracture Resistance, G_{IIc}, of Structural Adhesive Joints: an Effective Crack Length Approach", *Engineering Fracture Mechanics*, Vol. 72, 2005, pp. 877–897.
18. D. Alvarez, F.J. Guild, A.J. Kinloch, and B.R.K. Blackman, "Partitioning of Mixed-Mode Fracture in Adhesively-Bonded Joints: Experimental Studies", *Engineering Fracture Mechanics*, Vol. 203, 2018, pp. 224–239.
19. K. Tanaka, "Fatigue Crack Propagation from a Crack Inclined to the Cyclic Tensile Axis", *Engineering Fracture Mechanics*, Vol. 6, 1974, pp. 493–507.
20. X. Yan, S. Du, and Z. Zhang, "Mixed Mode Fatigue Crack Growth Prediction in Biaxially Stretched Sheets", *Engineering Fracture Mechanics*, Vol. 43, 1992, pp. 471–475.
21. B.A. Bilby, A.H. Cottrell, and K.H. Swinden, "The Spread of Plasticity from a Notch", *Proceedings of the Royal Society A*, Vol. 272, 1963, pp. 304–314.
22. J. Weertman, "Rate of Growth of Fatigue Cracks Calculated from the Theory of Infinitesimal Dislocation Distributed on a Plane", *International Journal of Fracture Mechanics*, Vol. 2, 1966, pp. 460–467.
23. J. Qian and A. Fatemi, "Mixed-Mode Fatigue Crack Growth: A Literature Survey", *Engineering Fracture Mechanics*, Vol. 55, 1996, pp. 969–990.
24. F. Fremy, S. Pommier, M. Poncelet, R. Bumedijen, E. Galenne, S. Courtin, and J.-C.L. Roux, "Load Path Effects on Fatigue Crack Propagation in I+II+III Mixed Mode Conditions—Part I: Experimental Investigations", *International Journal of Fatigue*, Vol. 62, 2014, pp. 102–112.
25. J. Mei, P. Dong, S. Kalnaus, Y. Jiang, and Z. Wei, "A Path-Dependent Fatigue Crack Propagation Model under Nonproportional Modes I and III Loading Conditions", *Engineering Fracture Mechanics*, Vol. 182, 2017, pp. 202–214.

HOMEWORK PROBLEMS

1. Explain why Mode I fracture mechanics analysis and test methods are limited for describing fracture in composite materials such as bones.
2. If we consider a material in which the fracture toughness is lower in shear (Mode II) than in Mode I and is given by:

$$G_c = 11.7 - 7.453\psi$$

   Repeat the calculations in Example Problem 8.1 and discuss the results.
3. If fracture toughness of bones, K_c in pure Mode I is defined by the value of K that results in 1 mm of stable crack extension, the values for a young, middle aged, and aged persons are $1.25, 2.0,$ and $2.4\,\text{MPa}\sqrt{m}$, respectively. Three-point bend specimens consisting of a span of 4 cm, width, $W = 1$ cm, thickness, $B = 1$ cm and edge crack, $a = 3$ mm, are made from the three bone samples. Calculate the load that will be required for stable crack extensions of 1 mm in each of those samples.
4. If the samples in Problem 2 are loaded in tension instead of bending, what is the tensile load that is needed for 1 mm of stable crack extension.

5. A crack in a wide bone of a young person is 3 mm in length and is at an angle of 30° from the horizontal. Using the fracture toughness data given in Figures 8.4 for bones in the transverse direction, compute the tensile stress, σ_y, needed to grow the crack by 1 mm. Assume that the K value in Mode I is approximately given by $K_I = 1.12\sigma\sqrt{\pi a}$ and for Mode II, $K_{II} = 1.12\tau\sqrt{\pi a}$, where σ is the stress normal to crack face (Mode I) and τ is the in-plane shear stress along the crack face (Mode II). From the stress–strain diagram in Figure 8.5, the elastic modulus of bones can be estimated as approximately 5 GPa. The Poisson's ratio is given as 0.25. Compare the stress required to grow the crack by 1 mm for a crack along the horizontal direction or pure Mode I.

6. A plate of steel contains a 4 mm long crack oriented at an angle $\beta = 30°$ from the direction of the applied uniform stress, σ. Using the maximum circumferential stress theory, calculate the value of critical stress for fracture assuming the following material properties: $K_{Ic} = 50\,\text{MPa}\sqrt{m}$, $E = 210\,\text{GPa}$, and $v = 0.3$.

9 Fracture and Crack Growth under Elastic/ Plastic Loading

9.1 INTRODUCTION

When using very ductile materials in applications that require strong assurance against fracture, the limitations of linear-elastic fracture mechanics become clear because in these materials, fracture is accompanied by significant plastic deformation. One such material is 304 austenitic stainless steel used extensively in load bearing applications in nuclear power-plants and in biomedical implants such braces and screws used to stabilize fractures in bones and other prosthetic devices. These materials are also resistant to oxidation and can remain microstructurally stable at elevated temperatures of up to 600°C. The microstructure in ferritic steels begins to change at these temperatures so, they are no longer reliable in such applications. Thus, these ductile materials are of high technological relevance and their fracture toughness and fatigue crack growth properties are of interest.

The estimated yield strength of 304 SS is 207 MPa and the estimated fracture toughness, K_{Ic}, is 300 MPa$\sqrt{m}$. Let us estimate the size of the smallest C(T) specimen required for measuring fracture toughness of 304 SS to illustrate our point about the limitations of linear elastic fracture mechanics (LEFM). To achieve a valid measure of fracture toughness, as described in Chapter 5, the $W-a$, a, and B values must all be larger than the following:

$$W - a, B, a \geq 2.5\left(\frac{300}{207}\right)^2 = 2.1 \text{ m}$$

This is a prohibitively large specimen for testing; the cost of material, the cost of specimen machining, the handling requirements in the laboratory, and the load capacity of test machines required for fracture toughness measurements are all extremely high. Further, these dimensional requirements for maintaining linear elastic conditions are higher than the section thicknesses of pressure vessels and piping components used in nuclear power-plants. Thus, LEFM is inadequate for assessing the integrity of these components during service. There is need for finding alternate analysis and test methods that allow the estimation of fracture toughness of such materials using conventional size specimens and assessments of the integrity of components during service that are made from these materials. Explorations in this field began in the 1960s in response to this need and have resulted in the development of the field of elastic-plastic-fracture-mechanics. An introductory overview of EPFM is provided in this chapter and a detailed treatment of the topic is available elsewhere [1].

DOI: 10.1201/9781003292296-9

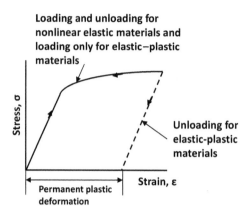

FIGURE 9.1   Nonlinear elastic and elastic-plastic stress–strain behavior.

9.2   RICE'S J-INTEGRAL

We introduce the concept of *J*-integral first proposed by Rice in 1968 [2] as an alternate crack tip parameter. The stress–strain behavior of nonlinear elastic materials and elastic-plastic materials are shown in Figure 9.1. An example of nonlinear elastic material is rubber that can stretch with a nonlinear stress–strain relationship when a load is applied but it returns to its original length when the load is returned to zero following the same path in the reverse as during the increasing load portion of the diagram. Examples of elastic-plastic materials are metals that sustain plastic deformation when stretched beyond their elastic limit resulting in a nonlinear stress–strain behavior but do not return to their original length after the load is removed. Instead, they unload following a linear stress–strain path with a slope equal to the elastic modulus and have a permanent plastic deformation of the magnitude shown in Figure 9.1. However, both materials have similar trends during the loading portion of the stress–strain diagram. The deformation theory of plasticity on which Rice's *J*-integral is based, ideally applies to materials exhibiting nonlinear elastic stress–strain behavior and it assumes that the stress–strain relationship is unique. In other words, for every value of strain there is a unique value for stress during both loading and unloading. The deformation theory of plasticity also applies to elastic-plastic materials if there is no load reversal in the loading history.

The stress–strain behavior in Figure 9.1 during the loading portion of the cycle is modeled using equation (9.1) known as the Ramberg and Osgood equation.

$$\varepsilon = \varepsilon_{el} + \varepsilon_{pl} = \frac{\sigma}{E} + \varepsilon_0 \alpha \left(\frac{\sigma}{\sigma_0} \right)^m \tag{9.1}$$

where $\varepsilon_0 = \dfrac{\sigma_0}{E}$; $\sigma_0 = 0.2\%$ yield strength of the material $= \sigma_{ys}$; α and m are regression constants derived from fitting experimental uniaxial tension stress–strain test data, and E = elastic modulus. ε_{el} and ε_{pl} are the elastic and plastic components of the strain, respectively. The function can be written in a different form with fewer constants as below.

$$\varepsilon = \frac{\sigma}{E} + \frac{\alpha\varepsilon_0}{\sigma_0^m}\sigma^m = \frac{\sigma}{E} + D\sigma^m \tag{9.2}$$

$$D = \frac{\alpha\varepsilon_0}{\sigma_0^m} \tag{9.3}$$

The deformation theory of plasticity assumes that the material is nonlinear elastic; in other words, the unloading path in the stress–strain behavior is identical to the loading path. Elastic-plastic metals behave differently but as mentioned earlier, during the loading portion both behaviors are the same. Therefore, when deformation theory of plasticity is applied to metals, no unloading is permitted. Another assumption implicit in the deformation theory of plasticity is that the resulting strains are small such that the strain-displacement relationships described in Chapter 2 are valid. These assumptions limit the applicability of the theory to conditions where the zone of large strains such as those that might be present in the immediate vicinity of the crack tip where fracture is about to occur, is limited in size.

Rice's J-integral is defined as contour integral, J_Γ or simply J in equation (9.4).

$$J_\Gamma = J = \int \left(W \sin\theta - T_i \frac{\partial u_i}{\partial x} \right) ds \tag{9.4}$$

where Γ is a contour that begins anywhere on the lower crack surface and traverses counterclockwise and ends on the upper crack surface as shown in Figure 9.2. u_i = component of the displacement vector, T_i = component of the traction vector given by equation (9.5) and W = strain energy density given by equation (9.6) and ds is an element along the contour Γ.

$$T_i = \sigma_{ij} \cdot n_j \tag{9.5}$$

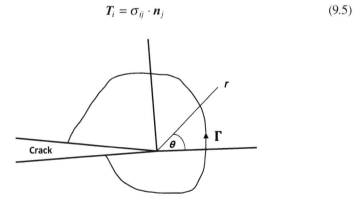

FIGURE 9.2 Crack tip contour used in the definition of the J-integral.

σ_{ij} are various components of the stress tensor, and n_j are the components of the unit direction vector.

$$W = \int_0^{\varepsilon_{ij}} \sigma_{ij} d\varepsilon_{ij} \tag{9.6}$$

Rice showed that J-integral is independent of the path chosen over which it is calculated. This means that the value of J remains the same if it is calculated along a path that is close to the crack tip or a path that reaches far away from the crack tip. He also showed mathematically that the value of J is equal to the strain energy release rate or the rate of change in strain energy with crack extension for nonlinear elastic materials. Thus, if strain energy is designated by U, then the following relationships were derived:

$$J = -\frac{1}{B}\left(\frac{dU}{da}\right)_\Delta = \frac{1}{B}\left(\frac{dU}{da}\right)_P \tag{9.7}$$

Under conditions of linear elasticity, we can show from this definition of J-integral that it is identical to the Griffith's crack extension force, G and we can write the following equation.

$$G \equiv J = \frac{K^2}{E'} \tag{9.8}$$

$E' = E$ for plane stress and is equal to $E/(1-v^2)$ for plane strain. Equation (9.8) makes J-integral a candidate fracture parameter that extends the Griffith's crack extension force concept to the nonlinear regime of material behavior. However, it should be noted that crack extension leads to unloading in the crack tip region, therefore, there must be limitations on the use of J-integral for characterizing crack growth. The use of J for characterizing fracture is limited to when the crack growth is small in comparison to the crack length and the planar dimensions of the specimen. Further, for elastic-plastic materials since the energy associated with plastic strains is not recoverable, the traditional interpretation of G as the crack driving force or the energy available for crack extension, no longer retains its significance. A third limitation comes from the condition that J-integral is only valid for the condition of small deformation. Thus, if the zone of intense deformation becomes significant, the validity of J as a fracture parameter comes into question. To overcome the limitations due unloading, the amount of allowable crack extension must be limited. We can enforce restrictions on the use J only when both zones of unloading and intense deformation are small in comparison to the region where the crack tip stress and strain fields are determined by J.

The crack tip stress fields are uniquely determined by J as shown by the following equations derived by Hutchinson [3] and independently by Rice and

Rosengren [4] and thus referred to in the literature as the HRR crack tip stress fields, equations (9.9).

$$\sigma_{ij} = \sigma_0 \left(\frac{J}{\alpha \sigma_0 \varepsilon_0 I_m r} \right)^{\frac{1}{1+m}} \hat{\sigma}_{ij}(\theta, m) \qquad (9.9a)$$

$$\varepsilon_{ij} = \alpha \varepsilon_0 \left(\frac{J}{\alpha \sigma_0 \varepsilon_0 I_m r} \right)^{\frac{m}{1+m}} \hat{\varepsilon}_{ij}(\theta, m) \qquad (9.9b)$$

In equations (9.9a) and (9.9b), I_m is a constant dependent on the value of the strain hardening parameter, m, and $\hat{\sigma}_{ij}(\theta, m)$ and $\hat{\varepsilon}_{ij}(\theta, m)$ are angular functions that depend on the angle θ and the value of m. Even though J does not characterize the energy available for crack extension, it does uniquely characterize the amplitude of the crack tip stress and strain singularities near the crack tip as per equations (9.9). Therefore, it is still an attractive crack tip parameter for characterizing fracture in elastic-plastic materials but, its use must be confined to the limitations of small amounts of crack extension and a small zone of intense deformation.

9.3 J-INTEGRAL AS A FRACTURE PARAMETER

In 1972, Begley and Landes [5] were first to propose J-integral as a parameter for fracture initiation and limited amounts of stable crack growth in a cracked body under elastic plastic and fully plastic conditions. To experimentally establish the point at which crack initiation occurs requires characterizing at least some amount of stable crack growth and its relationship to applied J. The crack initiation point can be then determined by extrapolation from a trend in which the stable crack growth is correlated to the magnitude of J; the value of J where the process of stable crack growth initiates can thus be determined. Landes and Begley first envisioned the physical process of ductile fracture [6] and is schematically illustrated in Figure 9.3.

The right side of Figure 9.3 shows the various stages of how the initial crack tip represented by Step 1 responds to increasing load by first blunting (Step 2) due to plastic deformation followed by the onset of crack extension (Step 3) and stable crack extension shown in Step 4. The corresponding J versus Δa relationship is shown on the left side of Figure 9.3. The J-resistance during crack blunting is seen in the initial linear portion of the curve with a higher slope and the J-controlled stable crack extension Step 4 is shown by the line with a smaller slope. The intersection between the 0.2 mm offset line from the initial linear portion, or the crack blunting portion, of the curve with the stable crack growth portion of the curve, is the J_{Ic} point. This is operationally defined as the onset of stable crack extension or ductile tearing. J_{Ic} is a material property, and its determination is standardized in ASTM Standard 1820 [7]. A magnified view of the crack tip region during loading is shown in Figure 9.4 to illustrate the crack tip opening displacement (CTOD) and its relationship to the blunting of the crack tip. This manifests itself in the form of crack extension due to stretching caused by plastic deformation.

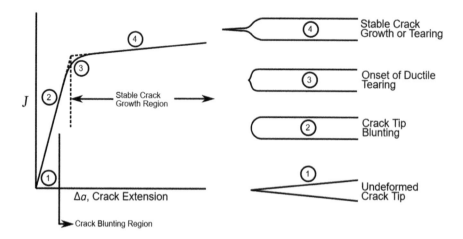

FIGURE 9.3 Schematic representation of crack tip blunting, onset of tearing or crack initiation and stable crack growth (right side of the diagram) the resulting J versus Δa relationship showing the various regions of crack growth and the J_{Ic} point (left side of the diagram).

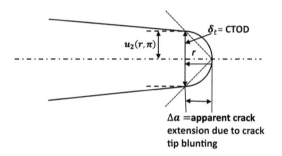

FIGURE 9.4 Magnified view of the crack tip deformation prior to ductile tearing.

From Figure 9.4, the following relationship between CTOD and Δa is derived.

$$\Delta a = \frac{1}{2}(CTOD) \approx \frac{J}{2\sigma_Y} \tag{9.10}$$

where $\sigma_Y = \frac{1}{2}(\sigma_{ys} + \sigma_{uts})$ and σ_{uts} is the ultimate strength of the material.

The determination of J_{Ic} from the crack blunting curve and the stable crack growth curve is shown in Figure 9.5 along with the construction of the 0.2 mm offset line. For a valid measurement of J_{Ic}, it is necessary to meet the conditions stipulated in equation (9.11) to ensure that the zone of large deformation is small in comparison to the specimen dimensions.

$$(W - a), B, a \geq 25 \frac{J}{\sigma_Y} \tag{9.11}$$

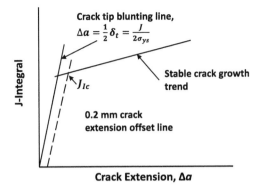

FIGURE 9.5 Schematic of the procedure for measuring J_{Ic} from measurements of J versus stable crack extension.

It is estimated that the zone of intense deformation is equal in magnitude to the CTOD thus the remaining ligament, crack size and the thickness must all exceed 25 times the CTOD. Also, since the crack extension in the blunting region is half of the CTOD, the three dimensions $W-a$, a, B will be approximately 50 times of the amount of crack growth. This method of determining J_{Ic} assures that we measure it when there is a small amount of crack extension and the zone of intense deformation at the crack tip is small in comparison to the region in which the stress field is controlled by J.

The relationship between K_{Jc} of the material and J_{Ic} is given by equation (9.12).

$$J_{Ic} \approx \frac{K_{Jc}^2}{E}\left(1-v^2\right) \tag{9.12}$$

This relationship allows one to estimate a value of K_{Ic} from tests conducted under significant plasticity conditions. It should be noted that K_{Jc} represents the initiation of stable crack growth while K_{Ic} is the onset of unstable crack growth. So, the equivalence of the two is only approximate, and typically K_{Jc} is less than K_{Ic} so, it is a conservative estimate of K_{Ic}.

9.4 EQUATIONS FOR DETERMINING J IN C(T) SPECIMENS

The compact type specimen C(T) is also the most common specimen geometry used in determining the J_{Ic} of the material. Thus, it is important to present an equation for estimating the value of J from the load versus displacement behavior. If we designate load-line displacement with Δ, then the value of J is given by equation (9.13) [8] below.

$$J = \frac{A}{B(W-a)}\left[2+0.522\frac{W-a}{W}\right] \tag{9.13}$$

where W=specimen width, B=thickness, and A=Area under the load versus load-line displacement, V, curve obtained from the test and shown in Figure 9.6.

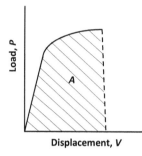

FIGURE 9.6 Schematic of the load and load-line displacement obtained from C(T) specimen during elastic-plastic fracture toughness test.

TABLE 9.1

Recorded Data and J Estimates Obtained from a C(T) Specimen from a Test Conducted on a Structural Steel

Load, P (kN)	Displacement, V (mm)	Crack Size, a (mm)	Δa (mm)	ΔA_i (kN·m)	Area, A_i (kN·m)	J (kJ/m²)
0.00	0.00	25	0	0.00	0.00	0.00
11.10	1.00	25.05	0.05	5.55	5.55	20.11
15.20	1.50	25.1	0.1	6.58	12.13	44.02
17.49	1.87	25.2	0.2	6.04	18.16	66.17
20.00	2.40	25.45	0.45	9.95	28.11	103.34
21.19	2.91	25.75	0.75	10.48	38.59	143.42
21.31	2.98	25.85	0.85	1.53	40.12	149.66
21.63	3.35	26.2	1.2	7.95	48.07	181.64
21.21	3.91	26.65	1.65	12.04	60.10	231.02
18.04	4.86	27	2	18.60	78.70	306.61

Example Problem 9.1

The load, displacement, and crack size data listed in Table 9.1 are recorded for a steel with $\sigma_{ys}=400$ MPa and $\sigma_{uts}=550$ MPa obtained from a test performed on a C(T) specimen with a thickness of 25 mm, a width of 50 mm, and an initial crack size of 25 mm. Estimate the J_R versus Δa curve and the J_{Ic} for the steel. What is the estimated K_{Jc} for the material if the elastic modulus is 207 GPa?

Solution

The load-displacement data recorded during the fracture test are plotted in Figure 9.7a. The load-displacement diagram is divided into m segments as shown in Figure 9.7a and points on the diagram are chosen for which the crack size is known or can be determined from the unloading compliance. The incremental areas, under each segment, ΔA_i, are determined using the following relationship as also shown in Figure 9.7a and listed in Table 9.1.

$$\Delta A_{i+1} \approx \frac{1}{2}(P_{i+1}+P_i)(V_{i+1}-V_i) \quad (\text{for } 0 \le i \le m)$$

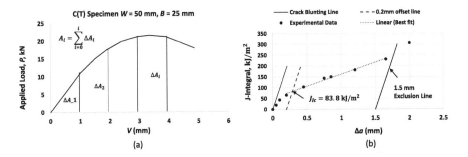

FIGURE 9.7 (a) The load displacement diagram from a fracture test on a C(T) specimen from a structural steel (b) stable crack growth relationship with J and determination of J_{Ic}.

The cumulative area under the load-displacement diagram can then be determined by the following equation:

$$A_i = \sum_{i=0}^{i} \Delta A_i \quad (\text{for } 0 \le i \le m)$$

From equation (9.13) the corresponding values of J at each of the points can be calculated and are listed in Table 9.1. Next, the J versus Δa points are plotted as seen in Figure 9.7b. This gives us the stable crack growth resistance curve from which we can estimate the J_{Ic} following the ASTM recommended procedure [7]. According to this procedure, several lines are constructed. First, we plot the crack blunting line using equation (9.10) recognizing that $\sigma_Y = \frac{1}{2}(\sigma_{ys} + \sigma_{uts}) = 475\,\text{MPa}$.

Next, we plot a line with a slope equal to the blunting line but offset with 0.2 mm crack growth. We then plot a line that is parallel to the blunting line but is offset by 1.5 mm of crack extension. This line is called the exclusion line because the data that lie to the right of this line are excluded from the estimation of J_{Ic}. We are interested in establishing and mathematically describing the relationship between J and Δa closer to the J_{Ic} than the one for larger amounts of crack growth; 1.5 mm of crack growth has been agreed upon by experts as that limit. Then, a straight line is fitted through the data to the right of the blunting line and the 1.5 mm exclusion line. That equation describes the stable crack growth resistance curve is as follows:

$$J = 51.965 + 110.48(\Delta a)$$

The intersection between the stable crack growth resistance curve and the 0.2 mm offset line is the J_{Ic} for the material. This can be obtained by solving the following equation:

$$J_{Ic} = 51.965 + 110.48\left(\frac{J_{Ic}}{2\sigma_Y} + 0.2\,\text{mm}\right)$$

This leads to a value of $J_{Ic} = 83.8$ kJ/m². We then assess if this value of J_{Ic} meets the requirements of equation (9.11). The value of $25(J_{Ic}/\sigma_Y) = 4.41$ mm. B, $W-a$, and a are all 25 mm that exceed the minimum value required for the test to be valid.

From equation (9.12), the equivalent value of K is $\sqrt{EJ_{Ic}/(1-v^2)} = 138\,MPa\sqrt{m}$, the K_{Jc} for the material. To measure a valid K_{Ic}, a specimen with following dimensions will be required:

B, $W-a$, and $a \geq 2.5\left(\dfrac{138}{400}\right)^2 = 0.119 \text{ m} = 119 \text{ mm}$ that is almost five times the size of the specimen used. In fact, a specimen much smaller than one with a $W = 50\,\text{mm}$ could yield valid measure of J_{Ic}. Note that a specimen with planar dimensions that are five times those of the 50 mm wide C(T) specimen is 125 times larger in volume and weight.

9.5 FATIGUE CRACK GROWTH UNDER GROSS PLASTICITY CONDITIONS

For low strength and high fracture toughness materials such as stainless steel or other materials that are used at elevated temperatures where considerable reduction in strength occurs, the linear elastic size requirements of equation (6.18),

$$(W - a) \geq \frac{4}{\pi}\left(\frac{K_{max}}{\sigma_{ys}}\right)^2,$$ used to ensure da/dN versus ΔK can be limiting. Thus, there

is a need for analytical concepts that can be used to characterize the fatigue crack growth behavior under large-scale plasticity.

Figure 9.8 shows the monotonic and cyclic stress–strain diagrams of a typical cyclic strain hardening material. The figure shows the definitions of various terms such as the yield strength and cyclic yield strength. This material undergoes considerable strain hardening during the cyclic loading and unloading when plastic deformation is encountered during each cycle. Other materials undergo softening during cyclic loading and their stress–strain diagrams will be lower than the monotonic stress–strain diagram.

The cyclic stress–strain behavior, equation (9.14) can be expressed by equations analogous to the Ramberg-Osgood relationship, equations (9.1) and (9.2), in which the stress is replaced by stress amplitude and strain by strain amplitude and the various constants are also analogous to their monotonic counterparts. To distinguish between the monotonic and cyclic constants, we add a prime as a superscript in the case of cyclic constants. Note that $\alpha\varepsilon_0$ has been combined into a single constant, α'.

$$\frac{\Delta\varepsilon}{2} = \frac{\Delta\sigma}{2E} + \alpha'\left(\frac{\Delta\sigma}{2\sigma_y^c}\right)^{m'} \tag{9.14}$$

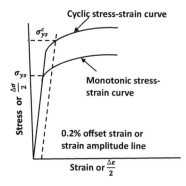

FIGURE 9.8 Monotonic and cyclic stress–strain diagrams illustrating the definitions of cyclic and monotonic yield strengths for cyclically hardening material.

Equation (9.14) can also be simplified as

$$\Delta\varepsilon = \frac{\Delta\sigma}{E} + D'(\Delta\sigma)^{m'} \tag{9.15}$$

For high strain hardening materials, the LEFM criterion for size is too conservative and a rationale can be made for revising it as in equation (9.16) by writing it in terms of ΔK and the stress levels required to cause reversed yielding, $2\sigma_{ys}^c$, within the crack tip plastic zone. Thus, we substitute $2\sigma_{ys}^c$ to replace σ_{ys} in equation (6.18).

$$W - a \geq \frac{4}{\pi}\left(\frac{\Delta K}{2\sigma_{ys}^c}\right)^2 \tag{9.16}$$

From a mathematical analogy to equation (9.4) and the definition of the J-integral, Lamba [9] defined a path-independent integral, ΔJ, as in equation (9.17).

$$\Delta J = \int\left(\Delta W \sin\theta - \Delta T_i \frac{\partial\Delta u_i}{\partial x}\right)ds \tag{9.17}$$

$$\text{Where,}\quad \Delta W = \int_0^{\Delta\varepsilon_{ij}} \Delta\sigma_{ij}\, d\Delta\varepsilon_{ij} \tag{9.18}$$

The cyclic J-integral defined in equation (9.17) also uniquely characterizes the crack tip stress and strain ranges under cyclic loading conditions analogous to equation (9.9a and b) for monotonic loading. For C(T) test specimens, the value of ΔJ was shown by Dowling and Begley [10] to be estimated by equation (9.19) below.

$$\Delta J = \frac{\Delta A}{B(W-a)}\left[2 + 0.522\frac{W-a}{W}\right] \tag{9.19}$$

ΔA is defined as the area under the loading portion of the load and load-line displacement diagram shown in Figure 9.9. In a slight variation than before in symbols, we

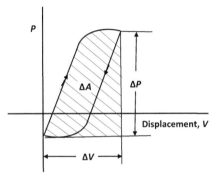

FIGURE 9.9 Schematic of the load versus displacement relationship during a cycle for C(T) specimen subjected to cyclic loading under plastic conditions in which the maximum and minimum displacements are fixed.

use V for displacement here and reserve Δ for its more conventional use to designate range of quantities such as load and displacement.

Since ΔJ is defined utilizing only the loading portion of the load-displacement diagram during a single load cycle as suggested by Dowling and Begley [10], there is no unloading involved in estimating ΔJ. Further, the amount of crack growth during one cycle is negligibly small from the point of view of causing unloading in the crack tip region. Because of these reasons, ΔJ is a valid parameter despite the unloading associated with the full loading cycle. Equations (9.20) and (9.21) then follow from these considerations.

$$\Delta J = \frac{\Delta K^2}{E} \tag{9.20}$$

$$\text{and} \quad \Delta J \neq J_{max} - J_{min} \tag{9.21}$$

Rather, ΔJ is the change in the value of J in going from minimum load to the maximum load during the fatigue cycle.

9.5.1   EXPERIMENTAL CORRELATION BETWEEN da/dN AND ΔJ

To demonstrate the validity of the ΔJ for characterizing fatigue crack growth behavior, Dowling and Begley [10] conducted several fatigue crack growth rate tests at various ΔV ranges in which the amount of plasticity was varied from dominantly linear-elastic conditions to the conditions of gross plasticity using C(T) and CCT specimens. Similarly, Brose and Dowling [11] conducted tests on stainless steels using C(T) specimens of several sizes with the smallest specimen having $W=25\,mm$ and the largest having $W=400\,mm$ to generate conditions that ranged from dominantly elastic to dominantly plastic. The results from their work are reproduced in Figure 9.10. The data are correlated with ΔK from large specimens in which dominant elastic conditions were maintained and with $\sqrt{E\Delta J}$ for small specimens in which plasticity was significant. The results clearly demonstrate the ability of ΔJ to normalize the effects of plasticity during fatigue loading. For a more detailed description of the topic and the various mathematical proofs of the equations presented, the readers are invited to consult another textbook [1].

Example Problem 9.2

The deformation of stainless steel at 600°C follows a relationship of the form given by equation (9.14) with the following constants: $E=135\,GPa$, $\sigma_y^c=200\,MPa$, $m'=5$, and $\alpha'=2\times10^3$. The cyclic yield strength of the material at the temperature is 200 MPa. A compact type specimen, 25 mm thick and 50 mm wide of this material is to be used to obtain crack growth rates for a ΔK up to 100 MPa$\sqrt{m}$. Assume that the maximum value of ΔK is reached when $a/W=0.6$. The relationship between the deflection range, ΔV and the load range, ΔP for $a/W=0.6$ for this material is given by the following equation, where ΔP is in MN, and ΔV is in meters.

$$\Delta V = \frac{100}{BE} + 585.4(\Delta P)^5$$

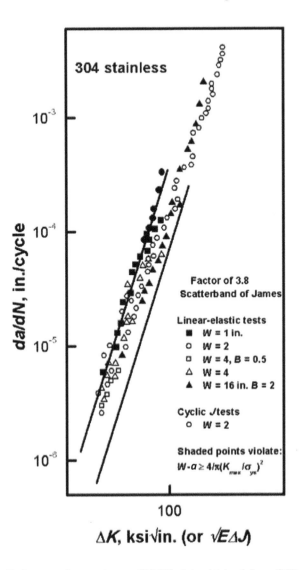

FIGURE 9.10   Fatigue crack growth rate (FCGR) data obtained from C(T) specimens of varying sizes to simulate conditions ranging from dominant elasticity to gross plasticity to validate ΔJ as a parameter for characterizing FCGR behavior over a range of plasticity (Figure taken from the work of Brose and Dowling [11] and reproduced with permission from ASTM, Conshohocken, PA).

Estimate the load and deflection ranges (load line) necessary to obtain this data. Estimate the size of specimen required to generate this data using LEFM procedures described in Chapter 6.

Solution

For 304 stainless steel at 600°C, $E = 135\,\text{GPa}$, $\sigma_0^c = 200$ MPa, $m' = 5$, $\alpha' = 2 \times 10^3$
   C(T) specimen $W = 50$ mm and $B = 25$ mm; data desired for $\Delta K = 100$ MPa$\sqrt{m}$ and lower with a maximum $a/W = 0.6$.

Equivalent value of ΔJ for ΔK of 100 MPa$\sqrt{m}$ is

$$\Delta J = \frac{\Delta K^2}{E} = \frac{100^2}{135 \times 10^3} = 0.074 \, \text{MJ/m}^2 = 74 \, \text{kJ/m}^2$$

$$\Delta J = \frac{\Delta A}{B(W-a)}\left(2 + 0.522\frac{W-a}{W}\right)$$

$$\Delta A = \int_0^{\Delta V}\left(\Delta P d(\Delta V)\right)$$

$$\Delta V = \frac{100}{0.025 \times 135 \times 10^3} + 585.4\Delta P^5 = 0.02963\Delta P + 585.4\Delta P^5$$

$$d(\Delta V) = \left(0.02963 + 2{,}927\Delta P^4\right)d(\Delta P)$$

$$\Delta A = \int_0^{\Delta P}\left(0.02963\Delta P + 2{,}927\Delta P^4\right)d(\Delta P)$$

$$\Delta A = 0.02963\frac{\Delta P^2}{2} + \frac{2{,}927}{6}\Delta P^6 = 0.0148\Delta P^2 + 487.8\Delta P^6$$

$$\Delta J = \frac{0.0148\Delta P^2 + 487.8\Delta P^6}{0.025 \times 0.02}\left[2 + 0.522\frac{0.02}{0.05}\right] = 65.35\Delta P^2 + 2.254 \times 10^6 \Delta P^6$$

The target value for ΔJ is 0.074 MJ/m². Thus, we need to find a value of ΔP based on the following equation and the relationship is plotted in Figure 9.11:

$$65.35\Delta P^2 + 2.254 \times 10^6 \Delta P^6 = 0.074$$

The value of ΔJ=0.074 MJ/m² is achieved with cyclic load of 0.0344 MN=34.4 kN. This corresponds to a value of displacement range of 0.00105 m or approximately 1 mm.

The minimum value of W required for obtaining the data at 100 MPa$\sqrt{m}$ in a test under dominantly linear elastic condition is given by equation (6.18). However, in this high strain hardening material, this criterion yields very conservative results. Substituting the values of the various terms in equation (9.16), we get that,

$$W - a \geq \frac{4}{\pi}\left(\frac{100}{2 \times 200}\right)^2 = 0.079 \text{ m} = 79 \text{ mm}$$

This implies that for a/W=0.6, the W value of 79/0.4=197 mm, significantly larger than the 50 mm wide specimen used here subjected to elastic-plastic loading. Compared to the specimen sizes required to assure linear elastic conditions during fracture testing, this is much smaller. LEFM conditions are known to extend much further during fatigue crack than for fracture testing.

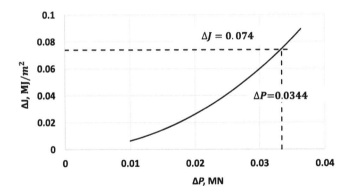

FIGURE 9.11 The relationship between ΔP and ΔJ for the C(T) specimen of 304 stainless steel at 600°C, $B = 25\,\mathrm{mm}$, $W = 50\,\mathrm{mm}$, $a = 30\,\mathrm{mm}$.

9.6 SUMMARY

In this chapter, the J-integral first introduced by Rice [2], was defined and its analytical basis for applicability as a fracture parameter for ductile materials was explored. It was shown that J-integral which is defined with the assumptions associated with the deformation theory of plasticity, can be used as a fracture parameter for limited amounts of crack extension. It was shown that, provided certain size requirements are met, and the amount of crack extension can be limited, the deformation fields ahead of the crack tip, are controlled by J, even in the presence of small amounts of crack growth. These findings pave the way, within certain restrictions, for J to be applied as a characterizing parameter for fracture in ductile materials for which LEFM is unsuitable.

The cyclic J-integral, ΔJ, was defined and shown to be a parameter that can unify fatigue crack growth data generated under dominantly small-scale-yielding condition to data obtained under conditions of gross plasticity.

REFERENCES

1. A. Saxena, *Advanced Fracture Mechanics and Structural Integrity*, CRC Press, Taylor and Francis Group, Boca Raton, FL, 2019, p. 307.
2. J.R. Rice, "A Path-Independent Integral and the Approximate Analysis of Strain Concentration by Notches and Cracks", *Journal of Applied Mechanics, Transactions ASME*, Vol. 35, 1968, pp. 379–386.
3. J.W. Hutchinson, "Singular Behavior at the End of a Tensile Crack in a Hardening Material", *Journal of Mechanics and Physics of Solids*, Vol. 16, 1968, pp. 13–131.
4. J.R. Rice and G.F. Rosengren, "Plane Strain Deformation Near a Crack Tip in a Power-Law Hardening Material", *Journal of Mechanics and Physics of Solids*, Vol. 16, 1968, pp. 1–12.
5. J.A. Begley and J.D. Landes, "The J-Integral as a Fracture Criterion", in *Fracture Toughness, ASTM STP 514*, 1972, American Society for Testing and Materials, Philadelphia, PA, pp. 1–23.

6. J.D. Landes and J.A. Begley, "The Effect of Specimen Geometry on J_{Ic}", in *Fracture Toughness, ASTM STP 514*, 1972, American Society for Testing and Materials, Philadelphia, PA, pp. 24–39.
7. Standard Method for Measurement of Fracture Toughness, E1820-15, American Society for Testing and Materials International, West Conshohocken, PA, 2015.
8. H.A. Ernst, P.C. Paris, and J.D. Landes, "Estimations of J-Integral and Tearing Modulus T from a Single Specimen Test Record", Fracture Mechanics: Thirteenth Conference, ASTM STP 743, American Society for Testing and Materials, 1981, pp. 476–502.
9. H.S. Lamba, "The J-Integral Applied to Cyclic Loading", *Engineering Fracture Mechanics*, Vol. 7, 1975, pp. 693–703.
10. N.E. Dowling and J.A. Begley, "Fatigue Crack Growth Under Gross Plasticity and J-Integral", in *Mechanics of Crack Growth,* J.R. Rice and P.C. Paris editors 1976, ASTM, Philadelphia, PA, pp. 82–103.
11. W.R. Brose and N.E. Dowling, "Fatigue Crack Growth Under High Stress Intensities in 304 Stainless Steel", in *Elastic-Plastic Fracture,* J.D. Landes. J.A. Begley, and G.A. Clarke, editors, 1979, American Society for Testing and Materials, Philadelphia, PA, pp. 720–735.

HOMEWORK PROBLEMS

1. What limitations of deformation theory of plasticity influence the validity of *J*-integral as a fracture parameter? How are these limitations overcome?
2. What is the significance of *J*-integral being path-independent?
3. What is J_{Ic} and how does it differ from K_{Ic}? If J_{Ic} of a structural steel is 35 KJ/m^2, estimate the K_{Ic} of the material. Comment on whether this estimate of K_{Ic} is conservative or optimistic and why.
4. Show that *J* for constant displacement conditions is given by.

$$J = -\frac{1}{B}\int_0^{\Delta}\left(\frac{\partial P}{\partial a}\right)_{\Delta} dV$$

and for constant load conditions, *J* is given by

$$J = \frac{1}{B}\int_0^{P}\left(\frac{\partial V}{\partial a}\right)_{P} dP$$

5. Show that for a center crack tension sample with a width of 2*W* and thickness, *B*, the *J* is given by

$$J = \frac{K^2}{E}(1-v^2) + \frac{PV_p}{2BW(1-a/W)}\left(\frac{m-1}{m+1}\right)$$

where V_p =plastic part of displacement, *P*=applied load. From dimensional analysis, we can write the following relationship between load and plastic part of displacement.

$$V_p = (W-a)f\left(\frac{P}{B(W-a)}\right)$$

For a power law hardening material, the above equation can be written as $\Delta_p = CP^m$, where C depends on the crack size.

6. Show that J-integral for linear elastic cracked bodies is equal to the Griffith's crack extension force, G.

7. What is the rationale for the crack blunting line given by equation (9.10) in connection to the measurement of J_{Ic}?

8. List reasons for why ΔJ can be used as a unifying crack tip parameter for characterizing fatigue crack growth behavior under conditions ranging from small scale yielding to large scale plasticity.

   Show that for plane stress conditions, $\Delta J = \dfrac{\Delta K^2}{E}$

10 Creep and Creep-Fatigue Crack Growth

10.1 INTRODUCTION

Several components of power-plants, chemical reactors, and land, air, and sea-based steam and gas turbines operate at temperatures where creep deformation and fracture are design concerns. Creep deformation occurs as a function of time due to sustained stress at elevated temperatures, and it becomes a design concern when service temperatures of the alloy exceed about 35% of the melting point in degrees Kelvin. Slow crack growth can occur in the presence of creep deformation and cause failure when the crack becomes critical size. This phenomenon is known as creep crack growth. When combined with cyclic loading, and creep and fatigue crack growth occur synergistically, the phenomenon is known as creep-fatigue crack growth. Environmental effects due to oxidation can further accelerate creep and creep-fatigue crack growth rates and must be considered as an additional factor within the realm of elevated temperature crack growth behavior. This chapter covers fracture mechanics approaches available for predicting elevated temperature crack growth behavior and the limitations of these approaches.

On April 17, 2018, a Boeing 737 aircraft operated by Southwest Airlines experienced a failure of its left turbofan engine [1]. Portions of the left engine inlet and fan cowl separated from the airplane and impacted the fuselage near a cabin window resulting in a rapid depressurization. The flight crew conducted an emergency landing and the airplane landed safely about 17 minutes after the engine failure occurred. Of the 144 passengers and 5 crewmembers aboard the airplane, 1 passenger received fatal injuries, and 8 passengers received minor injuries. The US National Transportation Safety Board (NTSB) made several new design safety recommendations to the Federal Aviation Administration (FAA) and to the airlines.

The fan blade fractured near the root and at multiple locations in the center of the airfoil. The fracture surface shown in Figure 10.1 had a smooth region with six well-defined curved lines, which are consistent with fractures caused by fatigue. These lines, referred to as crack arrest lines, generally represent changes in the stress state, temperature, or time interval associated with fatigue crack growth. Ratchet marks seen in the picture form when two adjacent fatigue cracks originate on different planes and join as shown in Figure 10.1. Ratchet marks are thus indicative of multiple fatigue origins.

Linear-elastic and elastic-plastic fracture mechanics (EPFM) concepts discussed in previous chapters are unable to predict crack growth in the presence of creep strains. This chapter will focus on developing concepts of time-dependent fracture mechanics (TDFM) in which creep deformation in the crack tip region is specifically included. In extending the fracture mechanics concepts to conditions where

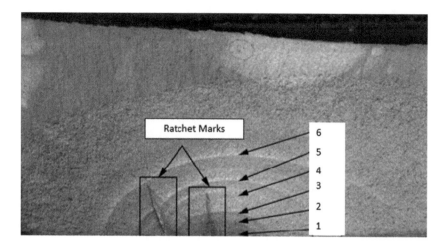

FIGURE 10.1 Fracture surface of a failed fan blade in an aircraft engine showing fatigue cracks with arrest lines and ratchet marks that form when multiple cracks grow and join [1].

time-dependent deformation is not a limitation, we will be taking advantage of mathematical analogies that exist between TDFM and the EPFM discussed in the previous chapter.

Figure 10.2 shows the creep deformation as a function of time in uniaxial specimens of 1 Cr-1Mo-0.25V steel subjected to a constant stress at 538 degrees C. The creep strains increase as time increases and the resulting creep deformation behavior is divided into three regions. The first region is called the primary creep region in which the strain rate continuously decreases with time until it reaches a steady-state value. This region is typically short lived but contributes significantly to the accumulated creep strain. During the second region, also called the steady-state creep region, the creep strain rate remains constant. In the tertiary creep region, which follows the steady-state region (not shown in Figure 10.2), the creep strain rates rise with time as necking develops and is followed by rupture. Since the objective of engineering designs is to avoid rupture, tertiary creep regime is not commonly encountered, so that condition is not common in components. Primary creep in several materials, as mentioned earlier, is short-lived, making steady-state creep region the most important in design considerations. Although the creep strain rates in this region remain constant with time but, they change significantly with stress and are described by a power-law, (also known as the Norton's law) equation (10.1), where $\dot{\varepsilon}_{ss}$ is the steady-state strain rate and $\sigma =$ applied stress.

$$\dot{\varepsilon}_{ss} = A\,\sigma^{n} \tag{10.1}$$

where A and n are constants derived from tests performed at different stress levels on uniaxially loaded specimens while experimentally measuring the creep strain rate. The measured steady-state strain rates are correlated to the applied stress levels to find A and n.

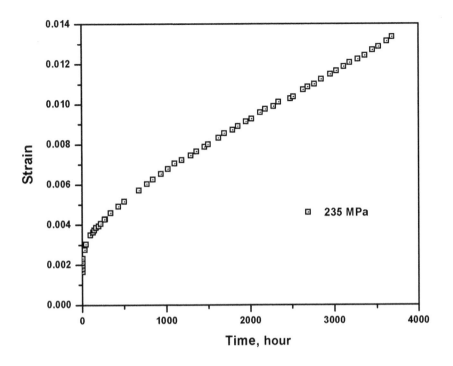

FIGURE 10.2   Creep deformation behavior of 1Cr-1Mo-0.25V steel at a stress level of 235 MPa at 538°C showing primary and secondary creep deformation. Tertiary creep is seen to begin at the end of the test that was interrupted prior to failure.

Some of the early studies using fracture mechanics to predict crack growth rates at elevated temperatures were reported by James [2] who extended the idea of using the cyclic stress intensity parameter, ΔK, to correlate fatigue crack growth behavior of 304 stainless steel at elevated temperatures over a wide range of loading frequencies. At about the same time, Siverns and Price [3] attempted to correlate creep crack growth rate to the stress intensity parameter, K. However, K uniquely characterizes the crack tip stress field only upon loading at time, $t = 0^+$. With elapse of time, stress redistribution occurs in the crack tip region due to creep deformation and there is no longer a unique relationship between K and the crack tip stress fields [4]. Thus, K is not expected to characterize the creep crack growth rates.

Under cyclic loading, the fatigue crack growth rates continue to be uniquely characterized by ΔK but only if the time parameters such as the frequency of loading, and the loading waveform are kept constant. It is also necessary that small-scale yielding and small-scale creep conditions are maintained for ΔK to be valid. The functional relationship between fatigue crack growth rate and ΔK can be written as in equation (10.2).

$$\frac{da}{dN} = f(\Delta K, v, R, T) \qquad (10.2)$$

where da/dN = fatigue crack growth rate, v = loading frequency, R = ratio of the minimum to maximum load during the cycle, and T = test temperature. In this chapter, we will first consider crack growth at high temperatures under static loading known as creep crack growth and then consider crack growth under cyclic loading at high temperatures resulting in creep-fatigue crack growth.

10.2 CREEP CRACK GROWTH

10.2.1 THE C*-INTEGRAL

In the mid-1970s, Landes and Begley [5] proposed a mathematical analogue of Rice's J-integral evaluated along a path, Γ, for cracked bodies subjected to wide-spread creep conditions and called it the C^*-integral defined in equation (10.3).

$$C^* = \int_\Gamma \left(W^* \sin\theta - T_i \frac{\partial \dot{u}_i}{\partial x} \right) ds \tag{10.3}$$

$$\text{Where,} \quad W^* = \int_0^{\dot{\varepsilon}_{ij}} \sigma_{ij} d\dot{\varepsilon}_{ij} \tag{10.4}$$

where T_i is a component of the traction vector, T, W^* is the stress-power density and $\dot{u}_i$ is a component of the displacement vector, $\dot{u}$. Γ is a contour that begins on the lower crack surface and travels counter clock-wise ending on the upper crack surface along which the line integral is evaluated. Note that in equations (10.3) and (10.4), all stress quantities in the definition of J are retained as stress but all the strain and displacement quantities are replaced by their rates. By analogy, C^* is capable of uniquely characterizing the crack tip stress fields, equation (10.5), under widespread secondary-stage creep conditions like the J-integral does for elastic-plastic conditions. C^* was also related to the stress-power dissipation rate (U^*) (rate of strain energy also referred to as stress-power) as shown in equation (10.6).

$$\sigma_{ij} = \left(\frac{C^*}{I_n A r} \right)^{\frac{1}{n+1}} \hat{\sigma}_{ij}(\theta, n) \tag{10.5}$$

$$C^* = -\frac{1}{B} \frac{dU^*}{da} \tag{10.6}$$

As mentioned before, A and n are constants in a Norton's power-law relationship, equation (10.1) between steady-state creep rate, $\dot{\varepsilon}_{ss}$, and the applied stress, σ, and r = distance from the crack tip, θ is the angular coordinate, $\hat{\sigma}_{ij}(\theta, n)$ is an angular function, B is thickness of the planar cracked body and a is the crack size. Landes and Begley [5] attempted to experimentally demonstrate the validity of C^* by conducting experiments on alloy A286 using both compact, C(T), specimens that are dominantly bend type specimens and center crack tension, CCT, specimens that are pure

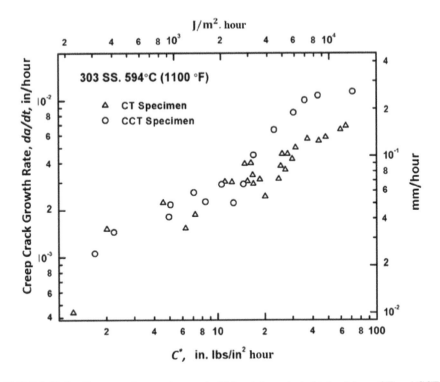

FIGURE 10.3 Creep crack growth rates in 304 stainless steel obtained from CT and CCT specimens, Saxena [4] (Figure reproduced with permission from ASTM, Conshohocken, PA).

tension specimens. Their results only met with partial success because the condition of widespread secondary creep was not met in their specimens for much of the duration of the tests. Consequently, the creep crack growth rates measured from C(T) and CCT specimens differed from each other. Saxena [4] showed that creep crack growth rates were uniquely correlated with C^*-integral for C(T) and CCT specimens from 304 stainless steel at 594°C because the necessary conditions of widespread steady-state creep were established shortly after loading in these tests, and the creep crack growth rate occurred primarily under these conditions. These results are reproduced in Figure 10.3.

From the analogy between J integral and C^*, the magnitude of C^* can be estimated from the expressions for J-integral by using applied load, P, and the measured load-line displacement rates in place of load-line displacement in the expression for J. For example, for C(T) specimens, C^* is given by equation (10.7).

$$C^* = \frac{P\dot{V}_{ss}}{B(W-a)}[2 + 0.522(1 - a/W)]\frac{n}{n+1} \qquad (10.7)$$

where $\dot{V}_{ss}$ = displacement rate measured during the test in which the C(T) specimen is loaded with a fixed force, P. Note also that the area term in the expression for J has been replaced by $P\dot{V}_{ss}\left(n/(n+1)\right)$ which in effect is the area underneath a P versus $\dot{V}_{ss}$ diagram.

The limitations on the use of C^* are that it only applies for conditions when the cracked body is under widespread steady-state creep deformation. This condition is commonly encountered in test specimens used for generating creep crack growth data but not as commonly in components that are designed to resist creep deformation. The components may be under small-scale creep or in the transition creep region, in-between small-scale to extensive creep. Other parameters that are used to characterize creep crack growth when small-scale-creep or transition creep conditions are needed.

10.2.2 $C(t)$ Integral and the C_t Parameter

With establishment of C^* as the crack tip parameter for characterizing creep crack growth rates under widespread steady-state creep conditions, the attention in the early 1980s turned to its limitations and search for parameters that could be used under small-scale creep conditions and would become identical to C^* when widespread steady-state creep conditions were established. The work of Riedel and Rice [6], Ohji, Ogura and Kubo [7], Bassani and McClintock [8], and Saxena [9] was key to unlocking the mechanics of this problem. Riedel and Rice [6] and Ohji et al. [7] analytically derived the crack tip stress fields ahead of the crack tip under small-scale creep conditions while Bassani and McClintock [8] numerically verified these analytical results and extended the analysis to conditions ranging from small-scale to wide-spread creep. They related the amplitude of the crack tip stress singularity from small-scale creep to widespread steady-state creep, equation (10.8) to a $C(t)$-integral. For widespread creep, $C(t)$ becomes identical to C^* and for small scale creep it is related to a line integral that is path-dependent whose value is determined along a contour very near the crack tip and is given by equation (10.9).

$$\sigma_{ij} = \left(\frac{C(t)}{I_n A r}\right)^{\frac{1}{n+1}} \hat{\sigma}_{ij}(\theta, n) \tag{10.8}$$

$$C(t) = \int_{\Gamma \to 0}\left[W^*\sin\theta - T_i\left(\frac{\partial \dot{u}_i}{\partial x_1}\right)\right]ds \tag{10.9}$$

The contour must be routed through a region where creep deformation dominates near the crack tip and thus the condition $\Gamma \to 0$. Under widespread creep, $C(t)$ becomes identical to C^*, by definition. Equation (10.10a–c) was then developed by Bassani and McClintock [7], to estimate $C(t)$ analytically as follows.

$$For \quad t \ll t_1, \quad C(t) \approx \frac{K^2\left(1-v^2\right)}{EA(n+1)t} \tag{10.10a}$$

$$For \quad t \gg t_1, \quad C(t) = C^* \tag{10.10b}$$

$$For\ 0 < \frac{t}{t_1} < \infty,\quad C(t) \approx \left[\frac{K^2\left(1-v^2\right)}{EA(n+1)t} + C^* \right] \tag{10.10c}$$

where the transition time t_1, between small-scale creep and widespread creep was obtained by equating equation (10.10a and b) for the two asymptotic conditions to derive equation (10.11) [5,6].

$$t_1 = \frac{K^2\left(1-v^2\right)}{EA(n+1)C^*} \tag{10.11}$$

In the small-scale creep region, the creep zone size, r_c, is given by the following equation obtained from finding the locus of points surrounding the crack tip where the accumulated creep strains equal the elastic strains.

$$r_c = B(EAt)^{\frac{2}{n-1}} \tag{10.12}$$

$$\text{where, } B \approx \frac{1}{2\pi}\left(\frac{(n+1)^2}{1.38n} \right)^{\frac{2}{n-1}} \tag{10.12a}$$

Based on equations (10.8) and (10.10a–c), $C(t)$ is an attractive parameter for characterizing creep crack growth rate under small-scale to extensive creep, but its disadvantage is that it cannot be measured at the loading pins during small-scale and transition creep. It also cannot be estimated along a contour far away from the crack tip in those regimes. These shortcomings are particularly a concern because its estimation is completely dependent on accurate constitutive equations which must include, in addition to elastic deformation and steady-state creep, the effects of instantaneous plasticity, primary creep, tertiary creep, 2-D versus 3-D stress fields in the crack tip region and the scale of creep deformation. All these deformation phenomena are operative in the crack tip region, so it is not sufficient to just account for elastic deformation and steady-state creep under planar stress/strain conditions as is assumed in deriving equations (10.10). By contrast the successful crack tip parameters, K/G, J, and C^* can all be evaluated in regions far away from the crack tip and in some form can be obtained from displacement or displacement rates measured at the loading pins.

Addressing the above shortcoming, Saxena [9] proposed a parameter, C_t, that is measurable at the load-line of the cracked body and thus includes contributions from all the different deformation mechanisms present at the crack tip. C_t generalizes the stress-power dissipation rate definition of C^*, equation (10.6) into the small-scale creep regime as follows:

$$(C_t)_{ssc} = -\frac{1}{B}\frac{\partial U_t^*\left(a,t,\dot{V}_c\right)}{\partial a} \tag{10.13}$$

where U_t^* is the instantaneous stress-power dissipation rate at any time, t, that is also a function of the crack size, a, and the deflection rate due to creep, $\dot{V}_c$. Note that $\dot{V}_c$ is also the instantaneous load-line displacement rate due to creep and it becomes equal to $\dot{V}_{ss}$ when steady-state conditions are reached in the specimen. Expressions were derived to measure the value of $(C_t)_{ssc}$ in specimens, equation (10.14) [9]. Subsequently, C_t was also related uniquely to the rate of expansion of the creep zone size at the crack tip, equation (10.15) [10].

$$(C_t)_{ssc} = \frac{P\dot{V}_c}{BW}\frac{F'}{F} \tag{10.14}$$

$$(C_t)_{ssc} = \frac{2K^2\left(1-v^2\right)}{EW}\frac{F'}{F}\beta\dot{r}_c \tag{10.15}$$

where W=width of the cracked body, E=elastic modulus, $\beta \approx \frac{1}{3}$, P=Load on the specimen, v=Poisson's ratio, F=K-calibration factor$=\left(\dfrac{K}{P}\right)BW^{\frac{1}{2}}$, $F' = dF/d\left(\dfrac{a}{W}\right)$, $\dot{r}_c$=rate of expansion of the creep zone size at the crack tip obtained from the derivative of equation (10.12) with time. C_t is also by definition identical to C^* or $C(t)$ under widespread steady-state creep conditions. However, under small-scale creep conditions $C_t \neq C(t)$ [10].

In creep-ductile materials, the region in which effects of growing crack are significant is small and the creep zone size evolves from being small to being comparable to the uncracked ligament size [11]. The creep crack growth rates have been shown in such instances to uniquely correlate with C_t as seen in Figure 10.4. In these tests conducted on C(T) specimens that were 254-mm wide, it was shown that very significant amounts of crack extension occurred prior to the establishment of widespread creep conditions and despite that, all creep crack growth rate data correlated well with C_t [12].

The experimental methods for measuring creep crack growth rates have been standardized in an American Society for Testing and Materials (ASTM) standard [13]. Thus, for long-term sustained loading conditions, the fracture mechanics methods for characterizing creep crack growth rates are well established.

The crack tip parameters C^*, $C(t)$, an C_t have all been formulated with the assumption that the crack is stationary; in other words, the crack growth rate is zero. However, creep crack growth rate cannot be zero, therefore, we must then ensure that the rate of crack growth is slow in comparison to the rates at which creep strains accumulate at the crack tip to justify using stationary crack parameters. This is done by comparing the change in load-line deflection due to crack growth, $\dot{V}_e$, with the total measured load-line deflection ($\dot{V}$) [14]. The correlations between da/dt and C^* and/or C_t should be expected only when $\dot{V} \gg \dot{V}_e$ using equation (10.16) [13].

$$\dot{V} \gg \dot{V}_e = \frac{2B\left(\dfrac{da}{dt}\right)}{P}\frac{K^2}{E}\left(1-v^2\right) \tag{10.16}$$

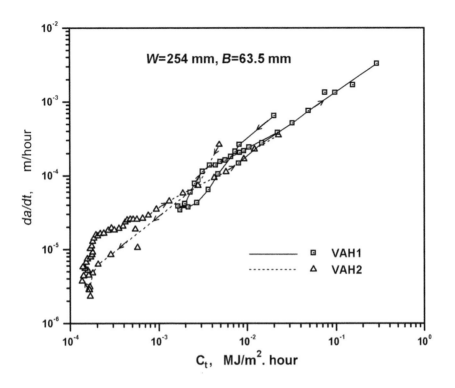

FIGURE 10.4 Creep crack growth rate as a function of C_t parameter in 1Cr-1Mo-0.25V steel at 538°C under small-scale creep to widespread creep in large C(T) specimens, Saxena, Yagi, and Tabuchi [12]. VAH1 and VAH 2 refer to two specimens tested at different load levels and the arrows point to crack growth rates as the test progressed from the beginning of the test going forward in time (Figure reproduced with permission from ASTM, Conshohocken, PA).

The materials in which the condition described in equation (10.16) is met, are called the creep-ductile materials. These are typically materials in which the rupture strain[1] during creep tests on smooth uncracked specimens exceeds 10%. Thus, there are limitations on the use of these crack tip parameters for correlating creep crack growth especially for creep-brittle materials, described next.

10.2.3 CREEP CRACK GROWTH IN CREEP-BRITTLE MATERIALS

Creep-brittle behavior during creep crack growth occurs when, $\dot{V}_e$, the rate of displacement increase due to crack extension in equation (10.16) becomes comparable to the experimentally measured value of $\dot{V}$. This signifies that the effects of crack growth begin to become as significant as the effects of creep deformation at the crack tip. Steady-state crack growth rate, da/dt, characterized uniquely by K can occur in such materials, but only under very limited conditions. Imagine an observer situated at the tip of a crack moving at a speed of da/dt while subjected

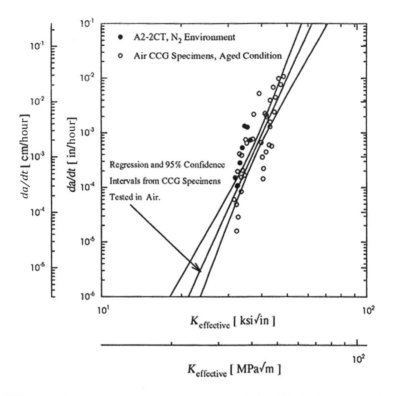

FIGURE 10.5   Creep crack growth behavior of a creep-brittle SA-106C steel at 360°C [11].

to a constant value of K. For steady-state crack growth to occur, the observer must see a crack tip stress field that does not change with time while K is held constant. In other words, the crack tip stress field must be characterized by equation (10.17).

$$\sigma_{ij} = \frac{K}{\sqrt{2\pi r}} f_{ij}(\theta) \qquad (10.17)$$

This can occur only if both the crack tip and creep zone boundary move at the same speed with respect to a fixed coordinate system, and the crack tip creep zone shape also does not change as the crack grows. These conditions are expected to occur only in materials that have low creep ductility, in other words creep-brittle materials. Small amounts of creep deformation must lead to crack extension and creep must not be allowed to accumulate at the crack tip so small-scale creep conditions can continue to persist as the test progresses. An example of creep crack growth rate characterized by K is found in Figure 10.5 for a carbon steel used in thick-wall piping that carries pressurized coolant in nuclear power plants and is subjected to a temperature of 360°C [15] during service. In this plot, the creep crack growth rates, da/dt, are plotted with the applied value of K that has been corrected for plasticity.

The tests were conducted in nitrogen and in air and showed similar results in both environments, confirming that the crack growth was in fact due to creep and not due to environment.

In several other creep-brittle materials, such as high strength aluminum alloys and Ni base alloys, steady-state conditions characterized by K require very substantial amounts of crack extension prior to reaching steady-state. It becomes necessary to use other crack tip parameters to account for the crack growth behavior under transient conditions in those cases. For more discussion on this topic, readers are referred elsewhere [11].

Example Problem 10.1

Why is $C(t)$ path-dependent in the small-scale and transition creep region? What is the implication of this path-dependency?

Solution

$C(t)$ integral as defined by equation (10.9) is path dependent for small-scale creep and transition creep because in these regimes, the stress redistribution occurs in the crack tip region. During the stress redistribution, elastic strains are being converted into permanent creep strains. Thus, the stress and strain-rate relationship, in uniaxial form, is given by the following equation:

$$\dot{\varepsilon} = \frac{\dot{\sigma}}{E} + A\sigma^n$$

Where, the dots indicate time rates. One of the assumptions in the proof that C^* is path independent is that the relationship between $\dot{\varepsilon}$ and σ is unique and thus single-valued. That is, for every value of σ, there is only one unique value for $\dot{\varepsilon}$. In the above equation, $\dot{\varepsilon}$ depends both on σ and on $\dot{\sigma}$. Thus, $C(t)$ is not path-independent under small-scale creep and transition creep. If no stress re-distribution was happening, the first term on the right-hand side of the equation will disappear and $\dot{\varepsilon}$ will only be a function of stress. So, for extensive creep conditions, $C(t) = C^*$ and becomes path independent. The implication of path-dependence is that $C(t)$ cannot be measured at the loading pins and it cannot be calculated along paths that are remote from the crack tip.

10.3 CRACK GROWTH UNDER CREEP-FATIGUE-ENVIRONMENT CONDITIONS

In earlier sections of this chapter, we built the analytical framework for considering time-dependent creep deformation in the crack tip region and used the analyses to define crack tip parameters for predicting crack growth that is accompanied by creep deformation. These analyses were under the conditions of sustained loading. Cyclic or fatigue loading is quite common in components that are operated at elevated temperatures, making crack growth behavior under creep-fatigue conditions important. Creep deformation can occur during both the sustained load portion and the loading portion of the fatigue cycle, depending on the loading rate.

In this section, we will discuss the analytical framework for addressing creep-fatigue crack growth behavior. We will first discuss the correlation between *da/dN* and ΔK for cycle times that are much smaller than the transition time given by equation (10.11). Subsequently, we will discuss creep-fatigue crack growth under longer cycle times.

10.3.1   *da/dN* VERSUS ΔK CORRELATIONS

During cyclic loading, the crack tip stress field during each cycle is expected to regenerate if small-scale creep and small-scale plastic deformation conditions are maintained and the crack tip creep zone is contained within the zone of reversed cyclic plasticity. The crack growth rate per cycle is characterized by ΔK in this instance if the loading waveform and cycle time are held constant. The full functional relationship between the crack growth rate per cycle, *da/dN*, and ΔK with frequency and waveform as explicit variables is expected to follow a functional form given in equation (10.18).

$$\frac{da}{dN} = f\left(t_r,\ t_d, t_h,\ R, T,\ \Delta K\right) \qquad (10.18)$$

In equation (10.18), $t_r =$ time for K to increase to K_{max} from K_{min}, during loading portion of a cycle; $t_d =$ the decay time for K to go from K_{max} to K_{min} during unloading, and t_h is the hold time for which K is held constant at K_{max}. $T =$ temperature at which testing is conducted, $R =$ load ratio, K_{min}/K_{max}, that is also held constant during fatigue crack growth testing. We define the loading frequency, v, such that $1/v = t_r + t_d + t_h$. If we hold t_r, t_d, and t_h constant, both the waveform and frequency are constant and if R is also held constant; it then follows from equation (10.18), that *da/dN* is only a function of ΔK. Further, if $t_r = t_d$ and $t_h = 0$, the cycle is called a balanced continuous cycle. We will limit our discussion here to the condition in which $t_r = t_d$ because in majority of the tests performed, this is common.

Some examples of correlations between *da/dN* and ΔK for different values of frequency and hold time are shown in Figure 10.6a and b. Figure 10.6a shows results generated on Inconel 718 alloy at 650°C at frequencies ranging from 0.01 to 1 Hz [16] for balanced fatigue cycles with no hold time. Figure 10.5b presents data for Astroloy at 655°C for hold time ranging from 0 to 15 minutes [17] for $t_r = t_d = 1$ second. We see a clear layering of the data with different loading frequencies or hold times validating the form of equation (10.18). Crack growth rates per cycle show an increasing trend with increasing hold time or increasing cycle time and a good correlation with ΔK.

A simple phenomenological model is described here that represents the time-dependent effects during high temperature fatigue crack growth behavior, namely the effects of frequency and hold time. Consider a cracked body loaded to a given value of K, as shown in Figure 10.7 [18]. Upon completion of a load cycle, a cyclic plastic zone forms around the crack tip. The magnitude of the cyclic plastic zone is proportional to $\left(\Delta K / 2\sigma_{ys}^{c}\right)^{2}$ as previously shown in equation (6.7). Two other zones begin to form at the crack tip with time, namely the creep zone in which creep strains

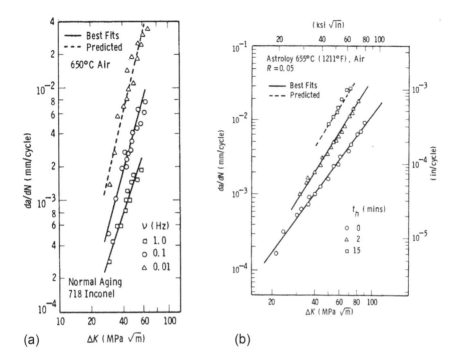

FIGURE 10.6 (a) Fatigue crack growth behavior of Inconel 718 at various loading frequencies [16] (b) for various hold times for Astroloy [17].

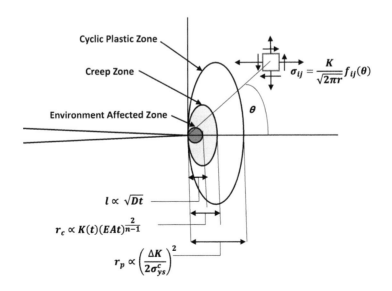

FIGURE 10.7 Schematic of the various crack tip zones, and their sizes as a function of time during cyclic loading (Adapted from Saxena [18]).

accumulate, and the environment-affected zone (EAZ) in which oxidation damage occurs. The size of the creep zone is governed by equation (10.12) and that of the EAZ, l, by equation (10.19) that is based on the well-known Fick's law for solid-state diffusion.

$$l \propto \sqrt{Dt} \qquad (10.19)$$

where D = diffusion coefficient of oxygen into the metal, and t = time. Oxygen is first adsorbed on the new crack surfaces, and it subsequently diffuses into the crack tip region aided by stress and the high temperatures. The diffused oxygen weakens the atomic bonds in the crack tip region causing the crack growth rates to be higher. In equation (10.19), it is assumed that the diffusion is the rate controlling step in the two-step process consisting of (a) adsorption and (b) diffusion of oxygen into the crack tip region. In other words, the adsorption kinetics are assumed to be faster than the diffusion kinetics, and the slower of the two steps in a sequential reaction is the rate controlling step. Degradation in the cohesive strength of atomic bonds presumably occurs in the high oxygen concentration region, previously termed as the EAZ, causing the crack growth rates to increase.

During high temperature fatigue crack growth, there is competition between three mechanisms of crack extension namely, (a) fatigue crack growth, (b) creep damage assisted crack growth, and (c) environment assisted crack growth due to oxidation. The latter two depend on time and have independent kinetics while the first depends on accumulated cycles. If the kinetics of creep deformation are rapid compared to oxidation, damage due to creep will prevail over oxidation, and vice versa. Thus, creep deformation kinetics compete with oxidation kinetics to determine the time-dependent portion of the crack growth rates.

Figure 10.8 shows a schematic of the three-region behavior between da/dN and the cycle time, $1/v$ at two values of ΔK. Region I is characterized by fast cycles in which the frequency is greater than a critical frequency v_0 for purely cycle dependent crack growth behavior. In this regime, the crack moves faster than the creep zone or the EAZ due to rapid accumulation of fatigue cycles. Thus, there is insufficient time for time-dependent damage in the form oxidation or creep to spread in the crack tip region. Region II is characterized by synergistic damage accumulation that includes damage due to fatigue cycling and due to creep and/or environment. Region III is characterized by time-dependent damage that can be either due to creep or environment, or both. Fatigue becomes less important in Region III. In the case of oxidation or environment dominant damage, da/dt is characterized by K but in materials where creep deformation and damage dominate, it is expected to be characterized by C_t or $C(t)$, because we assume small-scale creep conditions.

Assuming linear superposition of cycle and time dependent damage in Regions I and II, we can formulate equation (10.20) as follows:

$$\frac{da}{dN} = \left(\frac{da}{dN}\right)_0 + \int_{1/v_0}^{1/v} \left(\frac{da}{dt}\right)dt = C_0(\Delta K)^{n_0} + \int_{1/v_0}^{1/v} \left(\frac{da}{dt}\right)dt \qquad (10.20)$$

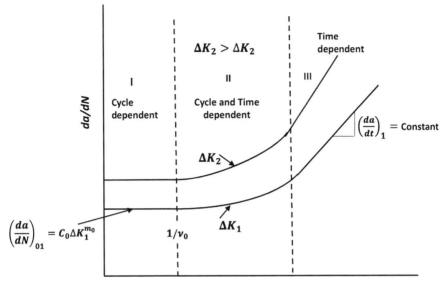

FIGURE 10.8 Schematic representation of high temperature fatigue crack growth behavior with cycle time (or 1/frequency).

If creep deformation/damage is rate-controlling, the time-dependent crack growth rate is given by equation (10.21). This equation is derived by noting that in equation (10.8a), the crack tip stress field is characterized by $\dfrac{K^2}{t}$ [19,20], essentially by $C(t)$.

$$\frac{da}{dt} = b\left(\frac{K_{max}^2}{t}\right)^p = b_1\left(\frac{\Delta K^2}{t}\right)^p \qquad (10.21)$$

For a balanced triangular waveform of loading, the crack growth per cycle is given by equation (10.22) as follows [20]:

$$\frac{da}{dN} = \left[C_0\Delta K^{n_0} + \int_{1/v_0}^{1/v}\left(\frac{da}{dt}\right)dt = C_0\Delta K^{n_0} + C\left(\Delta K^{2p}\right)\left(v^{p-1} - v_0^{p-1}\right)\right] \qquad (10.22)$$

For cycles with hold time, the crack growth rate per cycle can be written as in equation (10.23) [21]:

$$\frac{da}{dN} = \left[C_0\Delta K^{n_0} + \int_0^{t_h}b\left(\frac{\Delta K^2}{t}\right)^p dt = C_0\Delta K^{n_0} + C_1\left(\Delta K\right)^{2p}t_h^{1-p}\right] \qquad (10.23)$$

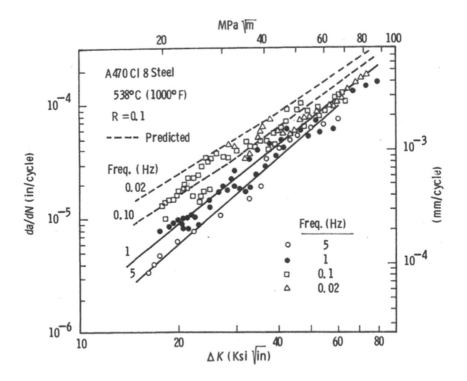

FIGURE 10.9 Fatigue crack growth rate behavior of 1Cr-1M0-0.25V steel at 538°C at several loading frequencies [19] (Figure reproduced with permission from John Wiley).

Figure 10.9 shows the fatigue crack growth data for a 1Cr-1Mo-0.25V high temperature ferritic steel obtained at various loading frequencies in the interval of $0.02 \le \nu \le 5$ Hz. The data clearly show the trend where the crack growth rate data at fixed ΔK increases with decreasing ν. In this case, $\nu_0 = 5$ Hz. The 5 Hz and the 1 Hz data are used to determine all the constants in equation (10.22) for which the resulting relationship is given by equation (10.24). The ΔK in this equation is in MPa$\sqrt{m}$ and da/dN is in mm/cycle and ν is in Hz.

$$\frac{da}{dN} = 6.8 \times 10^{-8} \Delta K^{2.51} + 2.2 \times 10^{-6} \Delta K^{1.454} \left(\nu^{-0.273} - 0.644 \right) \qquad (10.24)$$

The da/dN versus ΔK for test frequencies of 0.1 and 0.02 Hz are calculated from equation (10.24) and plotted in Figure 10.9 for comparison with experimental data. As can be seen, the model predictions are in good agreement with the experimental data. Similarly, there is data that supports the validity of the model for hold time effects represented by equation (10.23) [20].

Next, we consider an example of a material in which creep kinetics are slow and the time-dependent crack growth is dominated by oxidation. Nickel base alloys used in gas turbines are good examples of such materials. The IN718 and the Astroloy data shown in Figure 10.6a and b are examples that make this case. When environment damage is the dominant time-dependent mechanism, equation (10.25) is the characteristic equation [18].

$$\frac{da}{dt} \propto \frac{dl}{dt}\Delta K^q = \frac{c}{\sqrt{t}}\Delta K^q \qquad (10.25)$$

Substituting equation (10.25) into equation (10.20) yields the following result:

$$\frac{da}{dN} = C_0 \Delta K^{m_0} + C\Delta K^q \left(\sqrt{v} - \sqrt{v_0}\right) \qquad (10.26)$$

Note the difference between equations (10.22) and (10.26) is based on the choice of mechanism that dominates the time-dependent crack growth behavior. For Inconel 718, Figure 10.5a, the equation derived from regression analysis of the data for frequencies of 1 and 0.1 Hz is given in equation (10.27).

$$\frac{da}{dN} = 1.25 \times 10^{-8} \Delta K^3 + 8.08 \times 10^{-11} \left(\frac{1}{\sqrt{v}} - 1\right)\Delta K^{4.32} \qquad (10.27)$$

where v is in Hz, da/dN in mm/cycle and ΔK in MPa$\sqrt{m}$. Equation (10.27) is then used to predict the trend for 0.01 Hz and the predictions are seen to be in good agreement with the data validating the model.

For cycles containing hold times at maximum load in materials that are susceptible to time-dependent cracking due to oxidation, equation (10.28) can be derived. We assume in this derivation that t_r and t_d are both much smaller than t_h.

$$\frac{da}{dN} = \left(\frac{da}{dN}\right)_0 + A_2(\Delta K)^{2q}\sqrt{t_h} \qquad (10.28)$$

A_2, and q are regression constants. Equation (10.29) represents the behavior of Astroloy at 655°C for various hold times shown in Figure 10.6b. The solid lines represent the data used to estimate the regression constants and the dotted lines show predictions from the equation that are compared to the experimental data. The hold time is taken in hours, da/dN in mm/cycle and ΔK in MPa$\sqrt{m}$.

$$\frac{da}{dN} = 6.06 \times 10^{-8} \Delta K^{2.63} + 1.895 \times 10^{-8} \Delta K^{3.455}\sqrt{t_h} \qquad (10.29)$$

Comparing the 15-minute hold time data to the predicted trend in Figure 10.6b shows that the predictions from equation (10.29) are accurate.

Example Problem 10.2

50-mm wide and 12.5-mm thick C(T) specimens of a ferritic high strength low alloy steel were tested at 538°C at hold times of 0 and 5 seconds. The rise and decay times in both tests were 0.5 seconds. Regression fits were performed on the data for 0 and 5 seconds hold time tests, yielding the following equations.

$$\text{For } t_h = 0, \quad \frac{da}{dN} = 1.49 \times 10^{-7} \Delta K^{2.35}$$

$$\text{For } t_h = 5\,\text{sec} \quad \frac{da}{dN} = 5.7 \times 10^{-7} \Delta K^{2.09}$$

From these results and using the model in equation (10.23), plot the predicted crack growth behaviors for hold times of 0, 5, 50, and 500 seconds. The 0 and 5 second conditions are used to determine the constants in the model.

Solution

Figure 10.10 shows a plot of the da/dN versus ΔK trend for 0 and 5 second hold times from the equations given as part of the Example Problem. The values of C_0 and n_0 in equation (10.23) are 1.49×10^{-7} and 2.35, respectively. We then use the experimental data represented by the given equation for 5 second hold time to calculate the values of constants C_1 and p in equation (10.23) as follows. Rearranging equation (10.23), we get:

$$\left[\left(\frac{da}{dN} \right)_5 - \left(\frac{da}{dN} \right)_0 \right] = C_1 (\Delta K)^{2p} (5)^{1-p}$$

We then perform a linear regression between the log of the term on the left-hand side of the equation and log of ΔK to obtain the following values for constants C_1 and p:

$$C_1 (5)^{1-p} = 1 \times 10^{-6} \quad \text{and} \quad 2p = 1.5634 \quad \text{or} \quad p = 0.7817$$

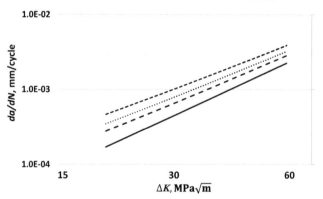

FIGURE 10.10   Predicted crack growth rates for conditions in Example Problem 10.2.

$$\text{Thus, } C_1 = \frac{1 \times 10^{-6}}{5^{0.2183}} = 7.037 \times 10^{-7}$$

The final equation from which the da/dN versus ΔK behavior for all the hold times can be calculated is as follows:

$$\frac{da}{dN} = 1.49 \times 10^{-7} \Delta K^{2.35} + 7.037 \times 10^{-7} \Delta K^{1.5634} t_h^{0.2183}$$

By substituting $t_h = 50$ and 500 seconds, the predicted crack growth rates are plotted in Figure 10.10.

10.3.2 CREEP-FATIGUE CRACK GROWTH RATES FOR LONG CYCLE TIMES

Let us consider creep ductile materials where high temperature crack growth occurs in the presence of creep deformation and the resulting damage is in the form of creep cavitation. If hold times are increased so that the creep zone becomes comparable in size to the cyclic plastic zone size, creep strains developed during the hold period are only partially reversed during unloading, and the crack growth rates per cycle cannot be expected to correlate with ΔK even for a constant hold time. Experimental studies performed on creep-ductile power plant materials using C(T) specimens have shown that TDFM parameters are needed for correlating creep-fatigue crack growth rates for wave forms with a hold time, t_h. In such cases, the time dependent crack growth, $(da/dt)_{avg}$, during the hold period is expressed as a function of the average magnitude of the C_t parameter, $(C_t)_{avg}$, defined in equation (10.30) [21].

$$(C_t)_{avg} = \frac{1}{t_h} \int_0^{t_h} C_t dt \tag{10.30}$$

$(da/dt)_{avg}$ is the average time-rate of crack growth during the hold time is defined by equation (10.31):

$$\left(\frac{da}{dt}\right)_{avg} = \frac{1}{t_h} \int_0^{t_h} \left(\frac{da}{dt}\right) dt \tag{10.31}$$

To estimate $(C_t)_{avg}$ for test specimens in which both load and load-line deflection behavior with time are measured, equation (10.32) may be used [22].

$$(C_t)_{avg} = \frac{\Delta P \Delta V_c}{BW t_h} (F'/F) \tag{10.32}$$

where ΔP is the applied load range, ΔV_c is the difference in load-line displacement between the end and start of the hold time, t_h, during a cycle. F'/F is given by equation (10.33) for C(T) specimens.

$$\frac{F'}{F} = \left[\left(\frac{1}{2 + a/W)} \right) + \left(\frac{3}{2\left(1 - \dfrac{a}{W}\right)} \right) \right] +$$

$$\left[\frac{\left(4.64 - 26.64\left(\frac{a}{W}\right) + 44.16\left(\frac{a}{W}\right)^2 - 22.4\left(\frac{a}{W}\right)^3\right)}{0.886 + 4.64\left(\frac{a}{W}\right) - 13.32\left(\frac{a}{W}\right)^2 + 14.72\left(\frac{a}{W}\right)^3 - 5.6\left(\frac{a}{W}\right)^4} \right] \quad (10.33)$$

The value of $(C_t)_{avg}$ from equation (10.32) is appropriate for the small-scale creep regime. For long hold times, $(C_t)_{avg} = C^*$ by definition. Thus, for very long hold times, the creep-fatigue crack growth rates blend with creep crack growth rates.

We consider that the cycle-dependent and the time-dependent crack growth rates during creep-fatigue are linearly additive. The cycle-dependent crack growth rate $(da/dN)_{cycle}$ is determined from fatigue tests without hold time and is expressed as a function ΔK. The average time rate of crack growth during hold time is expressed as a function of $(C_t)_{avg}$. The total crack growth rate during a cycle is given by:

$$\frac{da}{dN} = \left(\frac{da}{dN}\right)_{cycle} + \left(\frac{da}{dt}\right)_{avg} t_h \quad (10.34)$$

for a trapezoidal shaped creep-fatigue loading cycle consisting of increasing and decreasing loading portions and a hold time in-between, equation (10.34) can be rearranged to calculate $(da/dt)_{avg}$ as follows:

$$\left(\frac{da}{dt}\right)_{avg} = \frac{1}{t_h}\left[\left(\frac{da}{dN}\right) - \left(\frac{da}{dN}\right)_{cycle}\right] \quad (10.35)$$

Figure 10.11 shows a plot between the average crack growth rates estimated from equation (10.33) and $(C_t)_{avg}$ for a 1.25 Cr-0.5 Mo steel tested at 538°C for hold times ranging from 10 seconds to 24 hours with hold times of 98, 600, and 900 seconds in-between [22]. The data also include creep crack growth rate results. All the data from varying hold times and the creep crack growth rates collapse into a single trend. Thus, the following creep-fatigue model can be derived from these results.

$$\left(\frac{da}{dN}\right)_{C-F} = c_0 (\Delta K)^{n_0} + b_0 \left((C_t)_{avg}\right)^{\phi} t_h \quad (10.36)$$

where c_0 and n_0 are Paris equation constants derived from tests with zero hold time, b_0 and ϕ are constants derived from the $(da/dt)_{avg}$ and $(C_t)_{avg}$ relationship of the type shown in Figure 10.11. The following equation describes the actual data.

$$\frac{da}{dN} = 1.23 \times 10^{-5} \Delta K^{1.083} + 9.16 \times 10^{-3}\left[(C_t)_{avg}\right]^{0.825} \quad (10.37)$$

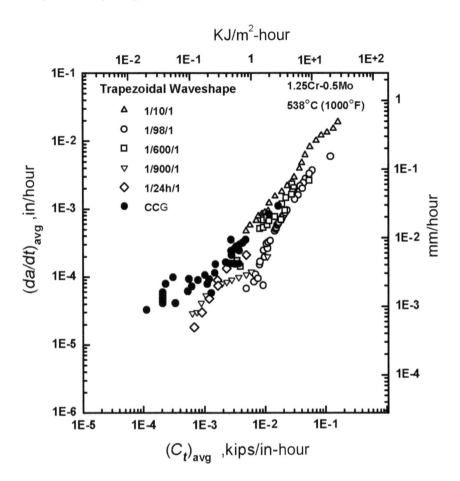

FIGURE 10.11   Average time rate of crack growth during hold-time as a function of $(C_t)_{avg}$ for hold times of 10 seconds to 24 hours for a 1.25 Cr-0.5 Mo steel [22] (Figure reproduced from reference [22] with permission from Springer).

where da/dN is in mm/cycle, ΔK in $MPa\sqrt{m}$, and $(C_t)_{avg}$ in kilojoules/m²-hour. The model is useful in estimating the lives of components made from this material and operated at 538°C containing cracks or crack-like defects. For more in-depth discussion of these parameters and how to estimate their value in components, the readers are referred to a more advanced text [11].

10.4   SUMMARY

In this chapter, the characteristic equations for evolution of crack tip stress fields as a function of time were derived under extensive and small-scale creep conditions. The creep deformation behavior was modeled with a power-law relationship between strain-rate and stress, ignoring primary creep and effects due to crack growth. C^*-integral was identified as the relevant crack tip parameter under extensive secondary creep conditions for characterizing creep crack growth rates. Methods for estimating C^* in commonly used test specimens were described.

When the additional effects due to stress redistribution during small-scale creep are considered, the C-integral was shown to characterize the amplitude of the crack tip stress field. However, the magnitude of C-integral cannot be measured at the loading pins or calculated along a path remote from the crack tip under small-scale creep and transition creep conditions. Therefore, an alternative crack tip parameter, C_t, that is measurable at the loading pins was defined and was seen to uniquely determine the rate of expansion of the crack tip creep zone size during small-scale creep. It was shown that under extensive creep conditions, C^*-integral, C-integral, and C_t are all identical and measurable at the loading pins. C_t was shown to uniquely characterize creep crack growth rates from small-scale to extensive creep conditions.

The effects of crack growth on the crack tip stress fields were considered and shown to be important in creep-brittle materials when subjected to small-scale creep conditions. It was theoretically argued that steady-state creep crack growth rates in these materials can be correlated with K if the crack tip and the creep zone boundary move at identical speeds and the shape of the creep zone also remains the same.

If small-scale yielding and small-scale creep conditions are maintained such as during loading frequencies at which creep effects are contained within the crack tip cyclic plastic zone, ΔK is shown to be the appropriate crack tip parameter for characterizing creep-fatigue crack growth rates for constant loading frequency and constant waveform. Phenomenological models were developed for extrapolating the effects of loading frequency and hold times on the crack growth behavior. As cycle times increase in creep-ductile materials, the validity of ΔK becomes questionable and TDFM parameters are needed. These parameters are also described and it was shown that $(C_t)_{avg}$ is the most successful crack tip parameter for representing hold-time effects during creep-fatigue crack growth.

REFERENCES

1. National Transportation and Safety Board Accident Report, "Left Engine Failure and Subsequent Depressurization in Southwest Airlines Flight 1380, Boeing 737-7H4, N772SW, Philadelphia, PA, April 17, 2018", NTSB/AAR-19/03, PB 2019-101439, November 2019.
2. L.A. James, "The Effect of Frequency Upon the Fatigue Crack Growth of Type 304 Stainless Steel at 10000F", in *Stress Analysis and Growth of Cracks, ASTM STP 513*, 1972, American Society for Testing and Materials, Philadelphia, PA, pp. 218–229.
3. M.J. Siverns and A.T. Price, "Crack Propagation under Creep Conditions in Quenched $2\frac{1}{4}$ Cr- 1 Mo Steel", *International Journal of Fracture*, Vol. 9, 1973, pp. 199–207.
4. A. Saxena, "Evaluation of C^* for Characterization of Creep Crack Growth Behavior of 304 Stainless Steel", Fracture Mechanics: Twelfth Conference, ASTM STP 700, American Society for Testing and Materials, Philadelphia, 1980, pp. 131–151.
5. J.D. Landes and J.A. Begley, "A Fracture Mechanics Approach to Creep Crack Growth", in *Mechanics of Crack Growth, ASTM STP 590*, 1976, American Society for Testing and Materials, Philadelphia, PA, pp. 128–148.
6. H. Riedel and J.R. Rice, "Tensile Cracks in Creeping Solids", Twelfth Conference, ASTM STP 700, American Society for Testing and Materials, Philadelphia, 1980, pp. 112–130.

7. K. Ohji, K. Ogura, and S. Kubo, "Stress-Strain Fields and Modified J-Integral in the Vicinity of the Crack Tip under Transient Creep Conditions", *Japan Society of Mechanical Engineering*, Vol. 790-13, 1979, pp. 18–20 (in Japanese).

8. J.L. Bassani and F.L. McClintock, "Creep Relaxation of Stress around a Crack Tip", *International Journal of Solids and Structures*, Vol. 17, 1981, pp. 79–89.

9. A. Saxena, "Creep Crack Growth under Non-Steady State Conditions", in *Fracture Mechanics: Seventeenth Volume, ASTM STP 905*, 1986, American Society for Testing and Materials, Philadelphia, PA, pp. 185–201.

10. J.L. Bassani, D.E. Hawk, and A. Saxena, "Evaluation of Ct Parameter for Characterizing Creep Crack Growth Rate in the Transient Regime", in *Nonlinear Fracture Mechanics: Time-Dependent Fracture Mechanics, ASTM STP 995*, 1989, American Society for Testing and Materials, Philadelphia, PA, pp. 7–29.

11. A. Saxena, *Advanced Fracture Mechanics and Structural Integrity*, CRC Press, London, England, 2019, p. 307.

12. A. Saxena, K. Yagi, and M. Tabuchi, "Creep Crack Growth under Small Scale and Transition Creep Conditions in Creep-Ductile Materials", in *Fracture Mechanics: Twenty Fourth Volume, ASTM STP 1207*, 1994, American Society for Testing and Materials, Philadelphia, PA, pp. 481–497.

13. ASTM Standard E1457-07, *Standard Test Method for Measurement of Creep Crack Growth Rates in Metals, ASTM Book of Standards*, American Society for Testing and Materials, West Conshohocken, PA, 2007.

14. A. Saxena, H.A. Ernst, and J.D. Landes, "Creep Crack Growth Behavior in 316 Stainless Steel", *International Journal of Fracture*, Vol. 23, 1983, pp. 245–257.

15. A. Saxena, *Nonlinear Fracture Mechanics for Engineers*, CRC Press, Boca Raton, FL, 1998.

16. S. Floreen and R.H. Kane, "An Investigation of Creep-Fatigue-Environment Interactions in Ni Base Superalloys", *Fatigue of Engineering Materials and Structures*, Vol. 2, 1980, pp. 401–412.

17. R.M. Pelloux and J.S. Huang, "Creep-Fatigue-Environment Interactions in Astroloy", in *Creep-Fatigue-Environment Interactions*, 1980, TMS-AIME, Warrendale, PA, pp. 151–164.

18. A. Saxena, "A Model for Predicting the Environment Enhanced Fatigue Crack Growth Behavior at High Temperature", in *Thermal and Environment Effects in Fatigue: Research—Design Interfaces—PVP* Vol. 71, Book No. H00257, 1984, ASME, New York, NY, pp 171-184.

19. A. Saxena, "A Model for Predicting the Effect of Frequency on Fatigue Crack Growth Behavior at Elevated Temperature? *Fatigue of Engineering Materials and Structures*, Vol. 3, 1980, pp. 247–255.

20. A. Saxena, R.S. Williams, and T.T. Shih, "A Model for Representing and Predicting the Influence of Hold Time on fatigue Crack Growth Behavior at Elevated Temperature", in Fracture Mechanics: Proceedings of the 13th National Symposium on Fracture Mechanics, ASIM STP 743, American Society for Testing and Materials, Philadelphia, PA, 1981, pp. 86–99.

21. A. Saxena and B. Gieseke, "Transients in Elevated Temperature Crack Growth" Proceedings of the MECAMAT International Seminar on High Temperature Fracture Mechanisms and Mechanics" EGF-6, 1987, pp. 19–36.

22. K.B. Yoon, A. Saxena, and P.K. Liaw, "Characterization of Creep-Fatigue Crack Growth Behavior under Trapezoidal Wave Shape Using C_t Parameter", *International Journal of Fracture*, Vol. 59, 1993, pp. 95–114.

HOMEWORK PROBLEMS

1. Describe the development of the crack tip stress fields for stationary crack subjected to constant load at elevated temperature. Assume that the initial response is in the regime of small-scale-creep.

2. Why is C^* path-dependent in the small-scale and transition creep region? What is the implication of the path-dependency?

3. Show that the transition time between small-scale and extensive creep is given by:

$$t_1 = \frac{K^2\left(1-v^2\right)}{E(n+1)C^*}$$

State the conditions for which the above equation is appropriate.

4. A 50 mm-wide and 25 mm-thick C(T) specimen of 304 stainless steel is subjected to a constant load of 18 kN at 594°C. If the A and n values of this material at 594°C are 2.13×10^{-18} and 6, respectively, for stress in MPa and strain rate in hour^{-1}, calculate the value of C^* when $a/W=0.5$ and the measured load-line deflection rate is 2.5×10^{-6} m/hour.

5. Show by analogy to J, a path-independent integral C^* can be defined for creeping cracked bodies. What are the assumptions for the validity of C^* and why is it suitable for characterizing creep crack growth? What is the significance of path-independence?

6. Show that

$$\dot{V}_e = \frac{2B\left(\dfrac{da}{dt}\right)}{P}\frac{K^2}{E}\left(1-v^2\right)$$

where $\dot{V}_e$ is the deflection rate associated with a change in elastic compliance resulting from crack growth at a rate of da/dt.

7. The fatigue crack growth rates in 304 stainless steel at 538°C at loading frequencies of 6.67 and 0.667 Hz are given by $\dfrac{da}{dN} = 9.9\times10^{-10}\,\Delta K^{3.83}$ and $\dfrac{da}{dN} = 9.17\times10^{-9}\,\Delta K^{3.26}$, respectively. The da/dN is in mm/cycle and ΔK in MPa$\sqrt{m}$. Derive the model constants that can be used to predict the effects of frequencies less than 0.667 Hz on the fatigue crack growth rate for 304 stainless steel.

8. The fatigue crack growth behavior of Astroloy at 650°C for a frequency of 1 Hz and $t_h=0$ is given by the following equation when da/dN is in mm/cycle and ΔK in MPa$\sqrt{m}$.

$$\frac{da}{dN} = 6.05\times10^{-8}\,\Delta K^{2.63}$$

Further, the fatigue crack growth rates for a hold time of 2 minutes superimposed on the above fatigue cycle yields a crack growth rate trend given by the following equation:

$$\frac{da}{dN} = 3.46 \times 10^{-9} \Delta K^{3.455}$$

Here also, *da/dN* is in mm/cycle and ΔK in MPa$\sqrt{m}$. Derive a phenomenologically based equation for extrapolating the effects of hold time on the fatigue crack growth behavior for this alloy at 650°C. You can assume that in this alloy the time-dependent crack growth is dominated by oxidation. Calculate the *da/dN* for a hold time of 15 minutes and compare it with data in Figure 10.5b for this material.

9. In your own words describe why is it necessary to formulate separate models for representing the effects of frequency and hold time on the high temperature fatigue crack growth rate behavior for materials in which time-dependent crack growth occurs due to creep damage and due to oxidation. Give examples of each type of materials.

NOTE

1 Rupture strain is measured as the strain at the time of rupture in a specimen with an initial smooth surface; frequently, cylindrical specimens with well-polished surfaces are tested to determine creep ductility of the material. A high rupture strain (>10%) implies a nominally creep-ductile material while a low ductility (<5%) implies a creep-brittle material.

11 Case Studies in Applications of Fracture Mechanics

11.1 INTRODUCTION

Fracture mechanics provides an analytical framework for developing a comprehensive methodology for predicting the potential for fracture in components. In this methodology, the magnitude and distribution of applied stresses, the resistance to fracture and crack growth of the material used in fabricating the component, and the size, location, and geometry of defects in the component can all be concurrently considered in predicting the potential for crack growth and fracture. This approach is also capable of determining the influence of deleterious service environment on the risk of fracture. In previous chapters of this book, we have described the physical and mathematical basis of these concepts. Here, we illustrate the use of those concepts by applying them to find solutions to real world example problems.

Applications of fracture mechanics can be classified in several categories, but the basic methodology is common even if the applications seem different. It is a matter of structuring the input/output variables to match the information/data that are available and what is desired as the answer. Figure 11.1 shows a flow chart of calculations

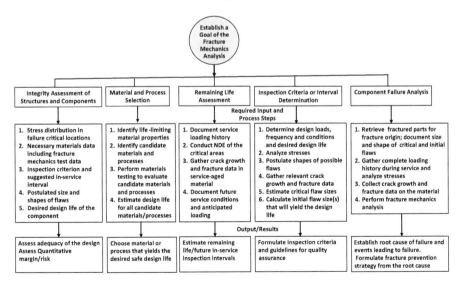

FIGURE 11.1 The various applications of fracture mechanics analysis in engineering applications and the associated steps involved in its implementation.

DOI: 10.1201/9781003292296-11

289

that can be performed for the various applications. Depending on the goal of the analysis and the input information available, equations can be set up to solve for the unknowns. The various classes of applications are as follows.

11.1.1 INTEGRITY ASSESSMENT OF STRUCTURES AND COMPONENTS

Preliminary and sometimes even final design of components is based on simple design rules which evolve from previous experience and because of their widespread use, become industry standards. They often result either in conservatism or in insufficient allowances for factors which contribute to fracture of components. For example, consider a design rule that the maximum allowable stress anywhere in a component should not exceed 75% of the 0.2% yield strength of the material. This design criterion can be over-conservative for a material with high fracture toughness but may have inadequate safety margin for low fracture toughness materials. In the latter case, the presence of even a small defect in a highly stressed region can cause catastrophic fracture. Fracture mechanics is used to assess the risk of fracture or the realistic safety margins in structures and components designed by such simple rules.

The risk (or safety margin) assessment consists of determining the flaw tolerance of the component using a fracture mechanics analysis. This flaw size is compared to ones that can realistically exist in the component due to the fabrication processes employed such as casting, welding, forging, etc., and/or to those that can be detected with 100% certainty during nondestructive evaluation (NDE). Such analyses can also be used for recommending a suitable in-service maintenance and inspection program for preventing catastrophic failures. Depending on the results of the analysis, the maximum allowable stresses can either be adjusted upwards to save material and weight or adjusted downwards to improve the safety margin. Thus, fracture mechanics analysis provides an interactive analytical tool for evaluating a preliminary design and achieving an optimized final design by balancing several structural integrity related requirements and cost.

11.1.2 MATERIAL AND PROCESS SELECTION

As described in the previous chapters, fracture mechanics provides a wide variety of test methods for evaluating materials and fabrication processes. Several of these methods have been standardized by international standards writing bodies. These include fracture toughness testing [1–6], fatigue crack growth testing [7], environment assisted cracking [8], dynamic fracture testing and crack arrest [9], creep crack growth testing [10], and creep-fatigue crack growth testing [11]. These methods collectively provide established techniques for testing and evaluating candidate materials and associated fabrication processes for fracture critical applications. However, these tests are expensive and occasionally prohibitively so. Therefore, fracture mechanics test methods are not replacements for the conventional test methods such as tensile test, hardness test, Charpy impact test, or other material screening tests that are often used in different industries. In fact, more than likely the preliminary material selection and process selections are made based on inexpensive tests and fracture mechanics methods for material testing are used to make final material selections.

11.1.3 DESIGN OR REMAINING LIFE PREDICTION

When a component is already in service and is approaching the end of its design life, engineers are often asked to assess how much longer the component is expected to last. From the knowledge of the magnitude and nature of the operating stresses and environment the component has experienced in service, the relevant fracture and crack growth properties and the nondestructive inspection data, fracture mechanics analysis is used to predict the remaining life of components already in service. Examples of such estimations will be discussed among the case studies described in the later sections of this chapter. The same methodology can be used to predict the design life of new components.

11.1.4 INSPECTION CRITERION AND INTERVAL DETERMINATION

Frequently, the nondestructive inspection techniques that are used during quality assurance (QA) or during service to detect cracks or crack-like defects in components are dictated by considerations such as the material, geometry, and size of the component, the accessibility of the critical areas of the component for placing transducers, and by economics. Inspections, especially during service, can be expensive because they are often time consuming and can cause long down-times involving expensive equipment, and require highly trained professionals to perform them and to interpret results. Conservative inspection criterion and intervals can result in high rejection rates and/ or costs. On the other hand, inadequate inspection criterion and insufficient inspection intervals can increase risk of catastrophic failures with devastating consequences. Fracture mechanics analysis provides an effective analytical tool for estimating realistic inspection intervals and accept/reject criteria. Several examples of such use of fracture mechanics will be illustrated in the case studies described in this chapter.

11.1.5 FAILURE ANALYSIS

Failure in machinery often occurs due to sudden fracture in a component of the machinery. Even in electronic components such as computers, failures are often related to thermal-fatigue fractures of joints that provide the electrical connections. In aerospace, automotive, and power-generation industries, the threat of failure caused by fracture of a component is always a crucial consideration. By conducting a thorough analysis of a failed part, important lessons can be learned about preventing similar failures in the future. Fracture mechanics is used extensively to explain the root cause of fracture.

11.2 GENERAL METHODOLOGY FOR FRACTURE MECHANICS ANALYSIS

The information necessary for conducting fracture mechanics analysis consists of:

- Identification of fracture critical regions which are sections of the component which are highly stressed (monotonic, sustained, or cyclic stresses) and/or subjected to aggressive environment.

- Operating conditions such as the service environment, loading parameters which determine the operating stress levels such as rotational speeds and power output, etc.
- Size, shape, and location of manufacturing or service generated cracks or crack-like defects,
- K-calibration expressions that combine stress magnitudes and flaw shapes/sizes for the pertinent component geometry.
- Material data including the crack growth rates under the appropriate environment and loading conditions that may include fracture toughness, tensile properties, fracture appearance transition temperature (FATT), material chemistry, fatigue crack growth rates (FCGRs), environment or creep assisted cracking rates etc., as applicable.
- Microstructure, chemistry, processing, and thermal exposure history of the component during manufacturing.
- For components already in service, an assessment is needed about the degradation in properties due to long-term exposure to service conditions.

After the information that is pertinent is assembled, the next step in the fracture mechanics analysis is to calculate the magnitude of the applicable crack tip parameters, K, J, C_t, C^* etc., as appropriate. These require the applied stress level and expressing them as a function of the characteristic crack dimension, a. From the fracture toughness and the expressions for estimating K and/or J, the final crack size, a_f, can be estimated. The crack growth rate data can then be used to estimate the crack size at inspection, a_i, to meet the design life requirement. Alternatively, if a_i is chosen as the largest crack that can go undetected by the chosen nondestructive inspection technique, the expected component life can be calculated. Often, a_i is chosen conservatively so the component life is also predicted conservatively. If that is the case, the predicted life should be treated as an inspection interval or the frequency at which inspections must be performed to detect cracks during service. If no cracks are detected during the in-service inspection that are greater than a detectable level, the component may be returned to service until the next inspection becomes due.

Fracture mechanics analysis is often performed with the aid of computers because once a computer program is developed several scenarios can be considered to identify optimum conditions. There are commercial computer packages available to perform fracture mechanics analysis but, engineers very often prefer to write their own computer programs which are specific to their needs and the components they most frequently work with. Component-specific computer programs are easier to use and maintain. The modern computing tools such as Excel[1] and MATLAB[2] are very useful in performing repetitive calculations.

11.3   CASE STUDIES

In this section, several examples are described to illustrate the use of fracture mechanics. The examples chosen are hypothetical but do have some resemblance to actual situations. In each case, the assumptions are discussed to identify the limitations of the results.

11.3.1 Optimizing Manufacturing Costs

Fracture mechanics provides a quantitative method for evaluating the influence of potential manufacturing defects/flaws in components and their effect on safe and reliable performance during service. This is done by selecting realistic inspection criteria, and choosing economic, but safe and reliable methods of NDE.

11.3.1.1 Problem Statement

Cylindrical shafts in large industrial centrifugal flow and axial flow fans are used for mounting hubs to which rows of blades are attached along the length of the shaft. These shafts can be as large as 10 m long and have a diameter of 580 mm and can weigh as much as 10–20 metric tons. These are large steel forgings that are produced as per ASTM specification A293 for plain carbon steel. The material is an American Iron and Steel Institute (AISI) and Society of Automotive Engineers (SAE) specification AISI/SAE 1045; see Tables 11.1 and 11.2 for chemical and mechanical properties of the steel. The company is experiencing new emerging competition and must take steps to reduce costs of their fans to retain a competitive market position. It was also noted that in the 50-year history of the company, no shaft failures during service were reported, indicating possible conservatism in design.

Primary fatigue stresses in the shaft are generated from gravity bending under its own weight and due to centrifugal stresses due to its own mass and by the mass of rows of attached blades. The fatigue stresses are fully reversed during each rotation of the shaft at a speed of 1,800 revolutions per minute (RPM). The centrifugal stresses fluctuate between the maximum value at full speed and 0 when the fan is shutdown. The QA procedure for the shaft forgings calls for (a) full ultrasonic inspection of the shaft to find forging bursts, if any, in the interior of the shaft close to its center and (b) magnetic particle inspection of the entire outside shaft surface to detect surface cracks. As a further precaution, 2 mm of the outer surface of the forging is machined to ensure that no shallow surface cracks remain. Due to these stringent and expensive QA requirements, one out of every three forgings is instead of are rejected, and the costs are passed on to the forgings that pass inspection and

TABLE 11.1
Nominal Chemistry of SAE/AISI 1,045 Steel (Wt%)

C	Mn	P	S	Fe
0.42–0.50	0.6–0.9	≤0.04	≤0.05	Balance

TABLE 11.2
Nominal Mechanical Properties of the Shaft Material

Brinell Hardness	Yield Strength (MPa)	Ultimate Strength (MPa)	% Elongation (50 mm Gage Length)	Reduction in Area (%)	E (GPa)	Poisson's Ratio, (v)
84	310	565	16.0	40.0	200	0.29

are approved for use. Safety of rotors cannot be compromised because they spin at very high rotational speeds. You are asked to design a test program and conduct a fracture mechanics analysis to support more realistic inspection requirements and accept/reject criteria for the shaft to reduce wastage and costs.

11.3.1.2 Approach

The stress analysis reveals that the magnitude of the normal stresses in the axial direction are due to gravity bending and are the highest on the surface of the shaft where they vary between ±52 MPa during each rotational cycle. The stresses are due to the weight of the shaft itself and due to the weight of the fan wheels attached to the shaft along its length. The neutral axis lies along the shaft axis where the axial stress is expected to be 0. The frequency of loading, as stated earlier is 30 Hz. The centrifugal stresses present in the shaft due to rotation are in the circumferential direction and are the highest at the center of the shaft. The magnitude of these stresses is also limited to 52 MPa by design and they fluctuate between that maximum value and 0 between starts and stops. These fans are estimated to undergo 10,000 start and stop cycles during their lifetime.

Growth of potential circumferential cracks on the surface of the shaft that lie on the circumferential-radial plane will be primarily driven by the gravity bending stresses that are normal to the shaft axis. Cracks in the central region of the shaft such as forging bursts will be driven by the centrifugal stresses. These cracks are schematically shown in Figure 11.2a and b, respectively.

The evolution of the shape of the surface cracks in cylindrical rods subjected to bending stresses have been studied by Forman and Shivakumar [12]. They tested cylindrical rods of Ti-6Al-4V alloy containing circumferential cracks and subjected them to reversed bending stresses. Figure 11.3 shows a picture of the evolving crack front from their work. The front of the growing crack appears to be in the form of a circular arc that intersects the outside surface of the shaft at an angle of 90°.

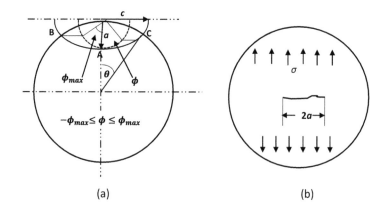

(a)   (b)

FIGURE 11.2 Postulated defect types that can cause failure in rotating shafts made from Society of Automotive Engineers (SAE) 1045 steel forgings used in large industrial fans (a) a semi-elliptical shaped defect in the radial-circumferential plane located on the OD of the shaft and (b) forging bursts that are likely present in the center of the shaft.

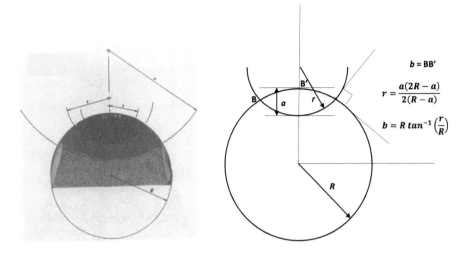

$b = BB'$

$r = \dfrac{a(2R - a)}{2(R - a)}$

$b = R\,tan^{-1}\left(\dfrac{r}{R}\right)$

FIGURE 11.3 (a) Picture of the evolving crack front in a circular shaft subject to pure bending and (b) schematic showing the various crack angles and dimensions (The picture is taken from Forman and Shivakumar [12] and reproduced with permission of ASTM, Conshohocken, PA).

Eventhough the initial crack in Figure 11.3a is machined and does not have the semi-circular shape, the crack grows quickly and stabilizes into a semi-circular crack front.

The next step is to assemble the expressions for estimating stress intensity factor, K or ΔK for the crack geometries in Figure 11.2. For the cracks shown in Figure 11.2a, the K-expressions were developed by Forman and Shivkumar [12] and are given as follows [12].

$$K = \sigma_b F_b\left(\frac{a}{2R}\right)\sqrt{\pi a} \qquad (11.1)$$

where σ_b = bending stress, a = crack size, R = radius of the shaft,

$$F_b(a/2R) = g(a/2R)\left(0.923 + 0.199\left(1 - \sin\frac{\pi a}{4R}\right)^4\right) \qquad (11.2)$$

$$g\left(\frac{a}{2r}\right) = \frac{0.92\left(\frac{2}{\pi}\right)\left[\dfrac{\tan\left(\pi a/4R\right)}{\cos\left(\pi a/4R\right)}\right]^{1/2}}{\cos\left(\pi a/4R\right)} \qquad (11.3)$$

$F_b(a/2R)$ is plotted in Figure 11.4.

The K-expression for a crack of the type shown in Figure 11.2b is conservatively approximated by equation (3.22) derived for a center cracked body subjected to uniform stress normal to the crack plane of finite width, $2W$.

$$K = \sigma\sqrt{\pi a}\left[\sec\left(\frac{\pi a}{2W}\right)\right]^{1/2}$$

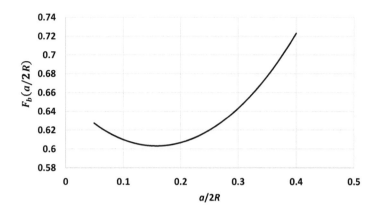

FIGURE 11.4 The K-calibration function for a circular shaft with a circumferential crack subjected to pure bending with a stress of σ_b [12].

In this case, σ = centrifugal stress at the center of the shaft and $W = R$, the radius of the shaft. The K values for the given stresses for the two flaw geometries are given in Table 11.3.

Next, we must assemble material properties needed to conduct the fracture mechanics analysis. The fatigue stresses due to gravity bending are low but the accumulated cycles during a 30-year life are estimated at 2.84×10^{10}. Thus, any crack that grows will propagate to failure. Based on this, the most relevant material property is the threshold value of ΔK below which fatigue crack growth does not occur. Since the bending stress reverses fully during each cycle, the load ratio under which FCGR data are generated must be −1. The central crack is expected to grow by fatigue loading from centrifugal stresses due to starts and stops but, such cycles accumulate to 10,000 during the lifetime of the shaft. The fatigue crack growth properties for AISI/SAE 1045 steel are summarized in Table 11.4. These are based on the constants for the wide-range FCGR model described by equation (6.6) and repeated below.

$$\frac{1}{da/dN} = \frac{1}{A_1 \Delta K^{m_1}} + \frac{1}{A_2}\left(\frac{1}{(\Delta K)^{n_2}}\right)$$

Compared to equation (6.6), the last term in the equation containing the critical K, K_c, is not included because the data in that regime were not available, and not needed for this application. We will assume a conservative value for K_{Ic} of 40 MPa$\sqrt{m}$ because fans could potentially be operating in extreme cold climates where there is risk of brittle fracture during a cold start. This value is representative of such conditions. Thus, fatigue life is considered to come to an end when K_{max} approaches K_{Ic}.

Note that FCGR behavior in Region II is identical at both load ratios, but different in Region 1. The FCGR trend for $R = -1.0$ is expected to be comparable to the trend at $R = 0.1$ and it was verified by conducting limited testing at $R = -1.0$ as described in reference [13]. The negative load ratio tests are typically harder to perform, so it was used just for verification testing.

TABLE 11.3

Stress Intensity Factor as a Function of Crack Size for a Circumferential Surface Crack in a Cylinder and for a Forging Burst Crack in the Center of the Shaft

Circumferential Surface Flaw on the OD		Forging Burst Flaw in the Center of the Shaft	
Depth, a (m)	$\Delta K \left(\text{MPa}\sqrt{m} \right)$	Half Crack Size (m)	$\Delta K \left(\text{MPa}\sqrt{m} \right)$
0.001	1.93	0.025	14.78
0.002	2.73	0.03	16.23
0.003	3.34	0.035	17.57
0.004	3.85	0.04	18.84
0.005	4.29	0.045	20.05
0.006	4.69	0.05	21.22
0.007	5.06	0.055	22.35
0.008	5.40	0.06	23.45
0.009	5.72	0.065	24.54
0.01	6.02	0.07	25.61
0.011	6.30	0.075	26.67
0.012	6.57	0.08	27.72
0.013	6.83	0.085	28.77
0.014	7.07	0.09	29.83
0.015	7.31	0.095	30.89
0.016	7.54	0.1	31.97
0.017	7.76	0.105	33.06
0.018	7.97	0.11	34.16
0.019	8.17	0.115	35.29
0.02	8.37	0.12	36.44
		0.125	37.62
		0.13	38.84
		0.135	40.09
		0.14	41.39
		0.145	42.73

Abbreviation: OD, outside diameter.

At 30 Hz, the shaft will accumulate 2.8×10^{10} fatigue cycles in 30 years. Thus, if the crack begins to grow, it will likely reach the critical size. Therefore, the maximum allowable crack size must be always such that the $\Delta K < \Delta K_{th}$. According to the conventional definition, ΔK_{th} is the value of ΔK corresponding to a da/dN of 10^{-7} mm/cycle. That is listed in Table 11.4 for load ratios of 0.1 and 0.5. We can also estimate the ΔK that corresponds to a da/dN that yields 1 mm of crack extension in 2.8×10^{10} cycles; in other words, $da/dN = 3.57 \times 10^{-11}$ mm/cycle or rounded to 1×10^{-11} mm/cycle. Those values are also listed in Table 11.4. That value of ΔK is 6.05 MPa$\sqrt{m}$. The crack depth at which that ΔK is reaches that value from Table 11.3 is 0.01 m = 10 mm as seen in Figure 11.5a. The corresponding surface length of the crack can be estimated from the following equations.

$$r = \frac{a(2R - a)}{2(R - a)} \tag{11.4}$$

TABLE 11.4

Constants in the Fatigue Crack Growth Equation for SAE/AISI 1,045 Steel Used in Fan Shafts

Load Ratio	A_1 (mm/MPa^{-n_1})	n_1	A_2 (mm/MPa^{-n_2})	n_2	ΔK_{th} $(\text{MPa}\sqrt{m})$	ΔK for $da/dN = 10^{-11}$ mm/cycle
0.1	8.85×10^{-28}	23	8.85×10^{-8}	3.5	9	6.05
0.5	1.14×10^{-22}	19	8.85×10^{-8}	3.5	6.25	3.83

The ΔK is in $\text{MPa}\sqrt{m}$ and da/dN in mm/cycle.

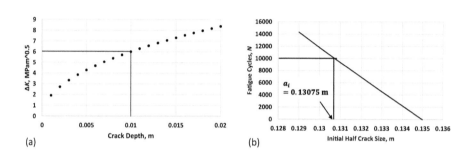

(a)   (b)

FIGURE 11.5   (a) The plot of ΔK as a function of crack depth for surface cracks showing that the crack depth, a, corresponding to a $\Delta K = 6.05$ MPa$\sqrt{m}$ is 10 mm. (b) The relationship between fatigue cycles and allowable initial flaw size showing that to achieve a cyclic life of 10,000 cycles, the maximum allowable half-flaw size should be 130.75 mm.

$$b = R \tan^{-1}\left(\frac{r}{R}\right) \tag{11.5}$$

Substituting $R = 0.29$ m, $a = 0.01$ m, we let $b = 0.01055$ m. Thus, $2b = 0.0211$ m $= 21.1$ mm. Thus, if magnetic particle inspection reveals a flaw that 23 mm long along the circumference, only then the shaft should be rejected. It is highly unlikely that a flaw of such a large size will be present in a shaft.

Next, we estimate the critical size of a forging burst type flaw. From Table 11.3, the K reaches a value of 40 MPa$\sqrt{m}$ at a half flaw size of 0.135 m $= 135$ mm. Thus, the total flaw size is 270 mm. Our inspection criterion should be based on an initial half flaw size that will grow to 135 mm in 10,000 start and stop cycles. At these ΔK levels, the da/dN versus ΔK relationship is given by equation (11.6), if da/dN is in m/cycle and ΔK in MPa$\sqrt{m}$:

$$\frac{da}{dN} = 1.13 \times 10^{-12} \Delta K^{3.5} \tag{11.6}$$

Thus, we can write the following governing equation for calculating the initial crack size at inspection, a_i as in equation (11.7).

$$\int_{0}^{10{,}000} dN = 10{,}000 = \int_{a_i}^{0.135} \frac{da}{\left(1.13 \times 10^{-12} \left(\sigma\sqrt{\pi a} \left[\sec\left(\frac{\pi a}{2R} \right) \right]^{1/2} \right)^{3.5} \right)} \tag{11.7}$$

Using excel to solve the above equation yields $a_i = 0.13075$ m $= 130.75$ mm as shown in Figure 11.5b.

The recommendation for the NDE criteria should be (a) the maximum permissible length of the circumferential flaw to be detected by magnetic particles is 21.1 mm and (b) the maximum permissible size of forging bursts should be 261.5 mm. Since both these sizes are large, very few shafts, if any, should have to be rejected. This will save the company considerable money.

11.3.2 RELIABILITY OF SERVICE-DEGRADED STEAM TURBINE ROTORS

In the early 1950s there was a surge in the world-wide demand for electric power when several countries were industrializing their economies. To produce power economically, it was necessary to develop turbines that could produce up to 200 MW of power. Figure 11.6 shows a cut out of a modern steam turbine showing the complexity of a machine in which a massive rotor spins at 3,600 rpm generating large amounts of kinetic energy. The rotor consists of a main shaft made from 1Cr-1Mo-0.25V steel forging, and several rows of blades mounted on hubs distributed along the shaft length. After several large turbines were built during the 1950s and were in use, it was discovered that 1Cr-1Mo-0.25V steels tend to embrittle when exposed to long periods of sustained temperatures in the range of 400°C to 600°C. This results in significant loss

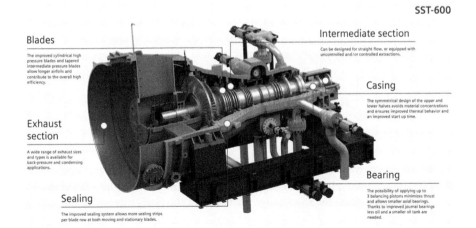

FIGURE 11.6 Picture of a cut-out of a steam turbine showing the rotor of the machine consisting of a shaft and blade mounted in rows along the rotor (Picture courtesy of Siemens Co. https://assets.siemens-energy.com/siemens/assets/api/uuid:1957c70c-6d96-4716-b239-428a 53fdff4b/se-brochure-industrial-steam-turbines-2021.pdf?ste_sid=a2b897357570bd382ee92 766475dbe8d).

in fracture toughness in the lower-shelf and the transition regions. The latter is characterized by a rise in the FATT. This phenomenon is known as temper embrittlement. In most cases, the hardness and tensile properties of the material do not change significantly, but the ductile to brittle transition temperature or the FATT can increase by as much as 100°C and the fracture toughness in the lower shelf can degrade by as much as 50%. The high-pressure (HP) and intermediate pressure (IP) regions of turbine rotors experience metal temperatures in the range of 427°C–538°C that is right in the range for temper embrittlement, causing significant concerns about the safety of rotors.

Temper embrittlement raises the risk of fracture significantly in the turbine shaft during cold starts[3] because under these conditions, the metal temperatures are near ambient to begin with, so in addition to low fracture toughness conditions, the rotor experiences high levels of thermal stresses due to transient temperature gradients between the outside diameter (OD) and the inside diameter (ID) of the rotor. Gallatin turbine rotor burst that occurred in June of 1974, is an example of the potential for such catastrophic fracture [14] that took the industry by surprise and generated considerable activity in risk assessment of turbines of the same vintage. Gallatin turbine had been in service for 17 years and the failure occurred suddenly without any warning during a cold start. Big metal pieces of the rotor were found far away from the location of the turbine. In the aftermath of the Gallatin turbine fracture, rotors of that vintage in service were evaluated for risk of fracture. The immediate response consisted of following steps:

- Ultrasonically inspect the rotors for presence of defects. Those with significant defect indications in the high stress areas were retired from service.
- Launch experimental programs to test materials from rotors that had been retired from service to quantify the extent of degradation in fracture toughness.
- Conduct fracture mechanics analyses to evaluate flaws and set up an accept/reject criterion.
- Investigate practical de-embrittlement heat treatments that could be applied in the field to regain either fully or partially the loss in fracture toughness.
- Initiate long-term programs to improve steel making processes to reduce impurity content such as sulfur (S) and phosphorus (P) to reduce the threat of temper embrittlement.
- Initiate basic research to develop time-dependent fracture mechanics approaches for predicting crack growth under creep and creep-fatigue conditions, relevant in high pressure and Intermediate pressure regions of the rotor.

The last two steps were directed toward addressing long-term issues and would only affect the future rotors, but not the ones in service at the time. In the following discussion, we will address the short-range responses of the types mentioned in the first four bullets above.

Two rotors that were retired from service after inspection revealing some flaw indications during ultrasonic testing were shipped to a laboratory for more inspections and testing. The results were reported in a paper by Argo and Seth [15]. Both rotors were produced from the same heat of steel with the chemical composition from the mill certificate as shown in Table 11.5.

TABLE 11.5
The Chemical Composition of the Steel Used in Fabricating the Retired 1Cr-1Mo-0.25V Rotors

C	Mn	P	S	Si	Ni	Cr	Mo	V	Fe
0.34	0.80	0.025	0.020	0.33	0.19	1.05	1.21	0.21	Bal

The following common observations are made from the before and after service data from the two rotors summarized in Table 11.6:

- There were small decreases in yield strength and tensile strength of the rotors during service consistent with classical characteristics of temper embrittlement.
- Significant decreases in percent elongation and percent reduction in areas at room temperature were recorded indicative of temper embrittlement.
- The Charpy impact energy at room temperature was in the lower shelf after exposure to service, and the FATT that represents 50% ductile and 50% brittle behavior, was high. No FATT measurements were available in the materials test records of the original rotors so that comparison was not possible.

TABLE 11.6
Tensile and Charpy Impact and Fracture Toughness Properties of the Original Rotors, after Service, and after Various Rejuvenation Heat-Treatments [15]

Heat Treatment	Rotor	0.2% Yield Strength (MPa)	Ultimate Tensile Strength (MPa)	% Elong	% Red in Area	RT CVN (J)	FAT (T°C)	RT K_{Ic} (MPa$\sqrt{m}$)
Original rotor	A	722	861.3	13.8	32.1	-	-	-
	B	646.4	820.4	15.16	38.3	-	-	-
Rotor after service	A	686.3	853.1	10.7	15.3	2.71	301	20[b]
	B	630.1	812.1	14.75	27.6	1.356	215.5	20[b]
1. 665°C for 18 hours	A	638.4	801.0	15.2	32.1	5.4	121.1	27.9[b]
1.	B	607.3	783.2	15.8	36.6	8.12	101.6	34.2[b]
2. 665°C for 32 hours/FC to 260°C/AC	A	632	805.4	15.8	38.8	6.78	121.1	44[a]
1.	B	608.7	786.4	17	42.5	5.4	126.7	52.8[a]
3. 954°C for 18 hours/ cooled at 37.7°C/ hours/676°C 18 hours	A	644.5	817.6	15.6	35.6	10.8	93.3	39.4[b]
1.	B	662.5	834.8	16.8	37.9	10.8	96.1	39.4[b]
4. 954°C for 18 hours/ cooled at 149°C/ hours/676°C 18 hours	A	648.7	817.6	16.0	37.9	8.12	96.1	34.2[b]
	B	664.6	836.2	14.5	30.5	8.12	96.1	34.2[b]

[a] Actual measured values.
[b] Estimated values from CVN energy.

The correlation between K_{Ic} and Charpy energy in the lower shelf is given by equation (5.11) repeated here for easy access.

$$K_{Ic} = \left[\left(12\sqrt{C_v} - 20 \right) \left(\frac{25}{B} \right)^{1/4} + 20 \right]$$

In the above equation, 20 MPa$\sqrt{m}$ is the fracture toughness in the lower shelf but even for that the required CVN energy is 2.8 Joules which is higher than the measured CVN energies of 2.71 and 1.357 J measured from the after-service rotor material. Nevertheless, we choose 20 MPa$\sqrt{m}$ as the lower shelf fracture toughness value for the embrittled material for use in the fracture mechanics analysis but will keep in mind that the fracture toughness could be even lower.

In the study reported by Argo and Seth [15], two de-embrittling and two complete rejuvenation heat treatments using specimens taken from the retired rotors were investigated. They characterized the mechanical properties after application of the heat-treatments. The results are summarized in Table 11.6 for comparison with original properties and that of rotors and after-service. The following heat treatments were used to recover the rotor properties:

1. A de-embrittling treatment consisting of exposure to 665°C for 18 hours and air cooled (AC).
2. A de-embrittling treatment consisting of exposure to 665°C for 32 hours and furnace cooled (FC) to 260°C, followed by air cooling.
3. Specimens were solution annealed at 954°C for 18 hours and then cooled at 37.7°C/hour and then tempered at 676°C for18 hours and then AC.
4. Specimens were solution annealed at 954°C for 18 hours, cooled at 149°C/hour and then tempered at 676°C for 18 hours.

Among the various heat treatments, the one most effective in raising the room temperature fracture toughness and lowering the FATT was the second heat treatment. FATT decreased by 180°C for Rotor A and by 114°C in Rotor B. Also, the measured room temperature fracture toughness values increased from an estimated value of 20 to measured values 44 and 52.8 MPa$\sqrt{m}$ for rotors A and B, respectively. Thus, this heat treatment is the most effective heat treatment for restoring the properties of the rotors. The yield and tensile strengths decreased some because of the heat treatment but substantial increases in percent reductions in area and in percent elongations that were observed were preferred as were the increases in room temperature values of K_{Ic}. Thus, this heat treatment that is also economical and less time consuming than treatments 3 and 4 is preferred. The practicality of subjecting the rotors to such treatment at the service location versus bringing them into a special shop to impart the treatment must be assessed separately for the rotors on a case-by-case basis.

11.3.2.1 Analysis of Stresses on the Rotor during Service

The primary steady-state stresses[4] in the rotor are in the circumferential direction, σ_θ, and are due to thermal gradients and due to centrifugal loads. These should be determined for the specific rotor. Here we assume that they vary with rotor RPM and with distance from the center of the rotor. At 3,600 rpm, we assume that the analysis leads to equation (11.8) below under steady-state conditions:

$$\sigma_\theta = 309.59 - 2.34(r - r_b) + 0.009(r - r_b)^2 \tag{11.8}$$

where r = distance from the center of the rotor axis, and r_b is the bore radius. The radial distance to the bore surface of the rotor or r_b is 76.2 mm, where the stresses are the highest. The stresses are calculated as a function of the rotational speed in rpm, and they are estimated in MPa. These values are plotted in Figure 11.7. In actual rotors, the stress distribution will be much more complex and will also vary in the axial direction because the radial distance to the tip of the blade varies along the rotor axis. In this exercise, we assume a rather simplistic stress distribution to illustrate the use of fracture mechanics.

11.3.2.2   Flaws in the Rotors and Their Evaluation

Flaws in rotors were detected and their size determined by ultrasonic inspections that were performed on rotors of the same vintage as the Gallatin rotor. Since the circumferential stress is the highest near the bore of the rotor, Figure 11.7, flaws that lie on the radial-axial plane at or near the bore of the rotor present the highest risk of initiating fracture. We assume that all flaws detected lie on the radial-axial plane and are of an elliptical shape with characteristic dimensions of a and c along the depth and length directions. This combination of flaw orientation and stress is the worst-case scenario. The dimension a is the flaw depth along the radial direction and c is the half flaw length along the axial direction. Since the ultrasonic technique yields an estimated flaw area, it is of interest to determine the a/c value for a fixed area that will yield the highest K. If the flaw is assumed to be in an infinite body, the value of K at the deepest point in the radial direction is given by equation (3.29) that is repeated below:

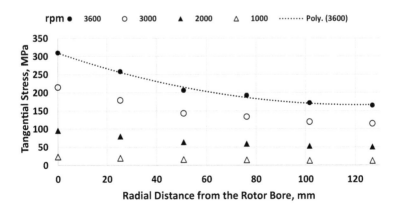

FIGURE 11.7   Centrifugal stress distribution in the rotor at various rotational speed as a function of distance from the bore surface. The radius of the bore that has the highest stresses is 76.2 mm.

$$K = \sigma\sqrt{\frac{\pi a}{Q}}; \quad Q = 1 + 1.464\left(\frac{a}{c}\right)^{1.65}$$

If the area of the flaw detected is A, it follows that

$$a = \frac{A}{\pi c} \quad \text{or} \quad a^2 = \frac{A}{\pi}\left(\frac{a}{c}\right)$$

Thus, $\quad \dfrac{K}{\sigma A^{1/4}} = \left[\dfrac{(a/c)^{1/2}}{1 + 1.464(a/c)^{1.65}}\right]^{1/2}$ \hfill (11.9)

Figure 11.8 plots the value of $\dfrac{K}{\sigma A^{1/4}}$ from the equation against the values of a/c and it is noted that the highest values of K are found when the a/c is 0.5. Thus, all selected flaws for evaluation are assumed to have an a/c of 0.5 in the interest of being conservative.

We consider three different types of flaws in the rotor as shown in Figure 11.9, including (a) a semi-elliptical surface flaw on the bore surface, (b) embedded elliptical flaw with its center located at a distance r from the rotor axis that is equal to a distance $r - r_b = t$ from the bore surface, where r_b is the bore radius, and (c) array co-planar elliptical flaws. We will then establish allowable sizes for these flaw types that can cause fracture for comparison with actual flaws detected during ultrasonic testing. A decision can then be made on whether the detected flaws pose a risk of fracture.

11.3.2.3 Semi-Elliptical Surface Flaw on the Bore

The circumferential stress is given by equation (11.8) and at the bore surface, $r = r_b$, it is estimated to be 309.59 MPa. We assume that the circumferential stress does not vary significantly along the radial direction over the crack face. From equation (11.8),

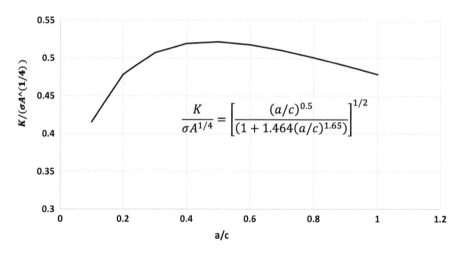

FIGURE 11.8 Variation in the value of nondimensional K as a function of a/c for a flaw with a given area A in the radial-axial plane of the rotor.

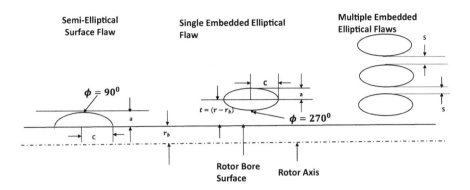

FIGURE 11.9   The three flaw geometries that pose a potential threat of fracture if present in size(s) large enough to cause fracture.

we see that the stress does vary radially but over the small distance across the depth a is assumed to be constant and characterized by its value at center of the ellipse. It follows then from the principle of linear superposition discussed in Chapter 3 that in equation (3.31) to calculate K, we can choose $H = 0$ and estimate K as follows:

$$K = \sigma_\theta \sqrt{\frac{\pi a}{Q}} f\left(\frac{a}{t}, \frac{a}{c}, \frac{c}{W}, \phi\right)$$

$$\text{Where,} \quad f = \left[M_1 + M_2 \left(\frac{a}{t}\right)^2 + M_3 \left(\frac{a}{t}\right)^4 \right] f_\phi \cdot g \cdot f_w$$

$$Q = 1 + 1.464 \left(\frac{a}{c}\right)^{1.65}$$

$$M_1 = 1.13 - 0.09 \left(\frac{a}{c}\right)$$

$$M_2 = -0.54 + \frac{0.89}{0.2 + a/c}$$

$$M_3 = 0.5 - \frac{1.0}{0.65 + \dfrac{a}{c}} + 14\left(1.0 - a/c\right)^{24}$$

$$f_\phi = \left[\left(\frac{a}{c}\right)^2 \cos^2\phi + \sin^2\phi \right]^{\frac{1}{4}} = 1 \quad \text{for} \quad \phi = 90°$$

$$g = 1 + \left[0.1 + 0.35\left(a/t\right)^2 \right]\left(1 - \sin\phi\right)^2 = 1 \quad \text{for} \quad \phi = 90°$$

$$f_w = 1 \quad \text{for} \quad \phi = 90$$

For $a/c = 0.5$, $Q = 1.466$, $M_1 = 1.085$, $M_2 = 0.731$, $M_3 = -0.369$

Thus, for a semi-elliptical flaw on the surface of the bore for $\phi = 90°$,

$$K = \sigma_\theta \sqrt{\frac{\pi a}{Q}} \left[1.085 + 0.731(a/t)^2 - 0.369(a/t)^4 \right]$$

For surface flaws that are small in comparison to the OD of the rotor, $a/t \approx 0$,

$$K \approx 1.085 \, \sigma_\theta \sqrt{\frac{\pi a}{1.466}} = 1.588 \sigma_\theta \sqrt{a}$$

At fracture, $K = K_{Ic}$, and $a = a_{cr}$ so:

$$a_{cr} = \left(\frac{K_{Ic}}{1.588 \sigma_\theta} \right)^2 = 4.1 \times 10^{-6} K_{Ic}^2 \text{ m}$$

For $K_{Ic} = 20$ MPa$\sqrt{m}$, $a_{cr} = 0.00164 = 1.64$ mm, and for $K_{Ic} = 40$ MPa$\sqrt{m}$, $a_{cr} = 6.5$ mm. Hence, by de-embrittling the rotor, the flaw tolerance can be increased substantially. It will also be prudent to increase the rotor rpm gradually allowing the metal temperature to increase above the FATT temperature before it gets to the full rpm of 3,600 thereby lowering the stresses as seen in Figure 11.7.

11.3.2.4 Single and Multiple Co-Planar Embedded Flaws

The schematic of a single and multiple embedded elliptical flaws is shown in Figure 11.9. In this case, the distance of the flaw from the rotor bore surface determines the stresses that the flaw experiences during service. The form of the K-expression is identical to the one for the semi-elliptical surface flaw except the various terms differ in detail as given below. In the case of a single embedded crack, the maximum value of K is expected at $\phi = 270°$ as shown in Figure 11.9.

$$M_1 = 1, \quad M_2 = \frac{0.05}{0.11 + (a/c)^{3/2}} \quad M_3 = \frac{0.29}{0.23 + (a/c)^{3/2}}$$

For $a/c = 0.5$, $M_1 = 1$, $M_2 = 0.108$, $M_3 = 0.497$, and $f_\phi = g = f_w = 1$, $Q = 1.466$.

$$\text{Thus, } K = \sigma_\theta \sqrt{\frac{\pi a}{1.466}} \left[1 + 0.108(a/t)^2 + 0.497(a/t)^4 \right]$$

Substituting equation (11.8) into the above equation, we get:

$$K = \left(309.59 - 2.34t + 0.009t^2 \right) \sqrt{\frac{\pi a}{1.466}} \left[1 + 0.108(a/t)^2 + 0.497(a/t)^4 \right] \quad (11.10)$$

For multiple cracks, we estimate a stress intensity magnification factor, M_K from the analysis of Hall and Kobayashi [16]. In this analysis, they estimated the magnification factor due to two and a linear array of embedded penny shaped cracks of diameter equal to $2a$ and located at a distance S apart from each other as shown in Figure 11.10. The equation describing the relationship is given in equation (11.11).

$$K = M_K \left(309.59 - 2.34t + 0.009t^2\right)\sqrt{\frac{\pi a}{1.466}}\left[1 + 0.108\left(a/t\right)^2 + 0.497\left(a/t\right)^4\right] \quad (11.11)$$

where,

$$M_K = 2.249 - 0.8647\left(\frac{S}{a}\right) + 0.252\left(\frac{S}{a}\right)^2 - 0.328\left(\frac{S}{a}\right)^3 + 0.001\left(S/a\right)^4 \quad (11.12)$$

Thus, K can be estimated for single cracks and for an array of multiple cracks for various values of a and S/a ratios. Note that the magnification factors have been estimated for penny-shaped cracks and to adopt that for elliptical cracks, we have assumed that M_K for an elliptical crack with crack depth of $2a$ is approximately equal to the M_K for a penny-shaped crack of diameter $2a$.

Assuming K_{Ic} of 20 MPa$\sqrt{m}$, the critical crack size, a_{cr}, needed for the K to reach a value of 20 MPa$\sqrt{m}$ can be estimated for cracks located at various distances from the bore surface and for various normalized spacing, S/a, as plotted in Figure 11.11. The case for a single crack is represented by the condition that $S/a > 6$. This indicates that for $S/a > 6$, there is no interaction between adjacent cracks, and they can be treated as single cracks. As expected, the critical crack size decreases for an array of co-planar cracks as the distance between the cracks is reduced.

The critical crack size, a_{cr}, for a surface crack on the bore was estimated earlier as 1.64 mm and compared to embedded cracks in the vicinity of the bore with an $S/a > 4$, represents a higher risk of fracture. However, for $S/a < 4$, the risk of fracture is higher

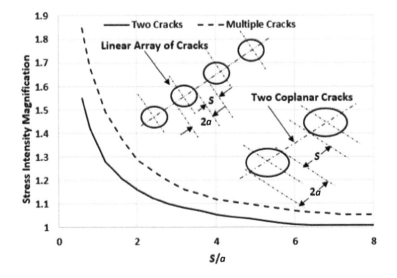

FIGURE 11.10 Stress intensity magnification factors for two and for an array of penny-shaped cracks of radius a relative to a single crack [15,16].

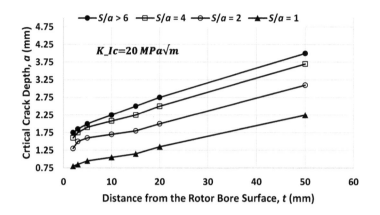

FIGURE 11.11 Critical crack depth of embedded flaws in the rotor located at various distances from the bore surface.

from the embedded flaws. Embedded single cracks of any size in the interior of the rotor away from the bore surface have reduced risks of fracture compared to surface flaws of the same depth located on the bore surface or near the bore surface.

In the study by Argo and Seth [15] all crack indications detected were in the range of $0.3 \leq a \leq 0.43$ mm. Thus, based on this analysis, these rotors were not expected to fracture, as was indeed the case when they were removed from service.

A program to inspect all 1Cr-1Mo-0.25V turbine rotors of the Gallatin rotor vintage for cracks in the bore region was necessary. All indications must be assessed using the methodology outlined here. The practicality of heat treating the rotors using a de-embrittling heat-treatment should be considered because of the considerable reduction in the risk of fracture that will result from it. If these steps are taken, it significantly reduces the risk of fracture. When rotors in service are inspected and the flaws detected have been all deemed to be acceptable or no flaws are detected, the question arises about the remaining safe life of the rotor and the frequency of in-service inspections going forward.

11.3.2.5 Remaining Life Assessment/Inspection Interval Calculations

Remaining life/inspection interval is the cycles/time required for cracks to grow from their current size to the critical crack size. The existing flaws are expected to grow by a relationship described by equation (11.13) taken from [17]:

$$\frac{da}{dN} = 1.41 \times 10^{-11} (\Delta K)^{2.7} \tag{11.13}$$

where da/dN is in m/cycle and ΔK is in MPa$\sqrt{m}$. This equation can be integrated between current crack size, a_0, to the critical crack size, a_{cr}, to determine the number of cycles to grow the crack from its initial assumed size to the critical size. It is prudent to choose a suitable safety margin when specifying an inspection interval or remaining life. Let us assume that the area of the largest crack located on the bore surface is 3.14 mm^2. Assuming $a/c = 0.5$, this translates into an $a_0 = 0.001$ m.

The $a_{cr} = 0.00164$ m, $\sigma_\theta = 309.59$ MPa. Thus, the cycles to failure, N_f, is given by the following equation.

$$N_f = \int_0^{N_f} dN = \int_{0.001}^{0.00164} \frac{da}{1.41 \times 10^{-11} \left(1.588\sigma_\theta \sqrt{a}\right)^{2.7}} = \int_{0.001}^{0.00164} \frac{da}{1.41 \times 10^{-11} \left(1.588\sigma_\theta \sqrt{a}\right)^{2.7}}$$

$$= 10945 \int_{0.001}^{0.00164} \frac{da}{a^{1.35}} = 10{,}945\left(0.001^{-.035} - 0.00164^{-0.35}\right) = 19{,}524 \text{ cycles}$$

The total anticipated number of start-up to shut-down cycles at a rate of approximately one per month for base loaded units is 120 cycles over 10 years. Comparing this to the calculated cycles, there is low risk of fracture if the rotors are thoroughly inspected using ultrasonic techniques, and there is assurance that no defects greater than 3.14 mm^2 in area are detected in the bore region of the rotor.

11.3.3 DESIGN OF VESSELS FOR STORING GASEOUS HYDROGEN AT VERY HIGH PRESSURES

Pressure vessels for storing hydrogen at pressures in the range of 300–1,000 bar (30–100 MPa) are a critical need for fueling stations for fuel-cell powered vehicles, for back-up power in residential and office buildings, and for fork-lifts in warehouses. These systems are also essential for producing clean power and reducing greenhouse gas emissions by accessing vast amounts of energy available from wind and sun. Excess energy during high generation periods can be stored in the form of compressed hydrogen produced by hydrolyzing water and can be made available on demand to power fuel cells during the low generation periods.

Figure 11.12 shows a picture of a 750-L hydrogen storage vessel capable of storing hydrogen at 100 MPa. High hydrogen gas pressures combined with potential for hydrogen embrittlement in ferritic steels used in fabricating these vessels create design challenges. Fracture mechanics experimental techniques and analyses are ideally suited for tackling these challenges. All steel pressure vessels are suitable for use up to 400 bar (40 MPa) hydrogen pressures. For higher pressures of up to 1,000 bar (100 MPa), vessels that are reinforced by a composite wrap made from fiberglass, carbon fiber, or high strength steel wires embedded in an epoxy resin are necessary.

We consider the design of a cylindrical vessel with an OD of 350 mm and a wall thickness of 31.75 mm, and a length of 10 m. The inner capacity of the vessel is measured by the amount of water it can hold and is approximately 589 L. The vessel is made from SA372 Grade J Class 70 steel commonly used for fabricating pressure vessels. The wall thickness chosen is the maximum possible to ensure microstructural consistency across the thickness from hardenability considerations. During service, the cylinder pressure is expected to cycle once every day between the maximum operating pressure (MOP) to 10% of MOP. The vessel must be designed for a life of 20 years with a safety margin of 2; in other words, it should have a calculated fatigue life of 40 years. You are asked to estimate the MOP for the vessel. The vessel after

FIGURE 11.12   Picture of wire-wound steel pressure vessel to store hydrogen at 100 MPa or 1,000 bar (Picture by courtesy of WireTough Cylinders, LLC reproduced with permission).

manufacturing will be subject to ultrasonic testing that is capable of detecting any flaw that has a depth, a, greater than 0.9525 mm and half length, c, on the surface of 2.38 mm.

11.3.3.1   *K*-Expressions for Cracked Pressurized Cylinders

A schematic of the radial-axial flaw that is oriented normal to the hoop stress in the cylinder is shown in Figure 11.13. Zheng et al. [18], Yang et al. [19], Raju and Newman [20], and Newman and Raju [21] have calculated the stress intensity parameter, K, along the crack profile for such cases. The following symbols are used in this section:

a = depth of the surface elliptical crack
c = half-length of surface elliptical crack
Q = flaw shape correction factor
ϕ = angle along the crack profile as shown in Figure 11.14; ϕ = 0 at the surface and is 90° at the deepest point in the flaw.
f_c = ratio of boundary correction factor with respect to a flat plate
t = cylinder wall thickness
R_i = inside radius of the cylinder

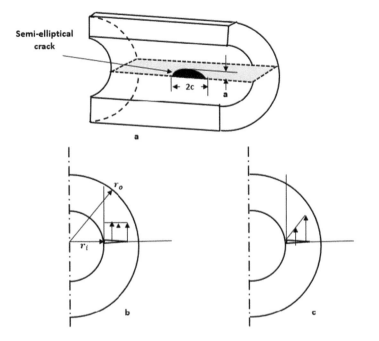

FIGURE 11.13 The cylinder model containing a semi-elliptical surface crack (a) the overall flaw shape and size, (b) crack surface subjected to uniform stress distribution, and (c) crack surface subjected to a linearly varying stress distribution.

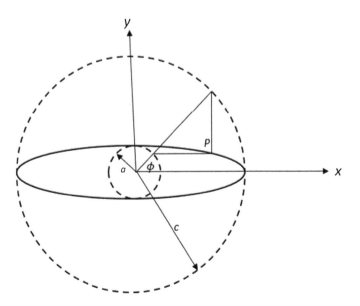

FIGURE 11.14 Definition of the angle ϕ at any point P on the boundary of an elliptical crack. The value of $\phi = 0$ on the surface and 90° at the deepest point.

p_{max} = maximum operating gas pressure
p_{min} = minimum operating gas pressure
$\sigma_{\theta i}$ = circumferential stress on the ID of the liner
$\sigma_{\theta 0}$ = circumferential stress on the OD of the liner
K_A = stress intensity parameter at the deepest point
K_B = stress intensity parameter at the surface point

The Newman-Raju [21] equations are used here. The equations provide K values at the deepest point in the flaw at $\phi = 90°$, and at the surface, $\phi = 0°$.

$$K = \frac{p}{Rt}\sqrt{\frac{\pi a}{Q}}F\left(\frac{a}{c}, \frac{a}{t}, \frac{t}{r}, \phi\right)$$
(11.14)

In equation (11.14), $\dfrac{p}{R_i t}$ is the average stress across the cylinder wall. The term F in equations (11.14) is written as follows:

$$F = F_1 f_c g(\phi) f(\phi) = \frac{K}{\dfrac{pr}{t}\sqrt{\dfrac{\pi a}{Q}}}$$
(11.15)

The various terms in equation (11.15) are given as:

$$Q = 1 + 1.464(a/c)^{1.65} \quad \text{for} \quad a/c < 1$$

$$F_1 = 0.97\left[M_1 + M_2(a/t)^2 + M_3(a/t)^4\right]$$

$$M_1 = 1.13 - 0.09\frac{a}{c}$$

$$M_2 = -0.54 + \frac{0.89}{0.2 + \dfrac{a}{c}}$$

$$M_3 = 0.5 - \frac{1}{0.65 + \dfrac{a}{c}} + 14\left(1 - \frac{a}{c}\right)^{24}$$

$$g(\phi) = 1 + \left[0.1 + 0.35\left(\frac{a}{t}\right)^2\right](1 - \sin\phi)^2$$

$$f(\phi) = \left[\sin^2\phi + \left(\frac{a}{c}\right)^2\cos^2\phi\right]^{1/4}$$

$$f_c = \left[\frac{R_0^2 + R_i^2}{R_0^2 - R_i^2} + 1 - 0.5\sqrt{\frac{a}{t}}\right]\frac{t}{R_i}$$

$$\text{Thus, } g(\phi = 0) = 1 + \left[0.1 + 0.35\left(\frac{a}{t}\right)^2\right]$$

$$g(\phi = \pi/2) = 1$$

$$f(\phi = 0) = \sqrt{\frac{a}{c}}$$

$$f(\phi = \pi/2) = 1$$

11.3.3.2 FCGR Properties of Pressure Vessel Steels in High Pressure Hydrogen

Design Code Committees of the American Society for Mechanical Engineers (ASME) formulate rules for pressure vessel design in the interest of public safety. As part of these rules, they specify materials and provide material data that is needed to design pressure vessels. FCGR behavior of SA 372 ferritic steels used to design of vessels for service in hydrogen was reported by San Marchi et al. [22]. The fatigue crack growth behavior of several low alloy ferritic steels including SA372 with ultimate tensile strength less than 945 MPa were very similar in high pressure hydrogen. The FCGR data of ferritic steels in hydrogen exhibit a knee in the da/dN versus ΔK trend at a value of ΔK_c that depends on the load ratio, R. The slope in the Paris equation changes from 6.5 at $\Delta K < \Delta K_c$ to 3.66 for $\Delta K > \Delta K_c$ as seen in Figure 11.15.

The FCGR behavior increases significantly with load ratio, R; thus, R must be explicitly accounted for in calculating crack growth rates in gaseous hydrogen. A relatively simple relationship, equation (11.16), conservatively represents the FCGR behavior of this class of steels for load ratio, R, $0.1 \le R \le 0.8$. These equations may be used for hydrogen pressures of up to 102 MPa and are as follows:

$$\frac{da}{dN} = C\frac{1 + C_H R}{1 - R}(\Delta K)^m \tag{11.16}$$

$$\Delta K_c = 9.35 + 4.484R - 1.865R^2 \tag{11.17}$$

where the values of C, C_H, and m are constants listed in Table 11.7 for ΔK less than or greater than ΔK_c.
For $R = 0.1$, $\Delta K_c = 9.78$ MPa$\sqrt{m}$; then

$$\frac{da}{dN} = 4.05 \times 10^{-11} \Delta K^{6.5} \quad \text{for} \quad \Delta K < 9.78 \, \text{MPa}\sqrt{m} \tag{11.18a}$$

$$\frac{da}{dN} = 2.0 \times 10^{-8} \Delta K^{3.66} \quad \text{for} \quad \Delta K > 9.78 \, \text{MPa}\sqrt{m} \tag{11.18b}$$

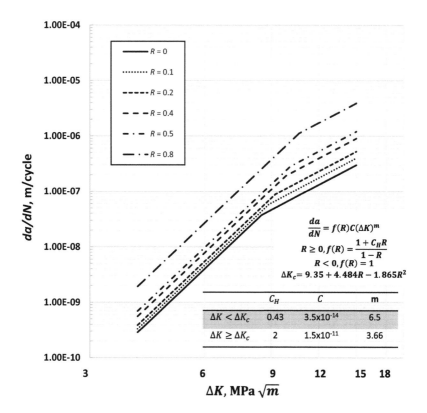

FIGURE 11.15 FCGR behavior of ferritic steels from the ASME code with ultimate tensile strengths less than 945 MPa [22]. ASME, American Society for Mechanical Engineers; FCGR, fatigue crack growth rate.

TABLE 11.7

FCGR Equation Constants in ASME Code for Ferritic Steels with an Ultimate Tensile Strength of Less Than 937 MPa in Gaseous Hydrogen Environment of Up To 102 MPa [22]

	$\Delta K \le \Delta K_c$ MPa$\sqrt{m}$	$\Delta K \ge \Delta K_c$ MPa$\sqrt{m}$
	da/dN in mm/cycle	da/dN in mm/cycle
C	3.5×10^{-11}	1.5×10^{-8}
M	6.5	3.66
C_H	0.43	2.0

The K_{IH} for this material is estimated to be 25 MPa$\sqrt{m}$. The ΔK corresponding to a K_{max} of 25 MPa$\sqrt{m}$ for $R = 0.1$ is $0.90 \times 25 = 22.5$ MPa$\sqrt{m}$. When ΔK_a reaches 22.5 MPa$\sqrt{m}$, the vessel will have reached its end of design life.

11.3.3.3 Design Life Calculations

To estimate the design life, we need the relationship between crack size and the number of fatigue cycles, N for various MOP values. We will then choose a MOP that meets the 20-year design life criterion with a safety margin of 2. The following steps are used to facilitate that calculation.

1. Using the initial crack depth, a, the half crack length, c, and the pressure, inner radius etc. in equation (11.15) we estimate the values of ΔK_a and ΔK_c at the deepest point in the crack and at the surface, respectively. All parameters needed to estimate ΔK_a and ΔK_c can be found in equation (11.15).
2. Next, we estimate the crack growth rates, da/dN, in the radial direction and the axial direction, dc/dN.
3. We calculate the number of cycles required to grow the crack by an incremental amount such as 0.254 mm and then determine the amount of crack growth in the c-direction in the same number of cycles as follows:

$$\Delta N_i = \frac{a_{i+1} - a_i}{\left(da/dN\right)_i} \quad \text{and} \quad \Delta c_i = \left(dc/dN\right)_i\left(\Delta N_i\right)$$

4. The number of fatigue cycles elapsed is given by the following equation

$$N_i = \sum_{i=2}^{n}\left(N_{i-1} + \Delta N_{i-1}\right); N_1 = 1$$

where n = the number of increments of crack depth needed to grow the crack from its initial size to final size.
6. Crack sizes a and c are tabulated in Table 11.8 and is plotted in Figure 11.16 for a MOP of 40.0 MPa.
7. The end of life is assumed to occur when maximum value of ΔK_a or ΔK_c approaches 22.5 MPa$\sqrt{m}$, the threshold K level for environment assisted cracking, K_{IH}. From Table 11.8, ΔK_c reaches that value first at a crack depth of 6.79 mm where the corresponding value of N is 13,926 cycles.
8. The design life in this case is $0.5 \times 13,926 = 6,963$ cycles or 19 years.
9. If we repeat the calculation with several values of MOP, we can derive a relationship between MOP and Design life as seen in Figure 11.17. To achieve a design life of 20 years or 7,300 cycles, we have to limit the MOP to 390 bar or 39 MPa.

11.4   SUMMARY

In this chapter, a general methodology for using linear elastic fracture mechanics concepts is illustrated through real-world practical examples. It is shown that fracture mechanics may be used effectively for assessing the integrity of structural components or for predicting the remaining life or design life of components. One of the most important applications of fracture mechanics is shown to be in selecting realistic inspection criterion and in-service inspection intervals for components designed to last a long-time for continued assurance against fracture.

TABLE 11.8

Results of Calculations for Design Fatigue Life of Vessels for Storage of Hydrogen at a Pressure of 40 MPa

i	a (mm)	c (mm)	ΔK_a (MPa$\sqrt{m}$)	ΔK_c (MPa$\sqrt{m}$)	da/dN (mm/cycle)	dc/dN (mm/cycle)	N (Cycles)
1	0.9525	2.38	11.52	8.02	1.51E−04	2.95E−05	1
2	1.2065	2.43	12.07	9.36	1.79E−04	8.06E−05	1,686
3	1.4605	2.55	12.54	10.45	2.06E−04	1.06E−04	3,107
4	1.7145	2.68	12.94	11.41	2.31E−04	1.46E−04	4,342
5	1.9685	2.84	13.34	12.25	2.58E−04	1.89E−04	5,442
6	2.2225	3.02	13.76	13.01	2.89E−04	2.35E−04	6,424
7	2.4765	3.23	14.20	13.71	3.24E−04	2.85E−04	7,303
8	2.7305	3.45	14.65	14.37	3.63E−04	3.39E−04	8,086
9	2.9845	3.69	15.11	14.99	4.07E−04	3.96E−04	8,785
10	3.2385	3.94	15.57	15.59	4.54E−04	4.57E−04	9,410
11	3.4925	4.19	16.04	16.17	5.06E−04	5.22E−04	9,969
12	3.7465	4.45	16.50	16.73	5.62E−04	5.91E−04	10,470
13	4.0005	4.72	16.96	17.27	6.22E−04	6.64E−04	10,922
14	4.2545	4.99	17.42	17.80	6.85E−04	7.41E−04	11,331
15	4.5085	5.27	17.87	18.31	7.53E−04	8.23E−04	11,701
16	4.7625	5.54	18.32	18.82	8.24E−04	9.09E−04	12,039
17	5.0165	5.82	18.76	19.31	8.99E−04	9.99E−04	12,347
18	5.2705	6.11	19.19	19.79	9.77E−04	1.09E−03	12,630
19	5.5245	6.39	19.62	20.27	1.06E−03	1.19E−03	12,890
20	5.7785	6.68	20.04	20.74	1.15E−03	1.30E−03	13,130
21	6.0325	6.96	20.46	21.20	1.24E−03	1.41E−03	13,351
22	6.2865	7.25	20.87	21.65	1.33E−03	1.52E−03	13,557
23	6.5405	7.54	21.28	22.10	1.43E−03	1.64E−03	13,748
24	6.7945	7.84	21.69	22.55	1.53E−03	1.76E−03	13,926
25	7.0485	8.13	22.08	22.99	1.63E−03	1.89E−03	14,093
26	7.3025	8.42	22.48	23.42	1.74E−03	2.03E−03	14,248
27	7.5565	8.72	22.87	23.86	1.86E−03	2.17E−03	14,394
28	7.8105	9.01	23.26	24.29	1.97E−03	2.31E−03	14,531
29	8.0645	9.31	23.64	24.71	2.10E−03	2.46E−03	14,659
30	8.3185	9.61	24.02	25.14	2.22E−03	2.62E−03	14,781
31	8.5725	9.91	24.40	25.56	2.35E−03	2.79E−03	14,895
32	8.8265	10.21	24.77	25.98	2.49E−03	2.96E−03	15,003

Three examples were chosen to illustrate the use of linear elastic fracture mechanics. These examples included inspection criterion and interval determination, failure analysis, risk assessment, and design or remaining life calculations. It is always good to compare predictions from the analyses to service experience. Such comparisons should be performed whenever possible because they provide realistic benchmarks for the validity of the analysis. This is even more important for complex components

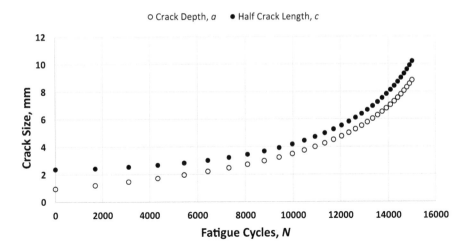

FIGURE 11.16 Plots of predicted crack depth, *a*, and half crack length, *c*, as a function applied fatigue cycles for cylinders at an MOP of 390 bar. MOP, maximum operating pressure.

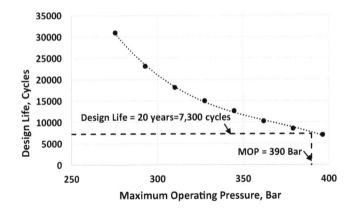

FIGURE 11.17 The relationship between MOP and design life of the pressure vessel. MOP, maximum operating pressure.

where several simplifying assumptions are frequently necessary to proceed with the analysis. Although we attempt to make conservative assumptions, as far as possible, it cannot always be guaranteed. Therefore, service experience, and lessons learned from field failures should be documented fully to avoid unexpected failures.

REFERENCES

1. E399-90, *Standard Test Method for Plane-Strain Fracture Toughness of Metallic Materials*, American Society for Testing and Materials, West Conshohocken, PA, 2020.
2. E1820-20, *Standard Method for Measurement of Fracture Toughness*, ASTM International, West Conshohocken, PA, 2020.

3. BS 5447:1974, *Methods of Testing for Plane Strain Fracture Toughness (KIC) of Metallic Materials*, British Standards Institution, London, England, 1974.
4. E561-94, "Standard Method for R-Curve Determination", in *Annual Book of ASTM Standards, Section 3*, 1994, American Society for Testing and Materials, Philadelphia, PA, pp. 489–501.
5. BS 5762:1979, *Methods for Crack Opening Displacement (COD) Testing*, British Standards Institution, London, England, 1979.
6. E1304-97, *Standard Test Method for Plane-Strain (Chevron-Notch) Fracture Toughness of Metallic Materials*, American Society for Testing and Materials, West Conshohocken, PA, 2014.
7. E-647-15, *Standard Test Method for Measurement of Fatigue Crack Growth Rates*, American Society for Testing and Materials, West Conshohocken, PA, 2015.
8. El681-03, *Standard Test Method for Determining a Threshold Stress Intensity Factor for Environment Assisted Cracking of Metallic Materials*, American Society for Testing and Materials, West Conshohocken, PA, 2013.
9. El221-12, *Standard Test Method for Determining Plane-Strain Crack Arrest Fracture Toughness, K_{la} of Ferritic Steels*, American Society for Testing and Materials, West Conshohocken, PA, 2012.
10. El457-15, *Standard Test Method for Measurement of Creep Crack Growth Rates in Metals*, American Society for Testing and Materials, West Conshohocken, PA, 2015.
11. E2760-16, *Standard Test Method for Creep-Fatigue Crack Growth Testing*, American Society for Testing and Materials, West Conshohocken, PA, 2016.
12. R.G. Forman and V. Shivakumar, "Growth Behavior of Surface Cracks in the Circumferential Plane of Solid and Hollow Cylinders", in *Fracture Mechanics: Seventh Volume, ASTM STP 905*, 1986, American Society for Testing and Materials, Philadelphia, PA, pp. 59–74.
13. A. Saxena and J. Opoku, "Evaluating Flaws in Metal Parts", Machine Design, Penton/IPC Publication, March 1980, pp. 1–8.
14. L.D. Kramer and D.D. Randolph, "Analysis of TVA Gallatin No. 2 Rotor Burst – Part I Metallurgical Considerations", 1976, ASME—MPC Symposium on Creep-Fatigue Interaction, MPC-3, 1976, pp. 25–40.
15. H.C. Argo and B.B. Seth, "Fracture Mechanics Analysis of Ultrasonic Indications in CrMoV Alloy Steel Turbine Rotors", in *Case Studies in Fracture Mechanics, Report AMMRC MS 77-5*, 1977, Army Materials and Mechanics Research Center, Watertown, MA, pp. 3.3.1–3.3.11.
16. L.R. Hall and A.S. Kobayashi, "On the Approximation of Maximum Stress Intensity Factors for Two Embedded Cracks", unpublished Boeing Memorandum, 1964.
17. G.A. Clarke, T.T. Shih, and L.D. Kramer, *Evaluation of Fracture Properties of Two 1950 Vintage CrMoV Steel Rotors, EPRI Contract Report- RP 502*, Westinghouse Research and Development Center, Pittsburgh, PA, 1978.
18. X.J. Zheng, A. Kiciak, and G. Glinka, "Weight Functions and Stress Intensity Factors for Internal Semi-Elliptical Surface Crack in Thick-Walled Cylinder", *Engineering Fracture Mechanics*, Vol. 58, 1997, pp. 207–221.
19. S.T. Yang, Y.L. Ni, and C.Q. Li, "Weight Function Method to Determine Stress Intensity Factor for Semi-Elliptical Crack with High Aspect Ratio in Cylindrical Vessels", *Engineering Fracture Mechanics*, Vol. 109, 2013, pp. 138–149.
20. I.S. Raju and J.C. Newman, Jr., "Stress Intensity Factors for Internal and External Surface Cracks in Cylindrical Vessels", *Journal of Pressure Vessel Technology*, Vol. 104, 1982, pp. 293–298.
21. J.C. Newman, Jr., and I.S. Raju, "Stress Intensity Factors for Internal Surface Cracks in Cylindrical Pressure Vessels", *Transactions of ASME*, Vol. 102, 1980, pp. 342–346.